I0709740

A Forest of Symbols

A Forest of Symbols

Art, Science, and Truth in the Long Nineteenth Century

Andrei Pop

ZONE BOOKS · NEW YORK

2019

ZONE BOOKS
633 Vanderbilt Street
Brooklyn, NY 11218

Printed in the United States of America.

Distributed by The MIT Press,
Cambridge, Massachusetts, and London, England

Library of Congress Cataloging-in-Publication Data
Names: Pop, Andrei, author.
Title: A forest of symbols : art, science, and truth in the long nineteenth
    century / Andrei Pop.
Description: Brooklyn, NY : Zone Books, 2019. |
    Includes bibliographical references and index.
Identifiers: LCCN 2019005766 (print) | LCCN 2019006411 (ebook) |
    ISBN 9781942130314 | ISBN 9781942130321 | ISBN 9781942130338 |
    ISBN 9781935408369 (alk. paper)
Subjects: LCSH: Symbolism (Art movement) | Art — Mathematics. |
    Science and the arts — History — 19th century.
Classification: LCC NX454.5.S9 (ebook) | LCC NX454.5.S9 P67 2019 (print)
    | DDC 700/.415 — dc23
LC record available at https://lccn.loc.gov/2019005766

Contents

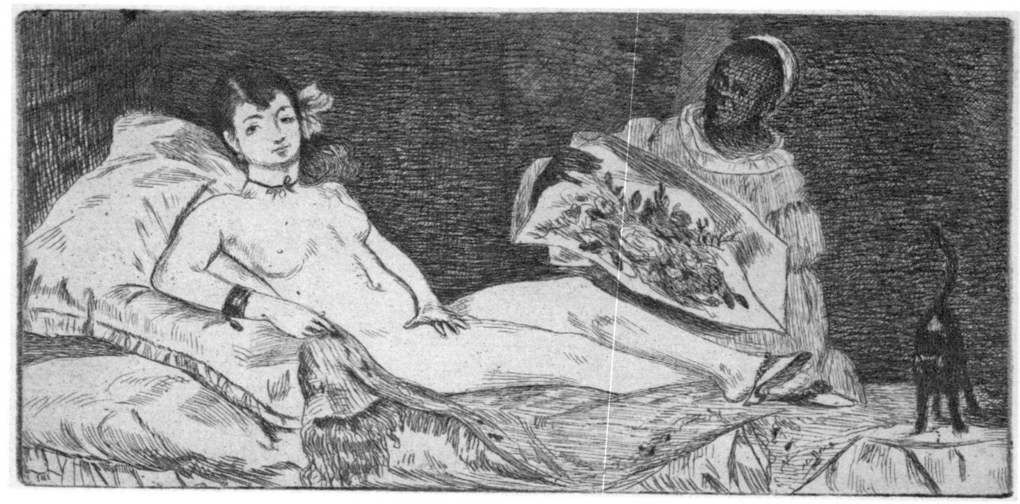

Figure P.1. Édouard Manet, *Olympia*, etching and aquatint (1865–67), New York Public Library. This plate was printed in Émile Zola's 1867 book on Manet, subverting its insistence that paint handling was everything to Manet, subject matter nothing.

Why Symbolism?

What is art for? The question might be asked by a child. Children love to paint or sing or sculpt with putty, but they are often puzzled by what grown-up artists do: sell their work ("Why? I like my pictures and want to keep them for *myself*"), or work hard to install it in an art space, only to see it dismantled after a few weeks or months ("Why? I want to keep my pictures *forever*"). Mostly, artists talk about their work interminably — to patrons, dealers, and art writers. But how do those people know what art is about?

> The most pervasive theory of the art object in art history . . . was its conception as a medium of communication or expression. The object was construed within its communicational or linguistic paradigm as a "vehicle" by means of which the intentions, values, attitudes, ideas, political or other messages, or the emotional state(s) of the maker — or by extension the maker's social and historical contexts — were conveyed, by design or chance.[1]

The author of these words does not think that art objects actually function as vehicles for what the artist has to say. But if they don't, this matters not only to the art writer. It also concerns the artist, and anyone encountering art, whether in a museum, on the street, or on the Internet. If artworks are not vehicles for meaning, what else could have caused the scandal over Édouard Manet's painting of the nude Olympia (1863)? Merely its appearance? The appearance of the oil painting is not that of the little etching Manet made for Zola's book about his art (Figure P.1). But how on earth are we to chart the correspondence, and the divergence, between the two without taking meaning into effect?

I do not attempt an interpretation of *Olympia* and its variants here. But it is plain that depriving art of meaning deprives it of its power. Not all of its power, surely: artworks have many uses and effects. The German critic Walter Benjamin once said, "Dada hit the spectator like a bullet," while Henri Matisse wanted his paintings to serve "like an armchair for the tired businessman." But notice the metaphorical "like" in both quotations: it is in virtue of its meaning, its sense, what we make of it, that a picture may hurt us like a bullet or comfort us like squashy furniture. In this book, I call art that *works mainly by virtue of its meaning* symbolist art. This is a conceptual and not a historical definition. Yet there were artists and writers at the end of the nineteenth century who called themselves symbolists, and whose unifying trait, for all their political and aesthetic differences, was a concern with how art gets its meaning. This book is about them.

The notion of art working in virtue of its meaning might seem either strange or banal. Is there really such a thing, and is it a distinct historical phenomenon? Consider a passage from Mark Twain's *Adventures of Huckleberry Finn*, published in 1884. Huck is trying, with little luck, to convince a skeptical Jim of the diversity of human languages:

> "Does a cat talk like a cow, or a cow talk like a cat?"
>
> "No, dey don't."
>
> "It's natural and right for 'em to talk different from each other, ain't it?"
>
> "'Course."
>
> "And ain't it natural and right for a cat and a cow to talk different from *us*?"
>
> "Why, mos' sholy it is."
>
> "Well, then, why ain't it natural and right for a *Frenchman* to talk different from us? You answer me that."
>
> "Is a cat a man, Huck?"
>
> "No."
>
> "Well, den, dey ain't no sense in a cat talkin' like a man. Is a cow a man? — er is a cow a cat?"
>
> "No, she ain't either of them."
>
> "Well, den, she ain't got no business to talk like either one er the yuther of 'em. Is a Frenchman a man?"

"Yes."

"*Well*, den! Dad blame it, why doan' he *talk* like a man? You answer me *dat!*"

Huck editorializes notoriously: "I see it warn't no use wasting words — you can't learn a nigger to argue."[2] Twain liked the exchange so much that he entitled public readings from the book after the second half of Huck's verdict. This ruffled feathers, more for the crudity of the epithet than for its racism.[3] But a close look at the text belies these apparent last words. Huck, who had just argued with Jim about the story of King Solomon's choice, thought he was offering logical argumentation: farmyard animals make different noises, so why shouldn't people? For his analogy to work, Americans and Frenchmen ought to differ as do cats and cows; more precisely, they ought to differ in the same respect, that of belonging to the same species. That is exactly what Jim sees, and what he tries to show Huck: neither a cat nor a cow is human, and just to be thorough, they are not of the same species either ("Is a cat a cow?"). A Frenchman, being a man, hasn't a cat's excuse for communicating differently. The Socratic conclusion is that a Frenchman doesn't speak differently from other people, in the respect that a cat or a cow does.[4]

There is call for belaboring the obvious point that Jim outsmarts Huck. For the moral and political insight that flows from the logical one, surely, is that Jim is as much a man as Huck. This is a lesson the book teaches, not tells. It does so by example, in the teeth of Huck's conventional resistance to it. This can only take place though with the careful uptake of what the book offers: action and humor and characterization, but also argument and allegory. The silly logical bout is at the same time a moving symbol of intellectual equality. And so *Huckleberry Finn* can be profitably thought of as a symbolist artwork, though it doesn't have the wan, moonlit ambiance of some of that art, an ambiance that Twain made fun of (Figure P.2).

Besides being art history, then, this book is a history of ideas, because the concerns of symbolist painters and poets were shared to a remarkable degree by theoretical scientists of the period, especially by mathematicians and logicians dissatisfied with the empiricism

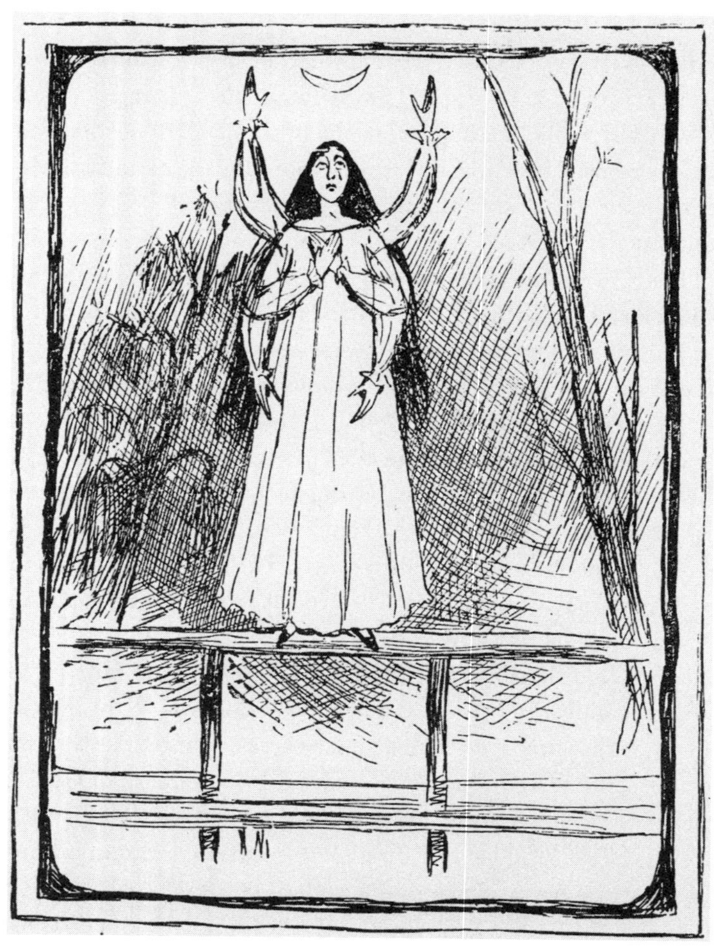

Figure P.2. E. W. Kemble, Emmeline Granger-
ford's last picture, from Mark Twain, *The
Adventures of Huckleberry Finn* (1884), ch. 17,
under the heading "Interior Decorations."
Twain captions the image with Huck's
comment on the multiplication of arms:
"It made her look spidery."

sweeping their disciplines. Above all, the assumption that all science consists of individually experienced observations seemed to leave no room for general laws of nature, opening up the possibility that different observers, differently trained and equipped, may not find any way to reconcile their divergent observations.[5] An analogous anxiety about achieving consensus, and the corresponding difficulty of attaining an understanding of the new, informed the most ambitious art of the era. Even if artists did not seek mind-independent truth, as did scientists, the very possibility of understanding an artwork was at stake, depending on some degree of agreement, both perceptual and conceptual, between artists and audiences.

This goal was not always within reach. It is worth dwelling briefly on an image whose dependence on its meaning for its effect is elusive (Figure P.3). Redon's drawing would be called symbolist even by historians reluctant to apply the term to a figure like Manet or Twain. They would rightly point to Redon's dreaminess, his effort to make visible a visionary or ideal state of affairs.[6] But what does that mean? Is there not plenty of solid observation in the rendering of the eye and its socket — a left eye, judging by the eyelid and tear duct? But this eye is placed smack in the middle of one surface of a cube! Are we looking at it or merely imagining it? As for the spheres orbiting the cube and the flat surface below (the bust resting on it suggests a table), is this an impossible landscape or just an unfamiliar one?

Without knowing the drawing's context of creation, we can still orient ourselves by the aesthetic tension present at its memorable center: a human body part, indeed a prime organ of perception, juxtaposed with a Platonic solid, which it seems to endow with awareness.[7] Can we go further? The Theosophical Society, founded in 1875 by Helena Petrovna Blavatsky, taught that man is "a mystic square" in the world of appearance and a cube in reality. Scientific-minded theosophists like architect Claude Bragdon interpreted this literally, drawing cross-sections of solids and citing higher-dimensional geometry as a vindication of their mysticism (Figure P.4).

It would be rash to equate Redon's seeing cube with the hypercubes of theosophy. The theosophists themselves got these ideas from

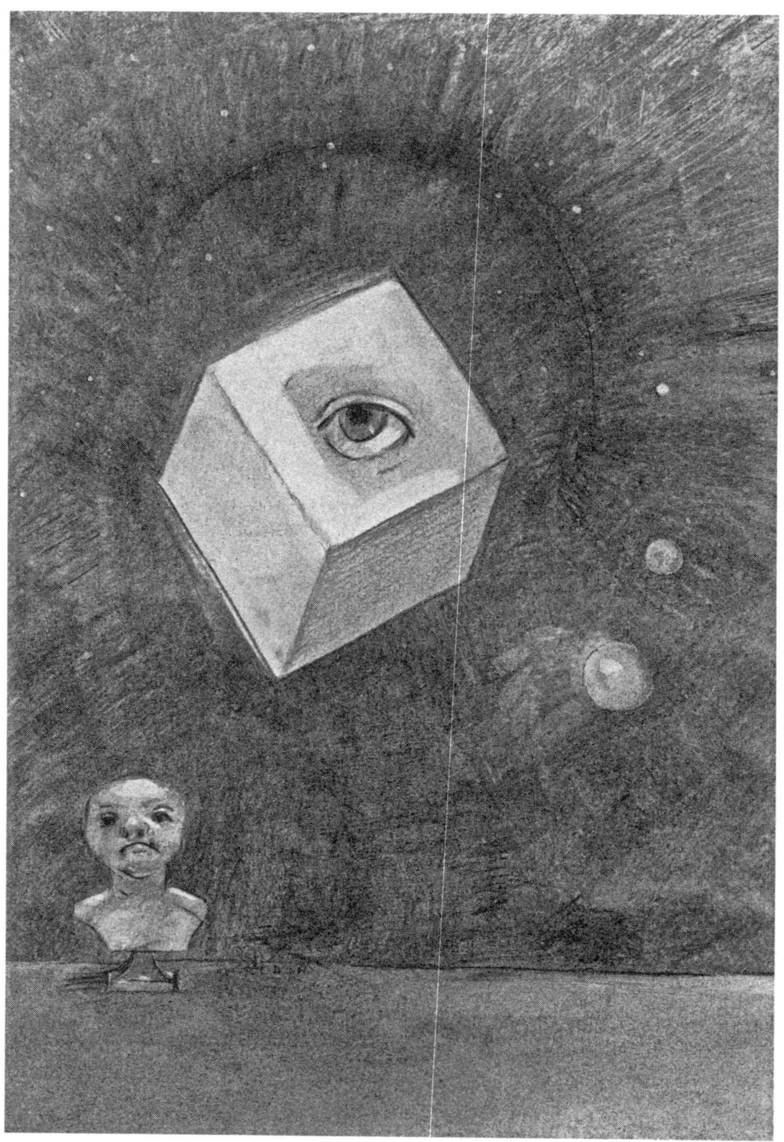

Figure P.3. Odilon Redon, *The Cube* (1880),
coal on paper, private collection. Note
the trace of a semicircular black outline
above the cube: a human head, or a figure
incommensurable with the square? Redon
explored the question in other drawings,
enlivening squares, triangles, and spheres.

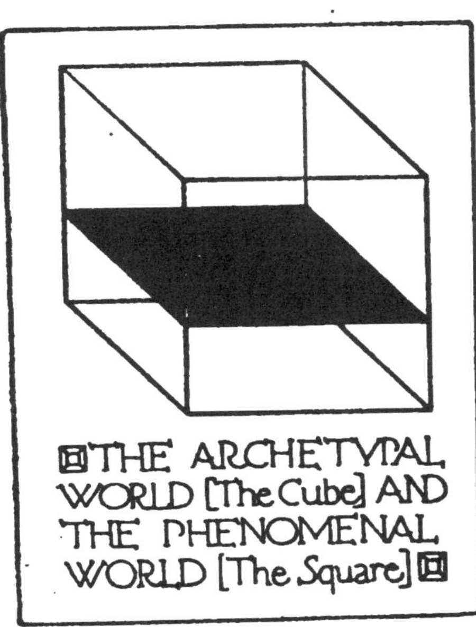

Figure P.4. Claude Bragdon, *Man the Square: A Higher-Space Parable* (Rochester, NY: Manas Press, 1912), pp. 11, 9. Bragdon's allusion to the riddle of Oedipus (what goes on four, then two, then three legs?) suggests that we do not in fact know everything about the fourth dimension but have only the geometrical analogy to guide us.

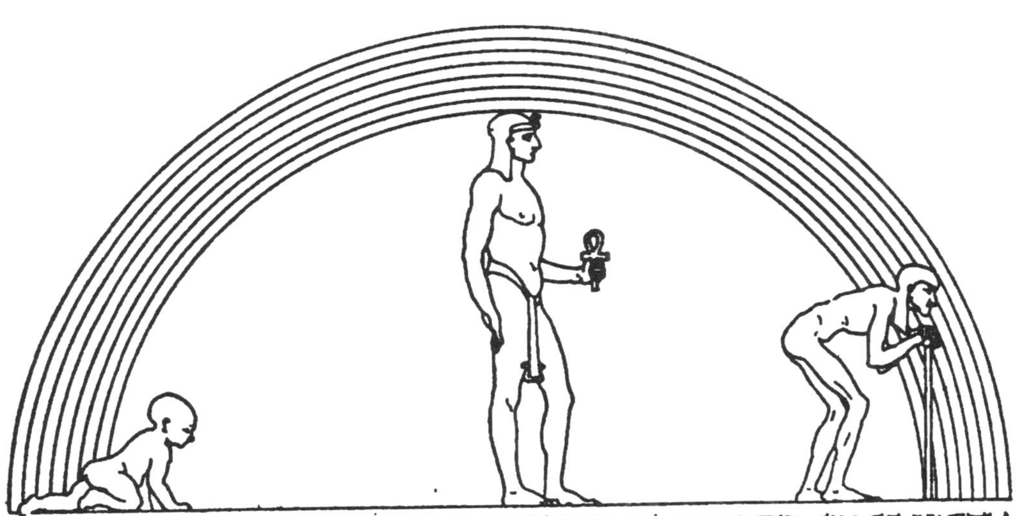

mathematicians like George Boole's son-in-law, George Hinton, who also inspired the multidimensional speculations of H. G. Wells. As Bragdon concedes, we do not picture four-dimensional space, but deduce it from our experience with lower-dimensional space: from the relation of a cube to a square, for instance. Redon himself claimed to proceed in much the same analogical manner, saying, "I think I have offered . . . in drawings and lithographs, varied human expressions; I have even, by permissible fantasy, placed them in a world of unlikelihood, in imaginary beings that I have tried to make logical with the logic of the structure of visible beings."[8]

Where is that logical structure of the visible in our drawing? Sticking to eye and cube, we might notice two peculiarities about them that tend in opposite directions. I have said that the eye is flush with one surface of the cube, but that is odd. That surface ought to be oblique to our angle of vision, since we can see two other surfaces of the solid, much as in *Man the Cube*. But those surfaces aren't foreshortened as if we were looking at a real die; they form parallelograms, just as if we were looking at a diagram of a cube — again like Bragdon's. The idea of a cube predominates over the visual experience of a cubic solid. As for the eye, which seems to emerge from the slightly rectangular upper face of the cube, its iris and pupil are turned upward against the lid, so that it stares at or through or past us. (The bust on the ground below looks up too — we can see its chin and both nostrils of its snub nose, as if it were meeting our gaze, centered on the cube.)

None of this necessarily exhausts the picture's meaning — only a silly formalism could make us think that pictures are self-evident in such a manner. But an analysis of this sort is clearly helpful not only in understanding what Redon was up to in an obscure chalk drawing, but also in such prominent works as his contributions to Stéphane Mallarmé's final book, *Un coup de dés* (*A Roll of the Dice*), which I will discuss in more detail in Chapter 2 (Figure P.5).

The presence of not one but two cubes in this lithograph goes far beyond its ostensible role of making visible the die, or dice, of the title: if the resting fist-sized die and the much larger airborne

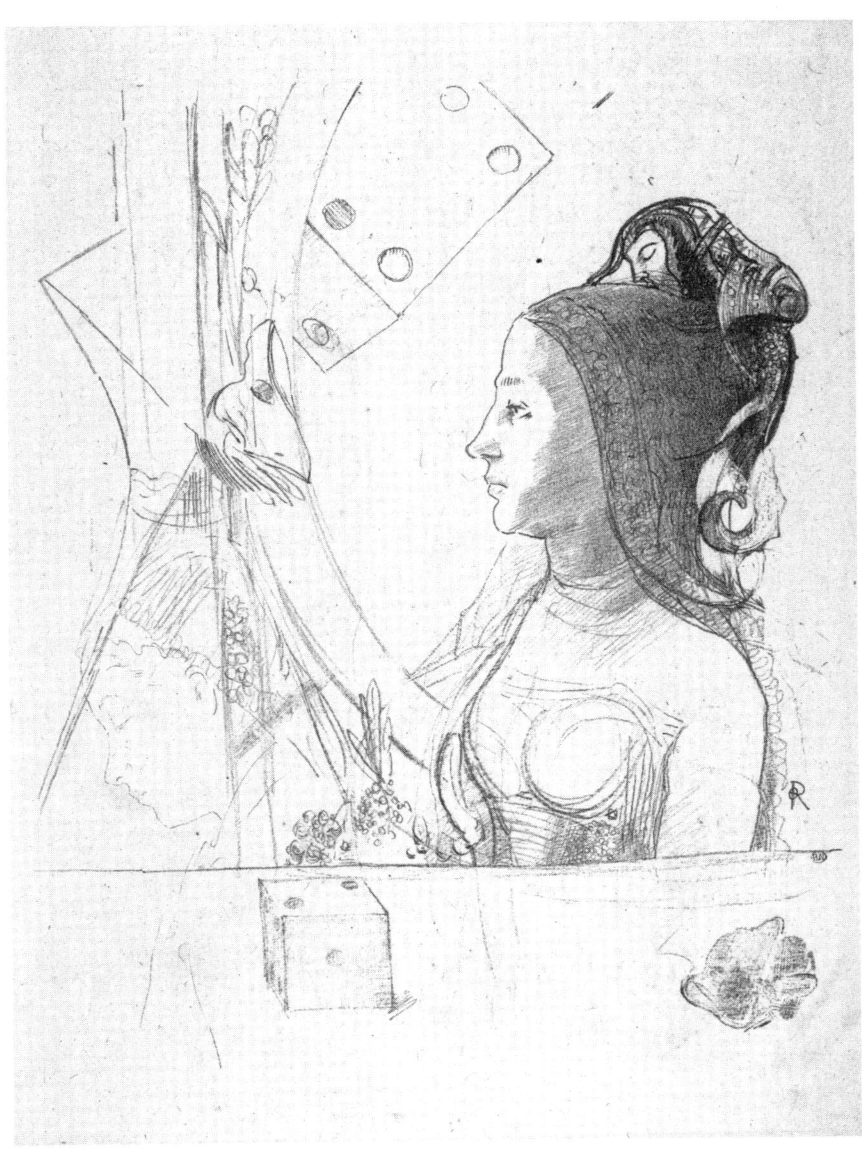

Figure P.5. Odilon Redon, illustration to Stéphane Mallarmé, *Un coup de dés* (1898), lithograph, Bibliothèque Nationale, Paris. The oblique face of the larger die, with its one spot in the corner, does not display a possible die configuration.

die represent something like the restless play of chance, they also correspond to the tangible, open-eyed profile bust of the woman below and the ghostly, closed-eye profile to the left. A robust dialectic of ideal and individual, the momentary and the eternal, is at play in Redon's "logical construction" of the fantastic. Again, meaning, though by no means easily available, is no by-product; on it rests the work's force, however elusive and subtle this may be.[9]

This logical structure of symbolism has often been neglected in favor of its links to the occult.[10] Scholars of the late nineteenth century have long been struck by the flourishing of esoteric doctrines, and their opposition to the positivist faith in progress that characterized the earlier half of the century.[11] But focusing on the irrational occludes the period's commitment to revolutionary science, which is central to its moral and political projects. It is not as a naïve rejection of reality, or of reform, that artists like Bragdon and Redon can be understood. What they have in common is rather the commitment to the concrete means, aesthetic and visual, but also logical and philosophical, by which reality, perceived or otherwise, is made accessible to more than one intellect.

To reflect this parallel but distinct project in art and science, my book will not dwell on what modern art borrowed from contemporaneous science, but on problems — some of them ancient, but gaining new urgency with industrial modernity — that afflicted both artists and researchers, and how these problems were addressed in both fields.[12] The problem of subjectivity in particular, of making oneself understood to others despite the privacy of one's consciousness, had to be met if collaboration in science or the sympathetic uptake of new art was to take place. To display the pervasive role of this problem, which might seem intimate to art but alien to science, in the culture of the fin de siècle, is the role of the next chapter.

Symbolisms in the Plural

Any reading of symbolism, however broad, must at the same time give some account of what lies beyond its boundaries. Take two balcony views of Paris, painted a decade apart (Figures 1.1 and 1.2). The paintings have much in common. From broad avenues teeming with hoof and foot traffic, to cast iron railings, modern "Haussmannized" Paris is their subject.[1] Its colorful human panorama framed by architecture looks as if made to be painted — even by the introvert Munch. The abruptly plunging perspective draws our gaze away from the featureless man and toward the crowd he is watching. Caillebotte's *Vue* is even more uncompromising: eliminating the onlooker he so often painted, it breaks up the view of street and kiosk, woman and hackney, following instead the floral curls of the iron railing, echoed by the simpler green shoots of a plant on the terrace.

We see substantially the same world in the two paintings, although the emphasis has shifted. The iron grate dominating both pictures, which so forcefully asserts the primacy of the painted surface in Caillebotte, is a stormy arabesque in Munch. Only the polychrome shadow it casts on the balcony floor can be easily traced by the viewer. This shadow, like the inarticulate gray one dripping from the man's left shoe, is naught but an absence of light, an optical consequence of the staccato illumination of the Rue Lafayette below. Shadows are literally nothing, gaps, physical absences, but in this painting, as so often in life, they assume legible form. What this form tells us — of the metal of the railing and its intricate design, of

Figure 1.1. Gustave Caillebotte, *Vue prise
à travers un balcon* (1880), oil on canvas,
Van Gogh Museum, Amsterdam. The desire
so typical of Caillebotte's voyeurs is muted
and generalized both here and in Munch;
if anything, what is desired is the whole visible world.

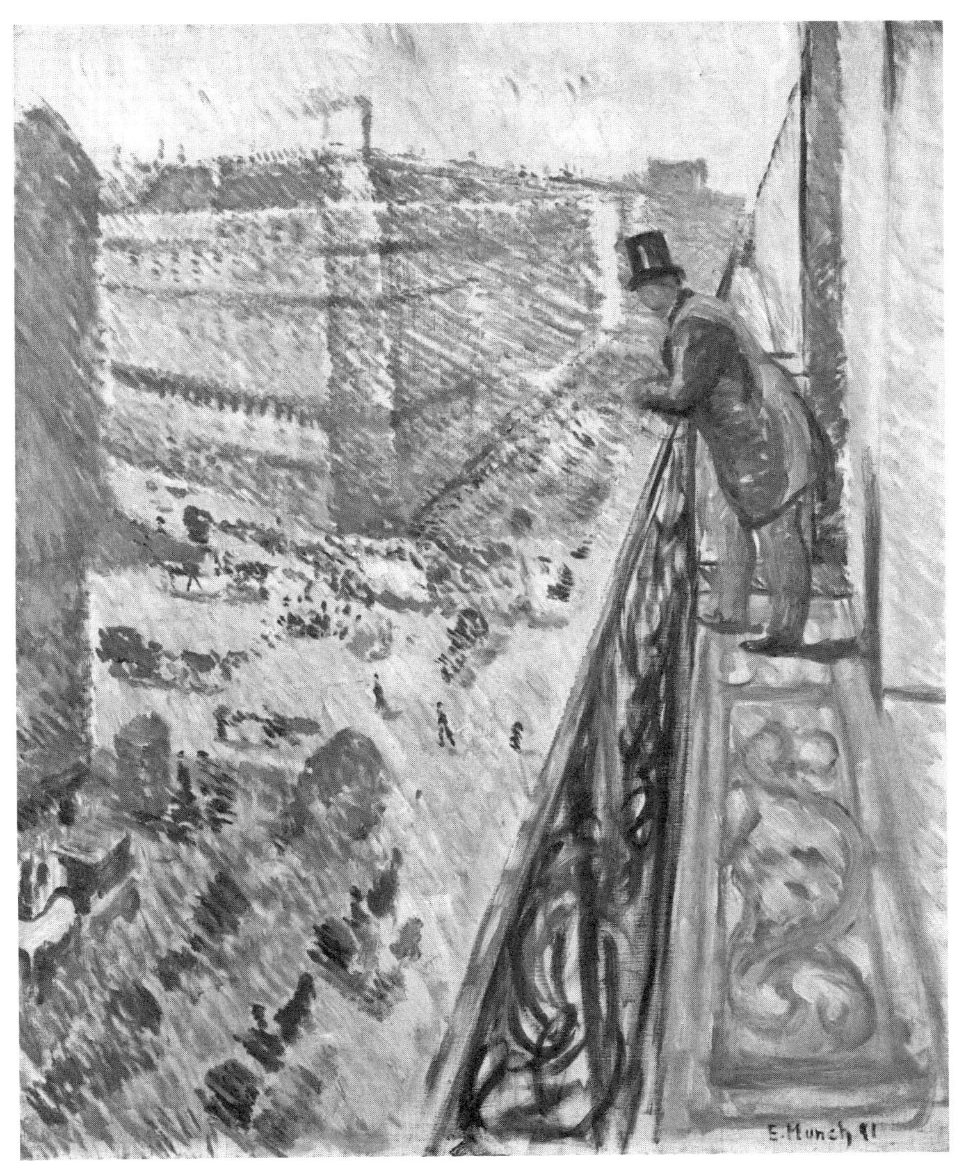

Figure 1.2. Edvard Munch, *Rue Lafayette* (1891), oil on canvas, Nasjonalmuseet, Oslo.

the more shadowy but no less convoluted thoughts and desires of the man leaning over it — it does indirectly, by dint of light and color and silence. Symbolism as this book sees it is not a turning away from the world, but a new way of looking at it, doing justice to what the daylit positivism of the industrial revolution had missed. To make sense of the world through such indirect means requires marshaling not just raw experience — what a radical empiricist like Caillebotte values most — but, more circuitously, our means of grasping it in word and image. But we cannot enter upon this master theme of symbolism without at least provisionally adopting its self-critical spirit. Symbolism in art and science is, not by coincidence, bound up with the rise of art historical writing. Art history is a symbolist undertaking in its search for the "how" and "why" of pictorial meaning. It shares with symbolism the difficulty of any self-reflexive practice: that of getting a clear view of itself.

Grasping Laocoon

To see how the birth of art history is entangled with symbolism, we must recall the artifact that first called forth passionate critical disagreement among artists and critics: the ancient Greek statue we call *Laocoon and His Sons*, dug up in Rome, on the Esquiline Hill, in January 1506, at the height of the Italian Renaissance. Two and a half centuries later, Johann Joachim Winckelmann proposed that the statue embodied classical decorum in depicting pain: the hero's contorted face and barely parted lips let out a sigh where Virgil had described a scream. This bold vision of Greek stoicism was challenged by the freethinking G. E. Lessing in 1766, only three years after Winckelmann restated it in his magnum opus, *The History of Ancient Art*. Lessing found the "sigh" wrong dramatically and theoretically: Laocoon is not reserved because it is manly to die without making a fuss, but because he is a creature of visual art, taking up space rather than time, as poetry, drama, and music do. Reserve is what visual art demands and all that it permits; an image of Virgil's scream would render it obscene.[2]

Winckelmann never responded to Lessing's critique, but he was pleased by the attention, noting to a correspondent that "as it is honorable to be praised by honorable people, it may also be honorable to be thought worthy of their censure."[3] Lessing himself was boastful one moment, dismissive the next, calling his *Laocoon* his best work but also "a mishmash of pedantry and whim." Other readings followed in rapid succession, from Johann Gottfried Herder's political reading of father and sons fighting for liberty to Goethe's pastoral tragedy: a trio of shepherds asleep in the wood, surprised by snakes. Artists also joined the fray. Anglo-Swiss painter Henry Fuseli rejected "the frigid fantasies of German critics," comparing the Laocoon formalistically to a "wave fluctuating in a storm"; and he drew images of a modern woman writhing in sympathy with the victim (Figure 1.3).[4] His friend William Blake covered an engraving of the statue he had prepared for Rees's *Cyclopaedia* with gnomic graffiti slogans like "Prayer is the Study of Art" and "For Every Pleasure Money is Useless."[5]

These facts suffice to let the problem arise on its own. What is this *Laocoon* that was argued over? Are Fuseli's wave, Goethe's sleeper, and Herder's fighter one and the same marble statue found in Rome in 1506? Obviously not, for a statue can neither sleep nor fight; but if it is its subject matter or imagined content or Greek myth or something of the kind, are we to conclude that there are "many Laocoons," as many as there are interpreters?[6] If there are, then there is no debate: the debaters didn't even find a common cause over which to argue. But that is clearly wrong, for there was a parallel and even longer-lived archaeological dispute about when the statue was made and how the fragments fit together. As early as 1863, Johann Jakob Bernoulli noted a "double dispute over the time of its making and the artistic motif."[7] The two sides of the dispute, material and conceptual, are not arbitrarily connected: what kind of entity we understand Laocoon to be has a bearing on when the sculpture was made, by whom, and how it was meant to be seen. The literary debate about meaning, about what we might call the Platonic idea of Laocoon, is intimately linked to archaeological and art historical questions. The opposite is also true. How the statue looks is significant to what it means, but art history has lost sight of this in attending to plural receptions.

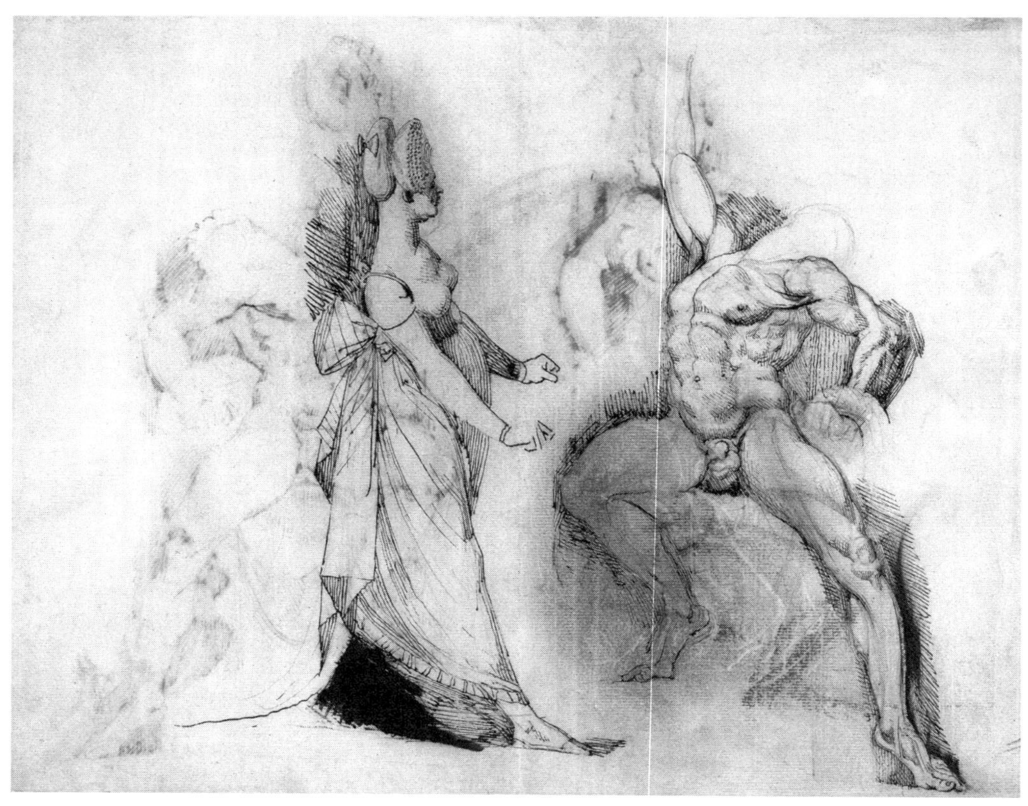

Figure 1.3. Henry Fuseli, *Woman before Laocoon* (1802), pen drawing, Kunsthaus Zürich. Fuseli drew this theme twice (on both sides of the same sheet of paper) during or after a visit to the Louvre in 1802, where Napoleon had brought art looted from Italy, including the Laocoon.

How did the Laocoon debate end? Interestingly, the matter was not one of diminishing returns, or of new fashionable controversies taking its place. The statue called forth impassioned argument deep into the nineteenth century — most notably from Anselm Feuerbach (the father of the German painter of the same name), who disputed Lessing's strictures on the expression of emotion in visual art, insisting that the Laocoon emits a bloodcurdling cry, and the anatomist Jakob Wilhelm Henke, who — on the basis of the chest musculature — saw in the Laocoon a case of arrested motion, which might issue in either a sigh or a scream, but only once the figure breathed out.[8] The discourse was certainly gaining in psychological sophistication — we are here far removed from Winckelmann's intuitive remarks about Laocoon's "great, resolute soul." But with this very sophistication, the end of the debate neared, and was announced boldly in the 1866 *Voyage en Italie* of the French historian, psychologist, and philosopher Hippolyte Taine, who saw the appeal of the Laocoon exclusively in its being "more than the others a neighbor to the modern style."[9] For Taine, the work of art is a mirror held before an epoch; the Laocoon interests moderns because it resembles their emotional distress. For the psychological critic, any objective content is effaced by the artwork's overwhelming subjective effect upon viewers. This tendency reaches an apex in what may be the most careful, if least-known, interpretation of the Laocoon, by the philosopher Hermann Lotze in his *History of Aesthetics in Germany* (1868). Agreeing with Henke that neither posture nor facial expression are consistent with intense pain, Lotze goes further, suggesting that the pulling of a tooth would hurt far more than death by snake. What we see instead in the statue is a "*psychic* process, [manifested] in the suddenly present hopelessness after prolonged stress and self-defense, and by no means in a physical pain against which the resilience of a great soul would be especially called for."[10]

Lotze's interpretation is substantial and lucidly argued. But subtly, almost imperceptibly, he has changed the subject, arguing from facial musculature and other physiological clues to a state of mind that could be expressed in other, less tragic forms. In fact, Lotze recalls a caricature that had appeared in the Parisian comic journal *Charivari* many years earlier, of a man in bed, awaking with a hangover and "precisely the attitude of Laocoon as he stretches and with the half-open, yawning mouth tries again to connect himself to miserable reality" (Figure 1.4).[11] The point is not the expendability of the snakes — an issue still passionately disputed.[12] Rather, it is that the artwork, for all its particularity, becomes only one of the many possible embodiments of a state of mind that may or may not be directly legible from the artwork (hence Henke's and Lotze's physiological and psychological arguments). The achievement of the *Laocoon* sculptors is thus merely that of having "known the ensemble of organic motions to the finest degree, in order to use them for the representation of a psychic process, which is not the only one capable of producing them, but which *also* may produce them, in which case it does so inevitably."[13]

Rethinking Objectivity

The end of the Laocoon debate is of consequence not just for psychophysical aesthetics, but for the very possibility of writing art history. In fixing attention ever more steadily on the subjective features of the mind that the artist expresses contingently in a physical object, Lotze is not just taking an ordinary risk of misinterpretation: the risk of missing his target. Instead he runs the risk of losing any grip on its meaning, indeed on any determinate meaning for this and other works of art. And he runs this risk not in spite of but *because* of his particular kind of scientific thoroughness. Consider a related difficulty canvased by Lotze's onetime pupil, Gottlob Frege:

> There is, let us suppose, a physiological psychologist. As is proper for a man of science, he is far from supposing the things he touches and sees to be his mental images. . . . Nerve fibers, ganglion cells he will never admit as contents of his consciousness, indeed he is rather inclined to consider his consciousness dependent on nerve fibers and ganglion cells. He confirms that light waves, refracted in the eye, stimulate the optic nerve. Some of that is conveyed by nerve fibers to ganglion cells. Further processes in the nervous system play their role, and color sensations form, which are connected perhaps to what we

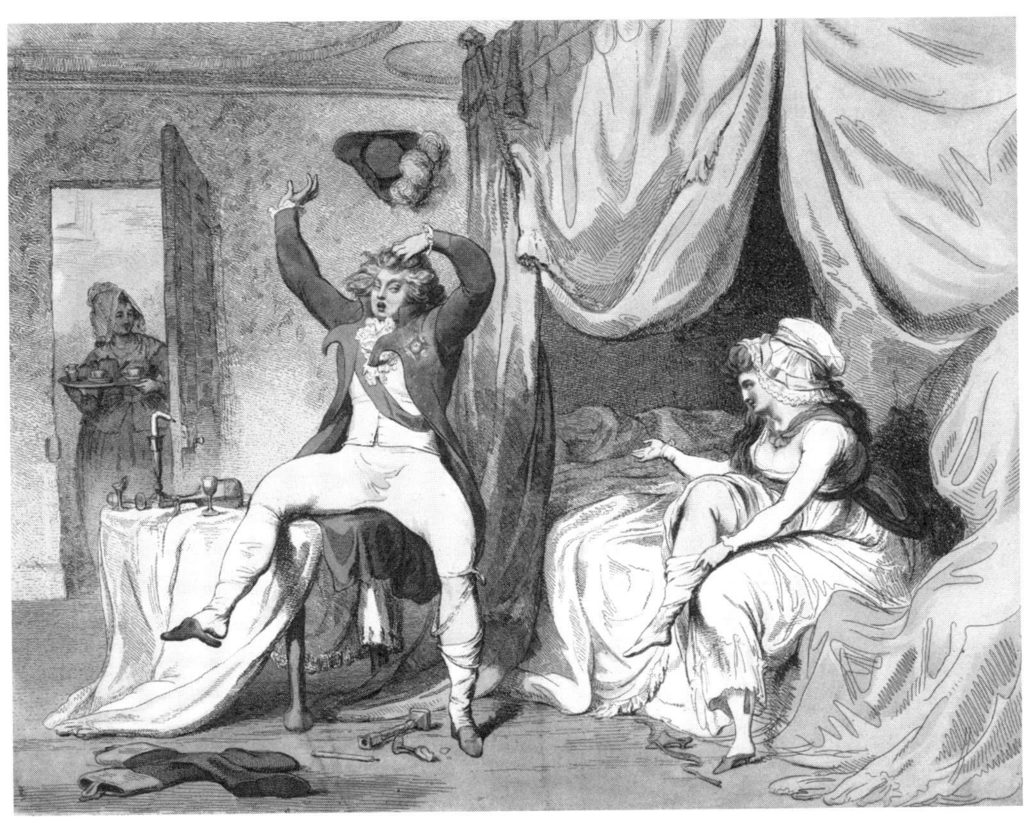

Figure 1.4. James Gillray, *The Morning after Marriage, or, A scene on the Continent* (1788), hand-colored etching, British Museum, London. An avid student of Fuseli (in a letter he speaks of "ye use to be made of him"), Gillray here cites the Laocoon more than a decade before Fuseli drew the statue. Note the witty transformation of the son trying to get the snake off his leg into the bride adjusting her stocking.

> call the image of a tree. . . . One can go a step further. . . . We think a thing independent of us stimulates a nerve and thus effects a sense impression; but speaking precisely, we experience only the end of this process as it intrudes into our consciousness. . . . If the researcher wishes to dispense with all assumptions, only images remain to him; all is dissolved in [mental] images, including the light waves, the nerve fibers and ganglion cells from which he began.[14]

This dramatic self-dissolution of the scientific attitude may sound like parody. But nineteenth-century psychological argumentation *was* this radical. Consider Salomon Stricker, a Viennese physiologist who, among other achievements, discovered the contractility of vascular walls. Stricker was fascinated by subliminal muscular sensations produced in speaking or singing. Extending this interest to mathematics, he reasoned that muscular sensations must accompany counting and all arithmetical operations. From this, he concluded that all thinking, from music to math, is nothing but inner perception of muscular movements. In turn, scientific argument, far from being binding, is but a species of persuasion, aimed at eliciting conversion to one's own views: "the rules of logic are called laws because all people who were in a position to express an opinion in the matter found those rules fitting for their own brains."[15]

The kind of philosophical psychology practiced by Stricker — a physical scientist tempted to give an empirical account of logical matters — might seem remote from the difficulties besetting modern art, now or in the nineteenth century. But the case of geometer and art historian Guido Hauck might change our mind. An expert in stereometry, Hauck wrote textbooks on perspective painting and even invented a "Perspektograph" to facilitate the drawing of perspectival views from plans and elevations.[16] This technical work, and some pioneering studies of eye musculature and movement, led him to suspect that the retina in fact produced curved images of straight lines, a previously ignored physiological "fact" that he thought explained the curvature of Doric columns (Figure 1.5).[17] Hauck's theory was disputed, but given that even as apparently uncontroversial aspects of visual art as straight lines might require psychological

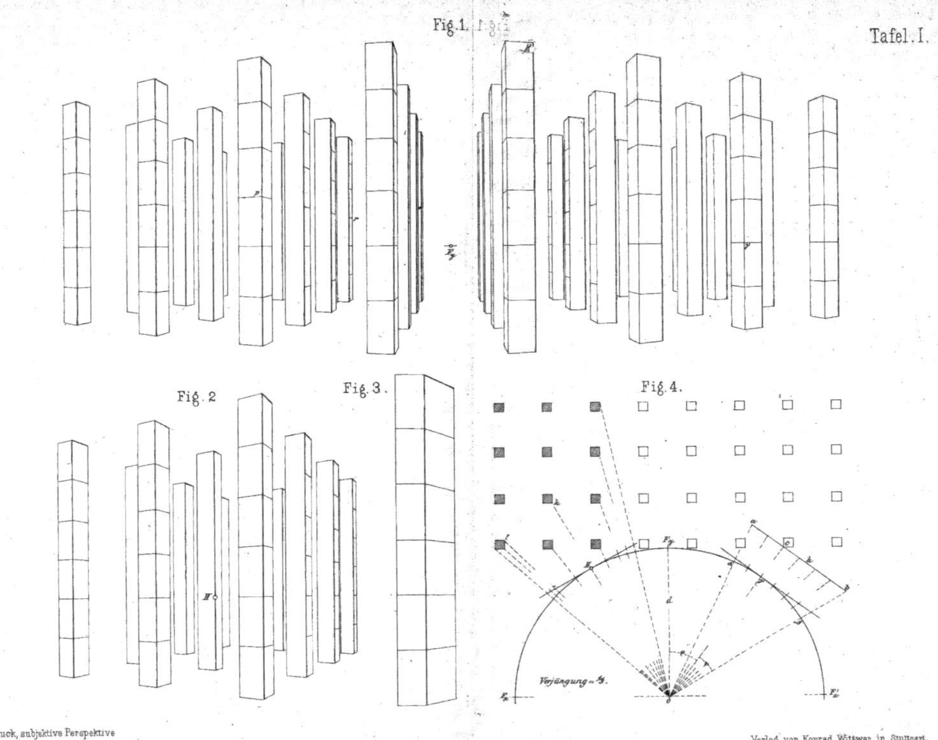

Figure 1.5. Guido Hauck, *Die subjektive Perspektive und die horizontalen Curvaturen des dorischen Styls* (1879), plate I. Erwin Panofsky believed that Hauck didn't go far enough in promoting curvilinear (Fig. 1) over linear perspective (Fig. 2): "Strictly speaking, even the verticals would have to submit to some bending (pace Guido Hauck)."

reinterpretation, it is clear that the kind of corrosive skepticism about our knowledge of the world articulated in Stricker's philosophy of muscle movements could become almost a kind of scientific platitude. One would have to prove that one *does* see straight lines rather than the curves that seemed more "natural" to the curved structure of the retina, and that despite the puzzling fact that, had we no experience of straight lines, it is not clear what could be meant by curved lines.[18]

Such skepticism is typical of the adoption of psychological method in disciplines other than psychology. It is a risk that must be taken whenever psychology really plays a decisive role in some field. Since hardly anyone can doubt this of art, interpreters and practitioners in the nineteenth century were especially under pressure. Ever more elaborate accounts of how art and poetry work upon the minds of spectators and readers appeared side by side with proclamations that the work of some artist or school, be it Courbet or Baudelaire, Manet or Mallarmé, is incomprehensible. And the attempt to explain the world through the workings of the human mind extended even farther, from mathematics to metaphysics, in the work of scholars ranging from the physiologist Stricker to the philosopher Edmund Husserl. Ernst Mach advocated the application of psychological method to physics, and other leading physicists, like Hermann Helmholtz, Gustav Fechner, and Wilhelm Wundt (also a logician), contributed to both disciplines.[19] It is striking that many of these figures were Germans or Austrians; but the French, Taine foremost among them, shaped this direction, as did John Stuart Mill and that generation of English logicians who thought of logic as revealing the "laws of thought."[20] Paradigmatic for the way they saw psychology at work in every intellectual domain might be Hermann Helmholtz's approach to the nature of numbers, which he derives from a human practice of counting: "Counting is a procedure that is based on the fact that we are able to keep in our memory the order of the sequence in which acts of consciousness temporally follow one another."[21]

Psychologism, as the tendency was called by its critics, is just this reduction of a scientific field to a particular human intellectual practice. But human practice is fallible, as are human minds. If truth is nothing *but* such practice, critique is useless: there is no elevated perspective from which its errors could be discovered or righted. The acute skepticism which results is expressed eloquently by Charles Darwin in a letter of 1881: "But then with me the horrid doubt always arises whether the convictions of man's mind, which has been developed from the mind of the lower animals, are of any value or at all trustworthy. Would any one trust in the convictions of a monkey's mind, if there are any convictions in such a mind?"[22]

This is a sobering confession: the inventor of the theory of evolution does not see how anyone could come up with scientific theories, if the theory of evolution is true. In consistently accepting this doctrine, it seems we must resign ourselves to a charge of deep irrationality.[23] Yet all is not so grim. This threat tacitly depends on the identification of reason with the working of actual minds. Had Darwin's question been whether monkeys or mollusks could ever grasp the rules of arithmetic, that would be an empirical inquiry into those creatures' intelligence and way of life. On the other hand, if reason is nothing *but* the thought processes of such beings, the question of success in reasoning shatters into a multitude of shards, as it did in the case of Laocoon. Would there be one reason for humans, another for monkeys, and yet another for mollusks? The difficulty is tied to the explanatory power of psychologism, and its application in evolutionary and physiological thought to reduce thinking to processes inside animal minds.

Darwin's doubt gets a cheerful twist in William James's *Principles of Psychology* (1890). There we are told that what matters in thinking is the starting point and conclusion, illustrated by a bundle of wayward paths:

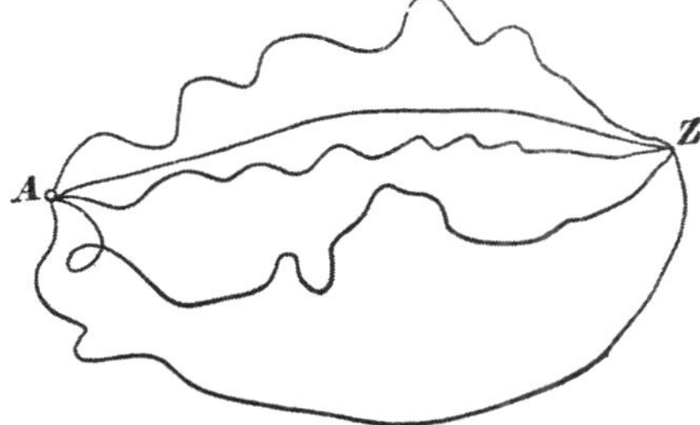

> Let A be some experience from which a number of thinkers start. Let Z be the
> practical conclusion rationally inferrible from it. One gets to the conclusion
> by one line, another by another; one follows a course of English, another of
> German, verbal imagery. With one, visual images predominate; with another,
> tactile. Some trains are tinged with emotions, others not; some are very
> abridged, synthetic and rapid, others, hesitating and broken into many steps.
> But when the penultimate terms of all the trains . . . finally shoot into the same
> conclusion, we say and rightly say, that all the thinkers have had substantially
> the same thought.[24]

Such biological relativism might seem harmless: let a hundred psychological flowers bloom, one might say. But it has its dark side,
both scientifically and politically. Some of its problematic features
are obvious in Darwin himself, who in the above-cited letter went
on to predict that "looking to the world at no very distant date . . . an
endless number of the lowest races will have been eliminated by the
higher civilized races throughout the world."[25] That the differences
James thought harmless could be marshaled as evidence to justify
social injustice was a fact not lost on Christine Ladd [Franklin], a
public school teacher, psychologist, and logician who completed the
requirements for a PhD at Johns Hopkins under Charles Sanders
Peirce in 1883, a degree granted to her only forty-four years later. In
1890, Ladd reflected on the dichotomy between deductive and intuitive reasoning, a question that "would have no more interest for the
general public than any other of the subjects which the metaphysician
exercises his ingenuity upon . . . were it not that there is an ancient
opinion to the effect that reason and intuition are marks respectively
of the manner of working of men's and of women's minds."[26] This she
disputed powerfully:

> It is not true that men's minds and women's minds have a different way of
> working; but it is true that upon certain occasions (and by far the greatest
> number of occasions) we all — men, women, and negroes alike — act from
> intuition, and that the circumstances of women's lives have hitherto been
> such as to make their interests lie somewhat more exclusively in those regions

in which conduct is intuitive than in those in which it is long thought out. It is not true that the Creator has made two separate kinds of mind for men and for women; but it is true that society, as at present constituted, offers two some-what separate fields of interest for men and for women, and that the nature of their conduct is of necessity determined by the character of the action which is demanded of them.[27]

Ladd's refutation of "feminine intuition" reads remarkably like Griselda Pollock's insistence a century later that in feminist art his-tory "difference is not essential but understood as a social structure which positions male and female people asymmetrically in relation to language, to social and economic power and to meaning."[28] The breadth of such questioning of empiricism in the late nineteenth century has been underestimated; it is the aim of this book to show that it is central, but not exclusive to, the development of symbolist art. Beyond establishing this matter of historical fact, the book aims to show the *utility* and *rightness* of this rationalist critique. Without working out the logical bases for our shared aesthetic, scientific, moral, and political projects, I do not see how we will overcome the tribalism overtaking twenty-first century life.

New Symbolisms

So far, I have tried to make plausible the hypothesis that the present crisis of the humanities, particularly those disciplines concerned with interpreting aesthetic objects, is akin to that nineteenth-century pre-dicament of the sciences whereby an exclusive emphasis on human cognition leaves the object of interpretation more mysterious and indeterminate than before, if it does not result in the outright impos-sibility of knowledge. In order to see how a fresh look at symbolist art can shed light on the state of the humanities today, it is useful to begin with the word, and those artists to whom it was first applied.

The very name, unlike that of the avant-gardes following in its wake, is retrospective; it is meant to recall the "forest of symbols that regard [man] with a familiar gaze" in Charles Baudelaire's 1857 poem "Correspondances."[29] This historical open-endedness will prove important in the next chapter. In any case, the term was first consistently applied in the manifesto *Le Symbolisme*, published in the Parisian daily *Le Figaro* on September 18, 1886, by a young admirer of Stéphane Mallarmé, the poet and critic Jean Moréas.[30] His account of what goes on in symbolist art was quite psychological: "The conception of the symbolist novel is polymorphous: sometimes a unique personage moves through milieus deformed by his own hallucinations, his temperament: in this deformation lies the sole *reality*."[31] Milieus — or media, *milieux* meaning both things — deformed by subjectivity, where that deformation, being objective, is the only thing *réel*: this is an analysis of subjectivity, not a flight from reality.

Symbolism as a subjective deformation of the objective is a good starting point for this study, but it is too vague to be of much immediate use.[32] How did symbolism differ from previous schools of art? Like his master Mallarmé, Moréas rejected Émile Zola's naturalist theory of art. But a more intimate and telling disagreement concerns the poets called Parnassians, in whose journal *Le Parnasse Contemporain* the leaders of the symbolist school, Paul Verlaine and Mallarmé, first published.[33] In a quite frank interview published in the newspaper *L'Écho de Paris* in 1891, which Mallarmé went on to incorporate into his collage-like theoretical text "Crise de vers," the poet drew a sharp line between symbolists and Parnassians:

> I think that at bottom the young [poets] are closer to the poetic ideal than the Parnassians, who still treated their subjects in the manner of the old philosophers and rhetoricians, presenting the object directly. I think that, on the contrary, there should be nothing but allusion. The contemplation of objects, the image taking flight on the reveries they bring about, these are [what makes] song: the Parnassians, they get hold of the thing as a whole and show it; because of this they lack mystery; they withdraw from the minds [of readers] this delicious joy of believing that they believe. *Naming* an object, that is suppressing three quarters of the delight of the poem, which is made of the joy of becoming a little at a time; to *suggest*, that is the dream.[34]

This celebrated interview, which says so much about both the symbolists' method and their hopes for a mentally alert public, has come to take on mythic proportions in the history of criticism: we inherit a triumphalist literary history according to which the Parnassians naïvely filled their pages with the names of things, hoping to summon them to life, while the symbolists understood that one may evoke the absent only by linguistic means. As Mallarmé had put it in his wry preface to René Ghil's *Traité du verbe* in 1886: "I say: a flower! And out of the oblivion to which my voice relegates every contour, which is something else than those chalices: up there rises musically an idea riotous or lofty, the Absent (*l'absente*) of all bouquets."[35] This intoxicating prose takes some deciphering: speech dispenses with our sense memories of specific flowers ("every contour"), so that in their stead the concept arises, which is radically different from, but logically complementary to (hence "the absent of") every collection of flowers. Even here, Mallarmé is elitist enough to distinguish the literal usage of the crowd, "facile and representative," from the "incantatory" art of the poet. A decade later, in "Crise de vers," this Platonist-sounding theory of meaning and a snobbish quest for purity had hardened into the slogan "hors du toute pierre," "outside (or away from) every [precious] stone." One may capture "on the subtle paper of the volume for instance the horror of the forest, or the mute thunder scattered among the foliage, but not the intrinsic and dense wood of the trees."[36]

This point is at bottom a reasonable one; as Mallarmé put the matter drily to his friend Degas, who was trying to write sonnets and finding the process very troublesome, "you can't make a poem with ideas; you make it with words."[37] You can no more make a poem with stones or pieces of wood than with ideas; language is the poem's body. But the theory of meaning in "Crise de vers" is not quite such bracing commonsense. Emotions like horror but also sensations like the "mute thunder" of the forest are supposed to be directly felt in writing, whereas the wood of the trees is not. Mallarmé seems tempted by the very mistake he imputes to the Parnassians of wanting to provide the thing itself in poetry by naming it. It is as if by a limitation of the subject matter of the poem to "the immediate givens of consciousness," as Henri Bergson would call them, he might bring forth *actual* fear or thunder, if not actual wood. Against it, we must insist, as Mallarmé did, that it is with *words*, whatever they mean, that one makes poems.

And one paints with paint. The lesson is valuable, but what it is not is a psychological recipe for making art. If we took it thus, we would arrive at the complacent view according to which symbolism is a catalyst for the development of abstract art, understood in turn as a vain exercise in serving up arbitrary signs. Symbolism on this influential view is something like a doomed if noble dead end, a falling silent of poetry in its efforts to transcend "the words of the tribe" with their real-life referents.[38] This might seem to do justice to a sonnet like "Le cygne," with its "white agony" that fits just as neatly a dying swan, frozen nature, and the ennui of the poet. But if colors like white and azure are so nimble in Mallarmé, it is thanks to their conceptual clarity and force, asserted over mere material bodies, and the change that haunts them.[39] The degree zero of literature is not silence so much as logic and sensuousness taken together.

As we have seen, Mallarmé might deserve some of the blame for the quietist reading of his theories and poems, but there is another, more charitable way of looking at both. As his friend and early interpreter Téodor de Wyzewa observed of his Parnassian poems, they are not distinguished by original imagery, nor are they musically on a par with "that native guitarist Verlaine." It is rather by the "conscious logic" with which Mallarmé builds up a theme, "with — but with nothing besides — its necessary expansion."[40] Wyzewa shows this, "indiscreetly," in discussing the poet's later marginal annotations to a printed copy of the early sonnet "À celle qui est tranquille." Where the "banal theme, borrowed from Baudelaire" ran, "And I am afraid of thinking when I lie alone," the mature poet put instead: "And I am afraid of dying when I lie alone."[41] Typically Mallarméan is not the melodrama, but rather the conscious development of a theme, which in his later poetry led to frequent charges of aloofness and abstruseness, charges that Wyzewa doesn't duck in speaking with tongue in cheek of the early period as "one of comprehensibility." There is here a real but gradual break with the Parnassians. Those poets wanted words to evoke objects vividly, like Gautier's "verse, marble, onyx, enamel."[42] Such poetry is in its economy itself richly suggestive of mood. Mallarmé updated Gautier's list in his "Salut" with his *"solitude, récif, etoile,"* retaining Parnassian conciseness, but stretching from hard material objects like the reef to states of mind and things at once material and ideal — the star. Symbolism does not aspire to a condition of music, but to analyze the conditions of meaning:

> *Previous poets made a pure music, seductive in itself:* M. Mallarmé thought that poetry should express something [or *some thing*], create an entire mode of life. To this new destination new means were suited: M. Mallarmé was thus led to consider what [things] poetry should signify, and by what means.[43]

The fundamental question of what means are necessary to express whatever art can express is not unique to poetry. How far symbolist art extends beyond poetry, and writing in general, is another question.

Symbolism and Impressionism

Symbolism in the visual arts is thought to be a creature of the 1890s, and something of a chimera: crowded between impressionism, or neoimpressionism, on the earlier end and the short-lived coteries of the nineties, from Maurice Denis's synthetism to Paul Sérusier's *Nabi* (Hebrew for prophet) group, which supplanted it, and whose members called themselves symbolists or collaborated with symbolist writers, notably at the Théâtre de l'Art and Lugné-Poe's Théâtre de l'Oeuvre.[44] The would-be symbolist painter might be imagined staring wistfully at the passing avant-garde pageant much like Munch's balcony-dweller, the hubbub of Paris swirling past, leaving the outsider with little but the shadow of the railing. That said, the groups did mix: the gallerist Le Barc de Boutteville held fifteen exhibitions of "Peintres Impressionnistes et Symbolistes" between 1892 and his death in 1897.[45] The catalogue writer of the second of these, Gabriel-Albert Aurier, complained about the excess of "-isms" in an article on Paul Gauguin in the *Mercure de France* for March 1891, notably called "Le Symbolisme en peinture":

> The public has, in sounding this word "impressionism," the vague notion of a program of some special realism; it expects works that are only the faithful transcript *with nothing besides* of an *exclusively sensorial impression*, of a sensation. If then, by chance, among the heterogeneous group of independent painters labeled thus there are found some artists engaged in different modes of art, contrary ones, the good public, that eternal and blessed worshipper of catalogues, will evidently not fail to lose its Latin, as they say, and already I see it shrugging its omnipotent shoulders and sneering: "What an idiot! . . . This impressionist paints me impressions that nobody could ever have felt!"[46]

Though gratuitous for the artist, a new term is needed to free the public of its preconceptions. Synthetist, ideist, symbolist — it's all the same to Aurier, as long as any misleading appeal to the realistic transcription of sensations is dropped.[47] Truth to sensation results in solipsism: "*the imitation of the material reality of things, as that reality is perceived by the divers temperaments of artists, presupposed to differ to infinity.*"[48] Who is to say what we are looking at in such a romantic chaos of competing views? A kind of "rudimentary" symbolism is required to make art intelligible, so that "there is never Art without symbolism."[49] Not that the symbolist painter is cut off from the world; but the reality that Moréas saw parsed through a subjectivity must, in effect, be reinterpreted through one: "This reality, which is supposed to be deformed by a temperament in the genesis of a work of art, what is, in the final analysis, the realized work, if not a *visible sign* of that temperament, what, if not a symbol of that temperament, the symbol of the eidetic and sensitive ensemble of the worker?"[50]

Symbolism's peculiarly international appeal, from Scandinavia to Russia to the Americas, is explained in part by this self-critical program. One did not need to study plein-air painting in Paris to participate in the new movement; it was enough to reflect critically on the psychic mechanisms that made such exercises of skill possible and apply them to a variety of subjects, from Mallarméan inner drama to the panoply of world myth and folk traditions. Doing so would turn into visible, intelligible signs the subjective distortions of reality that Aurier finds in all art. The growing legibility of these private experiences as signs of something else — the artist's awareness of sensation as a medium through which the natural world of objects first appears to thinkers, rather than being merely relayed — is a transposition of Mallarmé's account of symbolist verse to the more mimetic realm of painting. Represented objects, to avoid ambiguity between their nature and the artist's contribution, must "appear to him as *signs*. They are letters of an immense alphabet which genius alone knows how to enumerate."[51]

At his most bombastic, Aurier, like Nietzsche in *The Birth of Tragedy*, called for genius to pierce the shadows of the cave and espy the essences lurking in the forest of symbols.[52] Yet these symbols are not transcendent talismans or supersigns, but slivers of the visible reality of objects, selected and combined in ways that are conceptually articulate. As Antonin Proust would recall Manet exclaiming: "The Christ on the Cross, what a symbol! One could ransack the centuries without finding its equal."[53] This understanding of the symbol as a concept made visible is down-to-earth indeed compared to the romantic desire to express the inexpressible.

Aurier's vision of symbolism in visual art accorded well with the practice and occasional theoretical pronouncements of Paul Gauguin, Paul Sérusier, and Maurice Denis.[54] But more interesting is the friction that at times resulted between the most creative visual practitioners and the critic's pronouncements. Aurier's first manifesto was in fact his monographic article on Vincent van Gogh, under the romantic rubric "Les Isolés," "the isolated ones," in the January 1890 *Mercure de France*.[55] In the first printed response to Van Gogh, Aurier hailed the painter's colorism and his respect for "the reality of things," yet insisted that the use of recurring figures like the sower, the sun, and its botanical double the sunflower makes him a "symbolist who feels the continual need to clothe his ideas in precise, ponderable, tangible forms, in intensely sensual and material envelopes."[56] The childlike outline around the sun in Van Gogh's drawings seen by Aurier seemed to favor such intellectual emphasis (Figure 1.6).

"How else to explain [this] if one refuses to admit his persistent preoccupation with some vague and glorious heliomythic allegory?"[57]

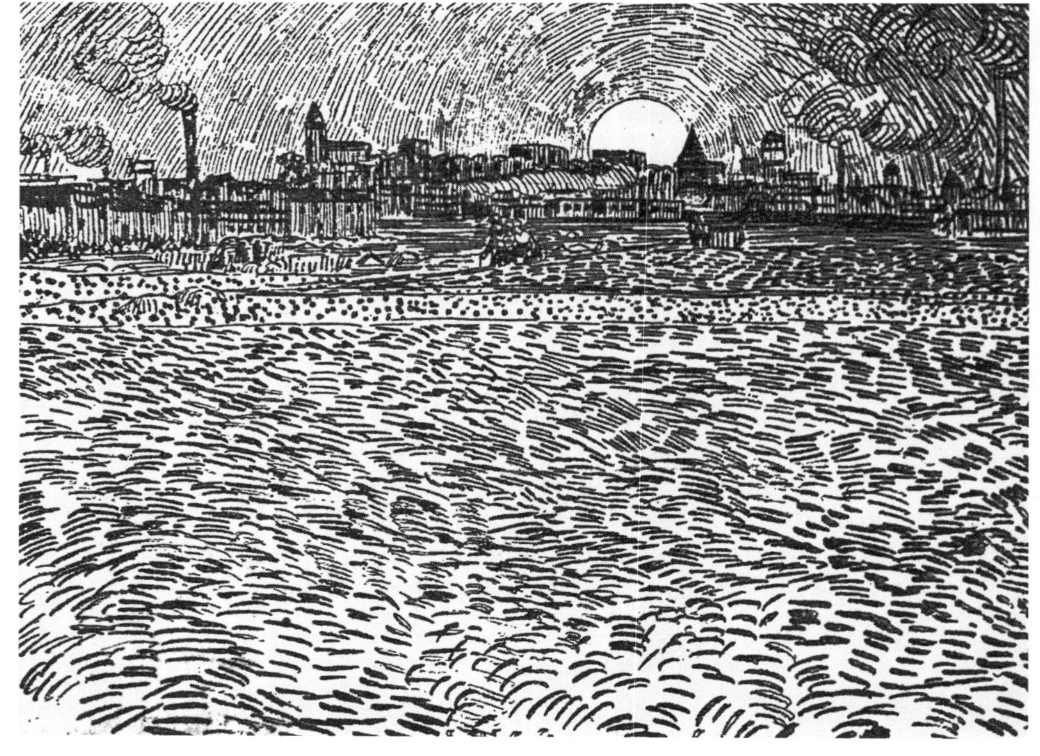

Figure 1.6. Vincent van Gogh drawing reproduced in G.-Albert Aurier, *Oeuvres posthumes* (1893), p. 203. A linear sun can also be found at p. 291, in a drawing by the symbolist Jeanne Jacquemin.

The art historical literature has not caught up with Aurier's question, but, strikingly, Van Gogh himself, then recovering from his psychotic episode in the St. Paul Asylum in Saint-Rémy, wrote Aurier one of his typically courteous, probing letters, modestly giving credit for his colorism to Delacroix and Monticelli, praising Gauguin and the academic painter Meissonier, whom Aurier had savaged.[58] Yet he was most exercised by Aurier's thesis concerning the symbolist unity organizing vision in his paintings:

> And then there is another question I want to ask you. Suppose that the two pictures of sunflowers, which are now at the Vingtistes' exhibition, have certain qualities of color, and that they also express an idea symbolizing "gratitude." Is this different from so many flower pieces, more skillfully painted, and which are not yet sufficiently appreciated, such as "Hollyhocks," "Yellow Irises," by Father Quost? The magnificent bouquets of peonies which Jeannin produces so abundantly? You see, it seems so difficult to me to make a distinction between impressionism and other things; I do not see the use of so much sectarian spirit as we have seen these last years, *but I fear the ridicule.*[59]

Van Gogh's subtle reservation is well-motivated: Aurier's insights, consistently applied, would certainly have cut across categories like impressionism and academicism. Nor is the most fruitful distinction a superficial sociological one between painters who signed one or another manifesto, between symbolists and Parnassians or impressionists. It is rather a matter of how one understands — and uses — the subjectively grasped external reality that is the main business of modern painting.

Symbolism and Modernity

To see where the forces of critique in symbolism tend, we must remain with Aurier for a moment. The first page of "Les Peintres symbolistes" rings with an epochal challenge:

> After having proclaimed the omnipotence of scientific observation and deduction for eighty years with childlike enthusiasm . . . the nineteenth century at last seems to perceive that its efforts have been in vain, and its boast puerile. Man is still walking in the midst of the same enigmas, in the same

formidable unknown, which has become even more obscure and disconcerting since its habitual neglect. A great many scientists and scholars today have come to a halt discouraged. They realize that this experimental science, of which they were so proud, is a thousand times less certain than the most bizarre theogony, the maddest metaphysical reverie, the least acceptable poet's dream, and they have a presentiment that this haughty science which they proudly used to call "positive" may perhaps be only a science of what is relative, of appearances, of "shadows" as Plato said, and that they themselves have nothing to put on old Olympus, from which they have removed the deities and unhooked the stars.[60]

The confrontation between positivist science and a range of human phenomena, from poetry to religion, is brought onstage here as dramatically as in the works of Nietzsche; equally familiar is Aurier's prophecy, in the next paragraph, that when the crisis comes, people will let go of "their catalogues and their algebras" to welcome poets and dreamers back to the polis. This sounds like the sort of Technicolor battle of good versus evil broad enough to admit on the side of the good all who are opposed to materialism, from idealists and mystics to theosophists and all manner of esoteric seers, in and out of art. Aurier himself was concerned with rehabilitating painters who had little in common theoretically, from Van Gogh and Gauguin to Monticelli, Puvis de Chavannes, Eugène Carrière, and J.-J. Henner. These were symbolists in the loose but politically and culturally significant sense that they rejected the positivist fixation on sensation as the sole concern of art. As Aurier was aware, however, he could not count on consensus even among his core of symbolist heroes, with Van Gogh expressing doubt about the wisdom of going beyond observation to paint "something like a music of tones," and calling Gauguin's use of traditional Christian themes "nightmares," "appalling" and "spurious."[61] It is wiser to take a step back and ask why science as presented here by Aurier should be opposed to symbolist art. Certainly his enthusiasm for Platonic philosophy indicates that it would be a mistake to lump Aurier's critique of positivism with any fashionable rejection of reason.[62]

As a matter of fact, there is a work of Aurier's, unpublished in his lifetime but printed in the *Oeuvres posthumes* of 1893, that clarifies his position on science, qualifying the demagoguery of "Les Peintres symbolistes." In the manuscript of his "Essai sur une nouvelle méthode de critique," which contains acute comments on Taine's *Philosophie de l'art* among other works by "materialist" authorities, Aurier is more circumspect in his critique of scientism:

> The peculiarity of the nineteenth century has been to wish to introduce sci- ence everywhere, even among those matters where it has least business; — and when I say "science," one should not understand mathematics, the only science properly speaking, but these obtuse bastards of science, the natural sciences. For the natural, or inexact, sciences, in contrast to the rational or exact sci- ences, being by definition insusceptible to absolute solutions, lead fatally to skepticism and fear of thinking. It is then right to accuse them of having made for us this society without faith, with its feet dragging on the ground (*terre à terre*), incapable of those thousands of intellectual or sentimental manifesta- tions which may be classed under the name of devotion.[63]

The distinction between mathematical and natural sciences makes better sense of what Aurier himself excused as the "rebarbative jar- gon" of "Les Peintres symbolistes."[64] Why claim that the best results of science are less certain than a theogony? Well, because by their own account the experimental sciences don't aim at and cannot guarantee certainty; the radical doubt and nihilism that rush in to fill the vacancies thus opened we have already examined in Darwin and Stricker. More interesting than this nostalgic, backward-look- ing strand in Aurier's cultural criticism, one shared with writers as diverse as Rainer Maria Rilke and Matthew Arnold, is the diagnosis of an alliance between art and the empirical sciences, with their a priori lack of absolute solutions, as the problem for culture. He does not pursue, like some early twentieth-century art theorists, a mysti- cism of art and mathematics. Although it is possible that he would have taken this familiar route had he lived to see the abstract painting of the second decade of the twentieth century, Aurier's own focus on logical analysis and purification of the artist's (subjective) means of communication fit far better with his somewhat earthy Platonism, according to which humans themselves are pale shadows of their Ideational selves, lost in the forest of symbols. Indeed, there would not be much moral or political interest in the critique of psychologism if we were not cognitively imperfect beings.[65] Our difficulties in making our own ideas clear to ourselves, to say nothing of making them clear to others, informs not only the project of philosophical critique Aurier is enthusiastic about, but connects it with the democratic or anarchistic politics of other symbolist critics, notably Gustave Kahn and Octave Mirbeau.[66]

Mallarmé himself had diagnosed the root of modern aesthetic individualism in the social differentiation and growing autonomy of the individual. In his 1876 essay on impressionism, which I will discuss at more length in the next chapter, he declared that "today the multitude demands to see with its *own* eyes"; by the time of his 1891 interview with Jules Huret he had abandoned all rhetoric of "the masses" and come to regard poets above all as social individuals, driven away from traditional rhyme and meter by the demands of a liberal modern society.[67]

The temptation to reply that, rather than destabilizing art, society ought to stabilize, can be felt in Aurier's complaints of the emptiness of the heavens and the death of faith. But for every Maurice Denis, who was a Catholic tertiary and flirted with the ultranationalist political organization l'Action Française, there were symbolists who agreed with Mallarmé that the modern world could not be reenchanted.[68] Aurier's more fine-grained critique of empirical science and its culture of positivism is a better guide to the relevance of symbolism today; after all, he did not become the priest in a homemade religion, as did some contemporaries like Joséphin Péladan, but sought to correct positivism through art and reflection on it.

Symbolism and Science

Aurier makes it plain that his critique of scientific skepticism and the nihilism that he believes follows from it does not apply to mathematics, "the sole science if we were to speak properly." His rhetoric and choice of epigraphs and metaphors (the cave, shadows, the "radiant heaven of Ideas") suggests that his authority for the critique of empiricism is no more recent than Plato.[69] As Albert Boime showed in his book on *Thomas Couture and the Eclectic Vision* (1980), the watered-down Platonism of academician and Plato translator Victor Cousin amounted to an official philosophy in nineteenth-century France, one well suited also to the continued primacy of ideal bodies in French painting and sculpture of the "*juste milieu*" (the "golden mean" suggests rather an Aristotelian tendency to this current of thought). Political and aesthetic radicals, it is true, were attracted rather to Proudhon and Comte, but in the context of a fin-de-siècle revolt against positivism, it is not surprising that an idealism only recently abandoned would inform the language of the critic, especially one, like Aurier, who had only recently finished his schooling.[70]

That is not all that is at work here. It was typical of the "eclectic" philosophy of Cousin and mid-nineteenth century France generally to try to reconcile natural science with the traditional trinity of "the true, the beautiful, and the good." Aurier's rationalism was of a more uncompromising stripe. In drawing a distinction between "objective ideas" and the "subjective ideas" through which the former "deform" the artist's perception of reality, he observes: "But to see this one must have a less materialist conception of the world, and not prefer Auguste Comte and Condillac to Plotinus or Plato."[71] Passages like this suggest that Aurier's Platonism is informed by actual developments not only in art but also in the philosophy and *science* of the second half of the nineteenth century. This development informed the theoretical work of French theoretical and experimental scientists. Thus the eminent physiologist Claude Bernard, justifying the pain inflicted on dissected animals: "The cowardly assassin, the hero and the soldier alike plunge the dagger in the heart of their fellow. What distinguishes them but the idea which directs their arms? The surgeon, the physiologist, and Nero alike mutilate living things. What distinguishes them, again, if not the idea?"[72] One could ask the same question of a striking image out of the portfolio of fifteen *Essays with X-rays* (*Versuche mit Röntgen-Strahlen*) published

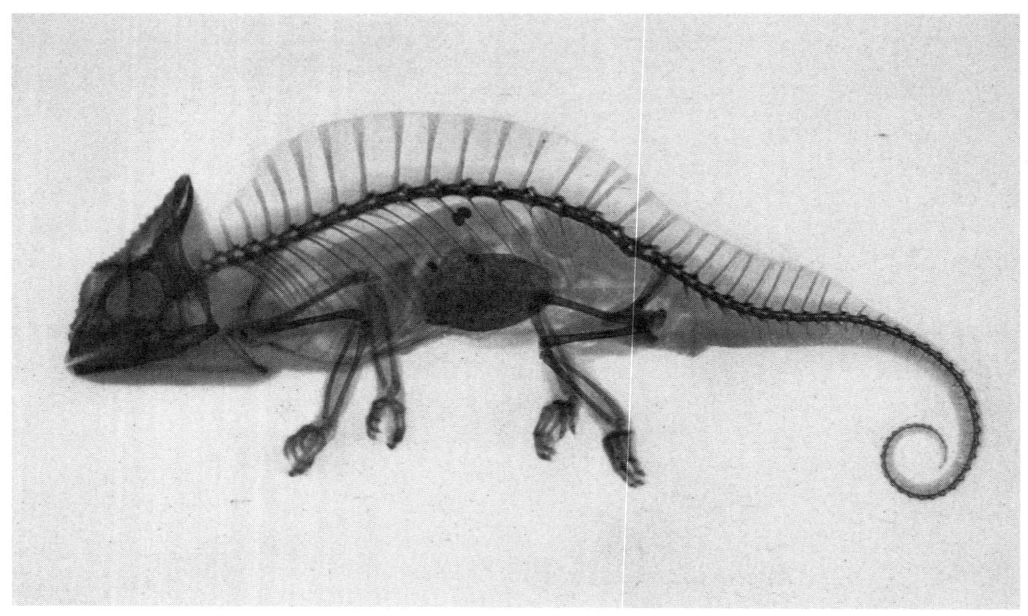

Figure 1.7. Josef Maria Eder and Eduard Valenta, photogravure from *Versuche mit Röntgen-Strahlen* (1896), Metropolitan Museum, New York. Four of the fifteen x-rays in Eder's portfolio were printed as positives. The dark particles in the chameleon's digestive track suggest it was still alive when exposed.

by Josef Maria Eder less than a month after the first publication concerning the newly discovered form of radiation.[73] Eder's mad rush to publish, and the odd mix of human, zoological, and artifactual samples scanned — besides human limbs and metal samples, three ancient cameos were also x-rayed — certainly bear witness to the breakthroughs in physical science and imaging technology of the fin de siècle. But foremost in the work's pristine aesthetic presentation, especially impressive in the positive prints (Figure 1.7), which clearly manifest various textures for various kinds of bone and soft tissue, is a conceptualist delight in finding in nature the order one had already learned to look for: equipping the fallible human sensorium with the tools to visualize Bernard's "animating idea."

An empirical Platonism of this passionate character did not, of course, remain without influence in the daily life of a rapidly industrializing France and Britain. But its stakes were perhaps clearest in the industrially less developed German and Habsburg empires. Indeed, the transformation appeared particularly sweeping to scientists working there:

> Material and intellectual intercourse is a sign of our time. Lands that until now stood remote from one another exchange their products by railroad and enter through the telegraph into rapid exchange of ideas. Sciences that have developed independently from one another, indeed to some extent have regarded one another with hostility, begin to take notice of one another, to intervene mutually and supportively in one another.[74]

The interdisciplinarity envisioned by Ernst Mach, to which he himself contributed in this first book by giving a physical interpretation of Hermann Helmholtz's psychological music theory in a popular style meant to be accessible to musicians, involved in the main the applied sciences, which had made prodigious advances in the quality of experimental equipment: optics, to which both Helmholtz and Mach contributed, is typical for the way it benefited from the contribution of physics, geometry, psychology, physiology, and even medicine.[75] It might seem to modern readers that the lingua franca of such work would be mathematical, but empirical researchers of the nineteenth century were suspicious at least of the a priori claims of traditional metaphysics, logic, and mathematics. Helmholtz, in a quite personal 1877 lecture "Thought in Medicine," wistfully recalled his own midcentury training as a surgeon, and how he had earned his doctorate with a lecture on the swelling of blood vessels, though he had never seen one operated on, much less operated himself.[76] But disaster intervened, and, being ill with typhus in Berlin's Charité Hospital in summer 1841, he was able to save up enough of his stipend to buy a microscope.[77] His career as an experimentalist took off from there, but Helmholtz's point was another; it was the dominance of theory, with its deductive method and its attribution of all human illness to one master principle, whether it be the life force (*vis vitalis*) or its modern replacement, nerve irritability (*Reizbarkeit*), that retarded the understanding of disease and treatment.[78]

Helmholtz's point was a sound one. The old theoretical pathologists "forgot that every deduction is only as certain as the premise from which it is deduced."[79] But his conclusion from this failure was sweeping: no science, not even Kant's rock of certainty, geometry, stood immune to experience, which will "confirm its axioms or perhaps disprove them."[80] Emboldened by non-Euclidean geometries studied by Gauss, Bolyai, Lobachevsky, and Riemann, Helmholtz went on to reduce the last bastion of abstraction, the numbers, to empirical laws of counting: "I regard arithmetic, or the study of pure numbers, as a method built upon purely psychological facts, teaching the logical application of a sign system (namely the numbers) of unbounded dimensions and unbounded potential for refinement."[81] This sign system, as we saw, is manipulated according to empirical laws: "*Counting* is a procedure that consists in the fact that we find ourselves capable of retaining in memory the sequence in which the acts of consciousness have followed one another."[82] There is no need to go into details of this arithmetical psychologism, but one thing should be clear: it pulls out the rug from under Aurier's distinction between (certain, rational) mathematical science and uncertain, ragged empirical science. Mathematicians, themselves caught up in a nearly century-long project to clarify and make rigorous the bases of their discipline, did not take the proposed reduction lightly: among other things, psychological acts could not account for infinite numbers, which no mortal mind would ever finish counting.[83] Georg Cantor, who first made the study of infinite numbers rigorous, in the process inventing modern set theory, was particularly critical of the empiricist tendency.[84] Pointing out its intolerance not only of the actual infinite but also of the "irrational numbers, recognized since Pythagoras and Plato," he made a prophecy: "So we see in Germany the currently dominant and powerful *academic-positivist skepticism*, which arose in reaction to the overextended idealism of Kant-Fichte-Hegel-Schelling, finally reach arithmetic too, where with utter and perhaps self-dooming consistency it draws the final conclusions available to it."[85]

Radical empiricism is self-defeating in its attempt to explain the most abstract and general laws of nature, number, and thinking. The assimilation of all theory to fallible experimental science was a matter of controversy in the late nineteenth century. In opposing it, Aurier and other artists achieved more than irrationalism: like Cantor or Frege, they wanted a sane apportioning of the domains of logic and subjectivity in the task of representation. That is the link between science, art, and even mathematics, and it is what made them such fruitful domains of philosophical reflection at the fin de siècle. In pursuing it, this book cannot rest content with the received history of science any more than that of art. Rather, it is their practice, especially their *pictorial* techniques, which will allow us to see commonalities among these realists. Cantor, for instance, did not just argue against empiricism or devise marvelously creative and speculative proofs, but printed with them images making intuitive how the two types of numbers he had distinguished, cardinal (one, two, three) and ordinal (first, second, third), may come apart. The arrangement of points in a two-dimensional matrix *shows* how different sets are derived from a certain sum of elements (Figure 1.8).

This isn't the symbolism of Mallarmé or Van Gogh. But like theirs, Cantor's image radically, and precisely, reconfigures the world through careful attention to the means of representation.

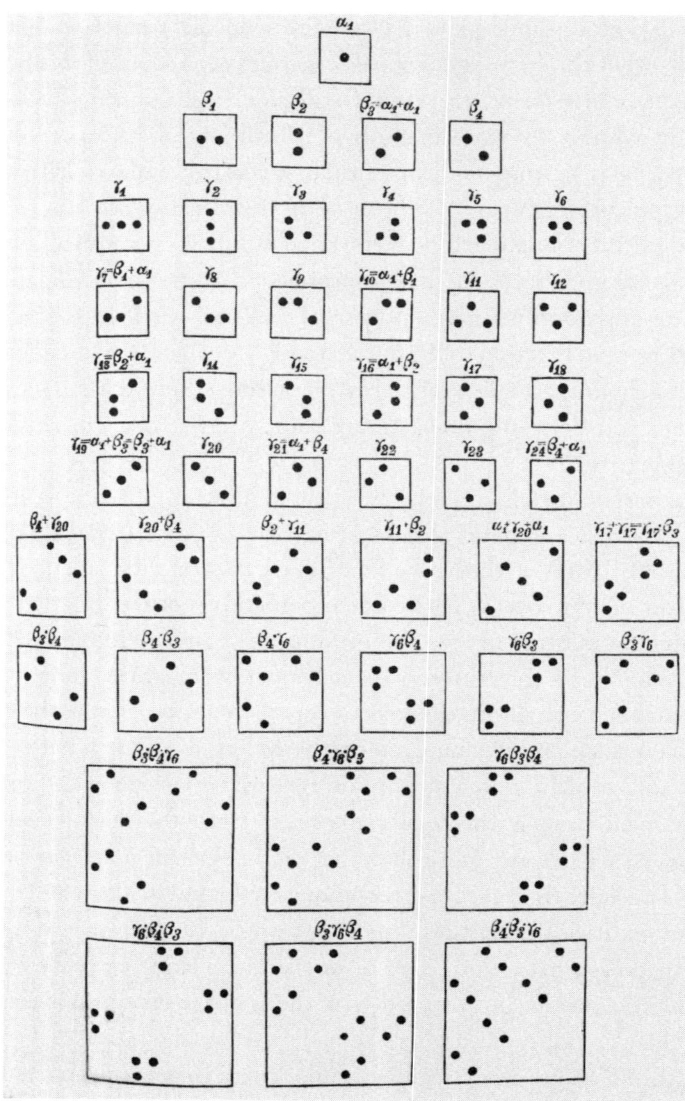

Figure 1.8. Diagram by H. Wiener in Georg Cantor, "Mitteilungen zur Lehre vom Transfiniten," *Zeitschrift für Philosophie und philosophische Kritik* 92 (1888), p. 255. The patterns at the bottom of the diagram are obtained by a process of combination of those at the top.

The Plan of This Book

To do justice to the variety of efforts to chart how meaning is possible in science and art, this book treats the work of several figures in detail, tracking problems and their attempted solution rather than sticking to biography and chronology. The first and most urgent question concerns the characteristic nineteenth-century division between subjective and objective, which persists in the endless debate between humanities and "hard" science. The second chapter attacks this question by inquiring whether there is any symbolist *method* in the arts, through a work of the 1870s that came to be regarded as a symbolist paradigm: Mallarmé's translation and Manet's illustration of Edgar Allan Poe's poem "The Raven." The difficulty of sharing subjective experience is not solved, but elegantly circumscribed in this complex artwork, made over a half century by an American poet, his French translator, and a painter associated with realism and impressionism, and their eventual rejection.

The third chapter reconstructs how art and thought first got into this impasse of the subjective and objective. Beginning with the phenomenon of the ineffability of color, which much fascinated the symbolist generation, from the mathematician Frege to the painter Van Gogh, the chapter traces the tradition of color subjectivity from symbolist monochromes to romantic ideas of ineffability to realist and impressionist efforts to paint, print, and photograph the world exactly as an individual subject perceives it. The paradoxical nature of such simulation, which *doubles* subjective experience in striving to capture it, is exposed.

The fourth chapter tackles this paradoxical mode of representation head on, in the form of "first-person" pictures made by impressionists like Caillebotte, symbolists like Redon, popular artists like Winsor McCay, and scientists like Ernst Mach and William James. Exploring the role of fiction in representation, and the threat that fiction poses to meaningful language and imaging, the chapter turns to a philosophical theory of pictures due to Frege and Wittgenstein, according to which there are logically articulated sensuous objects (pictures and writings and other works of art, e.g., sculpture and music) through which we can gain a shared understanding of the world and all of its strange denizens, from subjective experiences to fictions and mathematical entities.

The fifth chapter explores the consequences of this picture theory in symbolist and postimpressionist art, but also in the "pointillist" philosophy of Ernst Mach, William James, and Bertrand Russell. The reconciliation of logic and empiricism in their work, as well as in the canvases of Seurat and the practice of early film, is an important but neglected late achievement of symbolism. A brief conclusion returns to the wider stakes of concern with truth and objectivity.

CHAPTER TWO

Crises of Sense:

The French Take on Edgar Allan Poe

In 1889, the publisher and bookseller Léon Vanier, acting on the hunch that symbolism "is, with romanticism, the most serious manifestation of art in the nineteenth century," collected a bunch of Jean Moréas's 1886 manifestos and the counterblasts they provoked under the martial name of *The First Weapons of Symbolism*.[1] In the back of the book, a mail-order form invited readers to stock up on "symbolist and decadent publications" sold at "Bibliopole" Vanier's shop on the Quai Saint-Michel: among grab-bag offerings like Felix Fénéon's impressionist reviews and Plowert's *Petit glossaire des décadents et des symbolistes*, one could buy poetry by Jules Laforgue, Arthur Rimbaud, and especially Paul Verlaine and Stéphane Mallarmé. The latter was represented by an eclectic array of texts: the essay on his poetry by Téodor de Wyzewa, his translation of James McNeill Whistler's lecture on art for art's sake, his "Afternoon of a Faun" with decorations by Manet, available on Holland or Japanese paper, and an expensive (one-hundred-franc) edition of poems photolithographed from the autograph. This boasted a lurid frontispiece by Félicien Rops (doubling as an *ex libris*), showing the poet's disembodied hands strumming the harp of a nude muse, the classical ideal warring with a worn, gritty, sexually explicit reality (Figure 2.1). Almost lost among these appealing choices, Vanier also offered Mallarmé's prose translations of the entirety of Edgar Allan Poe's poems, "with portrait and

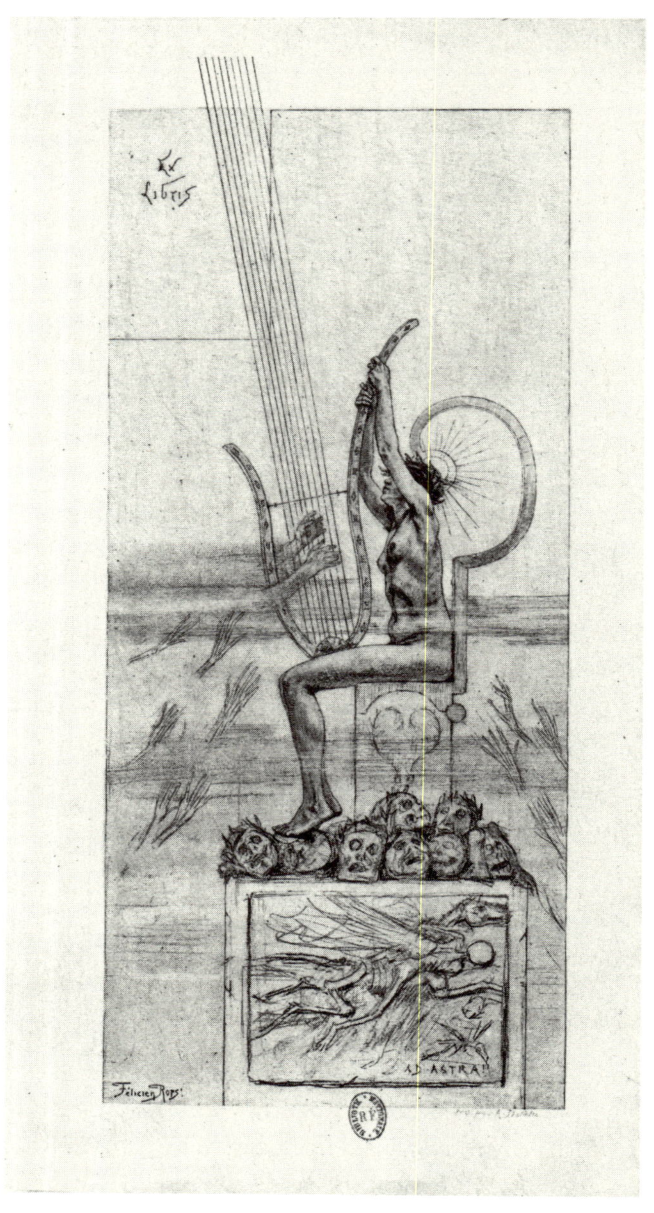

Figure 2.1. Félicien Rops, *Stéphane Mallarmé*, etching, in *Les Poésies de Stéphane Mallarmé* (Paris: Éditions de la Revue Indépendante, 1887). Note the sunken belly of the muse, and, on the frieze, the precariously mounted skeleton rider with his motto "To the stars!" It is as if Rops wished to combine ideal, reality, and obscene anti-ideal in one emblem.

illustrations by *Manet*, in magisterial octavo," and a large folio edition of *Le Corbeau*, Mallarmé's translation of Poe's "Raven," also illustrated by Manet. Unlike Poe's poems, brought out by Vanier in 1889, and the bevy of other fresh symbolist offerings, the *Corbeau* was already a decade and a half old.

That the Manet-Mallarmé edition of "The Raven," which had sold poorly for its original publisher, Richard Lesclide, would be offered by Vanier (at the original price of twenty-five francs) under the rubric of symbolism is not perhaps the greatest mystery in the annals of publishing. Yet we should pause a moment to let the oddness of the juxtaposition sink in. Poe's original poem, made famous in magazine form in 1845 and hastily collected by Putnam later that year in the collection *The Raven and Other Poems*, was widely read, praised, and ridiculed *before* the middle of the nineteenth century. What should a florid narrative poem, going on forty-five years old, be doing among the work of *enfants terribles* like Rimbaud, Laforgue, and other "accursed poets"? This seems a rusty canon indeed among the burnished new weapons of symbolism.

What's So Symbolist about "The Raven"?

Understanding the symbolist fascination with Poe's "Raven" requires rethinking some clichés about the forward-looking nature of avant-gardes. The symbolist artists were not just introverts: they were what we might call retroverts. The stories recounted by Mallarmé's pupils of his attempting to teach them English by reading Shakespeare aloud are hardly uncharacteristic. Like their slightly older English contemporaries, the Pre-Raphaelite Brotherhood, the symbolists revered masters of allegory like Dante, and myths, ceremonies, and religious rituals taken from a variety of cultures, ranging from China to Tahiti. This nostalgia for the ornate and the formal coexisted with brisk liberal or anarchist political convictions, so it cannot be chalked up to pure conservatism. Like Eliot (one of their great exegetes) half a century later, they thought stealing a better tribute than imitation. This approach to older art and culture is distinctly modern: it envisions the commerce with the past more as a collision with kindred minds situated in alien cultures than a harmonious learning process within a settled tradition.[2] And the tradition that they were most anxious to annex was Edgar Allan Poe's. Not by accident did Paul Gauguin decorate his 1891 portrait of Mallarmé with the head of a raven (Figure 2.2), recalling Manet's design for the book wrappers of the 1875 edition: it lurks behind Mallarmé's skull, as if arising from his murky brain. In the final state, Gauguin darkened the skein of musical notes, or rests, comprising the background, the better to bring out the raven's rapacious beak.[3] (The beak will prove important later.) For the moment, what we want to know is this: Why did Poe play this role for the symbolists? And if it had to be Poe, why, among all his works, "The Raven"?

The choice had little to do with the poem's American popularity, which is hard to square with the myth of the neglected genius.[4] Nor did stories of Poe's rock-'n'-roll-like dissolution — though it might have impressed Verlaine or Rops — cut much ice with the professorial Mallarmé.[5] More compelling was Poe's singling out of "The Raven" as a paradigm of his poetics, and of art in general. A year after the poem's success, Poe printed "The Philosophy of Composition," perhaps the most outspoken aesthetic manifesto of the century, disguised as a line-by-line account of the composition of "The Raven." The theory consists essentially, as a hostile critic put it, in asserting that "a poem is a metrical composition without ideas."[6] This requires, however, a most intellectually poised performance on the part of the poet, who wishes to attain a particular and powerful effect on the mind of the reader. Poe tells with tongue in cheek how, in pursuit of a melancholy tone he came by "ordinary induction" to hit upon the refrain as a particularly felicitous mechanism, and especially on a one-word refrain both sonorous and sad: "Nevermore." To make the extended repetition plausible, he needed a nonhuman speaker: "and, very naturally, a parrot, in the first instance, suggested itself, but was superseded forthwith by a Raven, as equally capable of speech, and infinitely more in keeping with the intended *tone*."[7]

The account continues in this vein, from the identification of

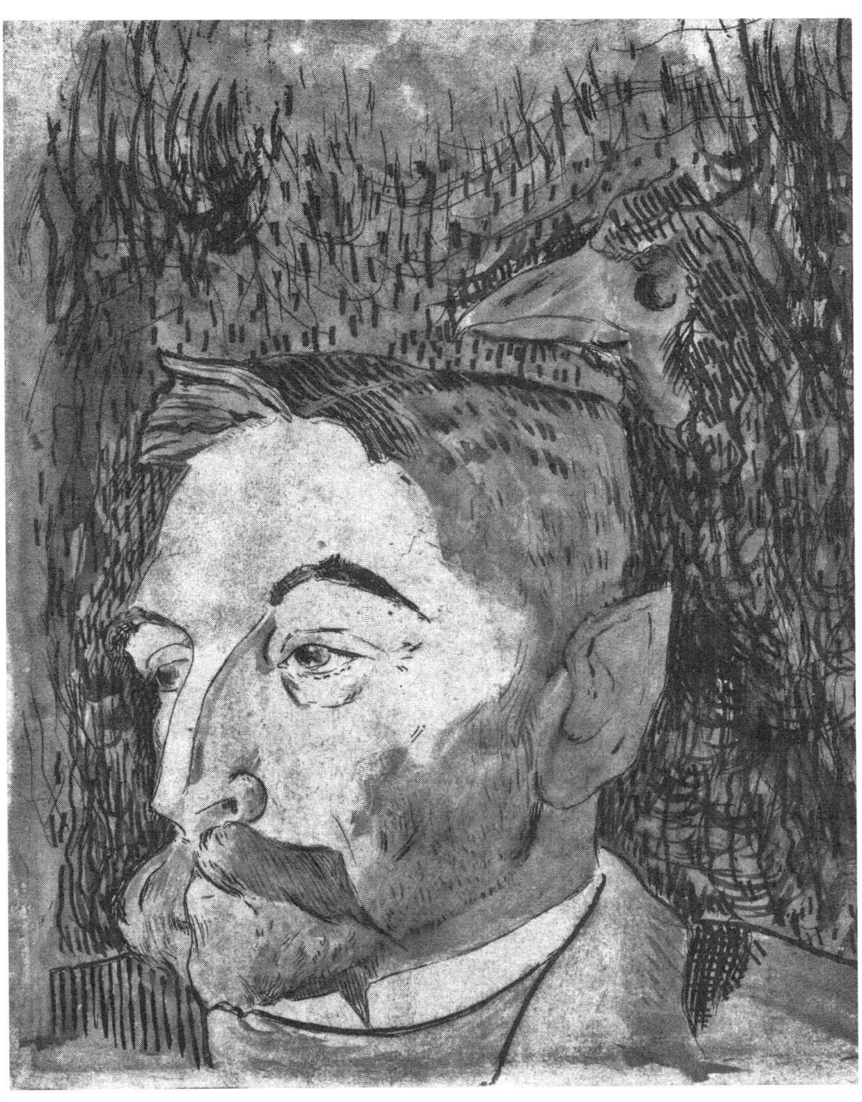

Figure 2.2. Paul Gauguin, *Portrait of Stéphane Mallarmé* (1891), etching with pen and black and brown ink and brush and black and gray wash, first state, Art Institute of Chicago. Note the comic sharpness of the poet's features (the elfin ear) and the strategic positioning of the raven's head just over his own, as if the bird were both figment and totem.

death—especially that of a woman—as the perfect tragic subject matter, to details of imagery and versification. This tale of the birth of "The Raven" has struck critics as incredible, though symbolists seem to have taken it literally. Gauguin for one wrote to his thirteen-year-old son Émile, apropos artistic intention, of "the raven on top of the head of Pallas which is there rather than a parrot because of the artist's choice, a calculated choice."[8] Mallarmé's position, as usual, was subtle: citing a letter by an acquaintance of Poe's claiming the whole exercise a joke, Mallarmé gently replies that, even if that were so, "what is thought, *is* [so]" (*Ce qui est pensé l'est*). Even if it didn't actually occur, the procedure sketched by Poe is legitimate. "The eternal wing-beat does not exclude a lucid gaze scrutinizing the spaces devoured by flight."[9] For Mallarmé, it is in Poe's aesthetic theory, his logical reflection on artmaking, that poem, poet, and bird become one.

Manet's Path to Poe

The symbolist appeal of Poe's "Raven," then, is both theoretical and aesthetic, but for the financial risk of a deluxe book like *Le Corbeau* of 1875, there had to be some commercial appeal too. True, there was the general French enthusiasm for Poe, launched by Baudelaire. This enthusiasm had supposedly moved Mallarmé, in adult life an English teacher, to learn the language as a boy "simply in order to better read Poe."[10] This enthusiasm might have moved Richard Lesclide, publisher and founder of the Librairie de l'Eau-Forte, who had already worked with Manet, to hope for a popular *and* critical success in *Le Corbeau*, a manuscript that Mallarmé's previous editor had rejected as "a pack of insanities."[11] Alas, the printing of the book, despite Lesclide's enthusiasm, was plagued by delays, skyrocketing costs, and changes of mind, especially on the part of Manet, whose multiple proofs nearly derailed the project. Shops did not get their copies on time, and as Lesclide predicted gloomily, as a result they did not reorder. But as far as the quality of the book itself is concerned, it is hard to imagine a more uncompromising luxury object, from the format "two feet in height" to Manet's four bursts of transfer lithography (Figures 2.3–2.6), breaking the flow of Poe's poem and Mallarmé's

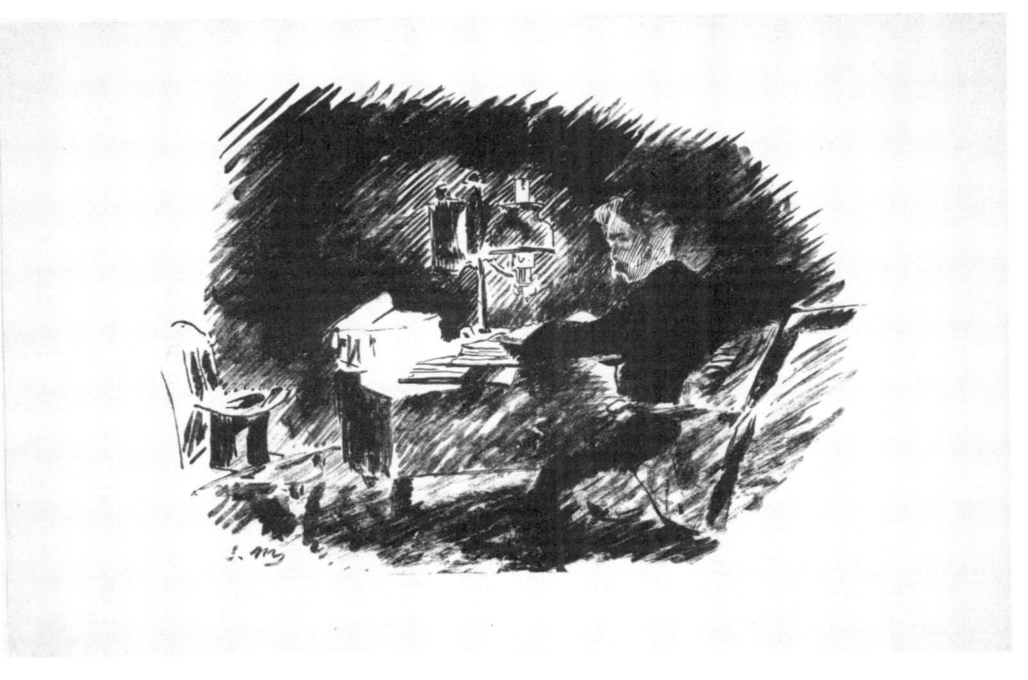

Figure 2.3. Édouard Manet, *Le Corbeau* (1875), first image (*Le Bureau*), transfer lithograph, British Museum, London. In parentheses are the names used by Lesclide in his correspondence with Manet and Mallarmé: in this case, "The Desk."

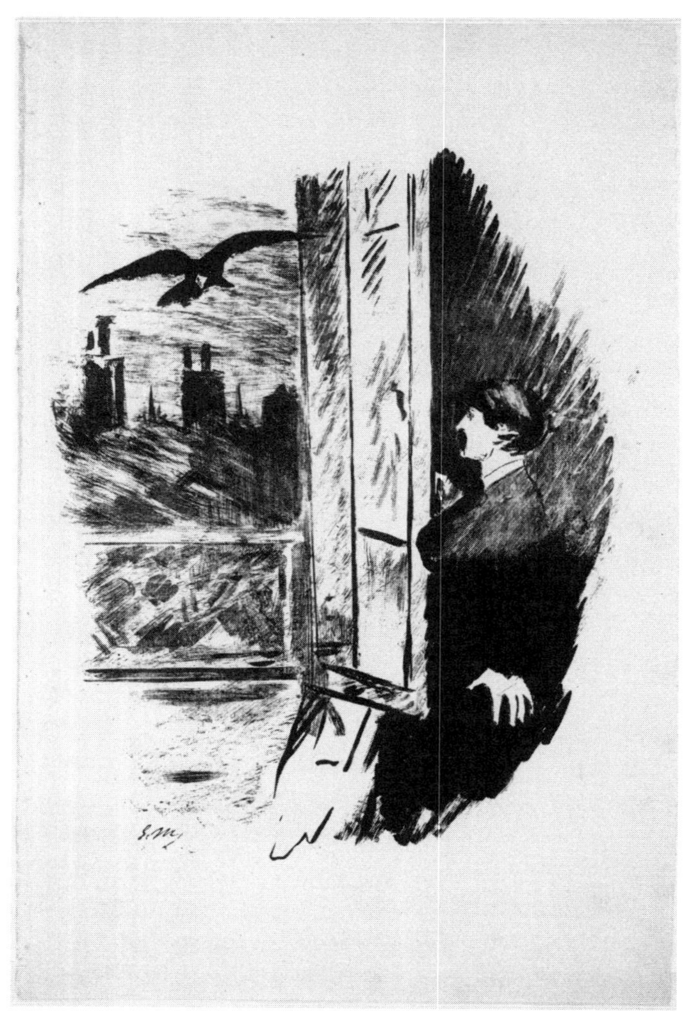

Figure 2.4. Édouard Manet, *Le Corbeau* (1875), second image (*La Croisée*), transfer lithograph, British Museum, London. *La Croisée* means casement window (with its crossbars), but also "The Crossing," evoking the crossed paths of narrator and bird

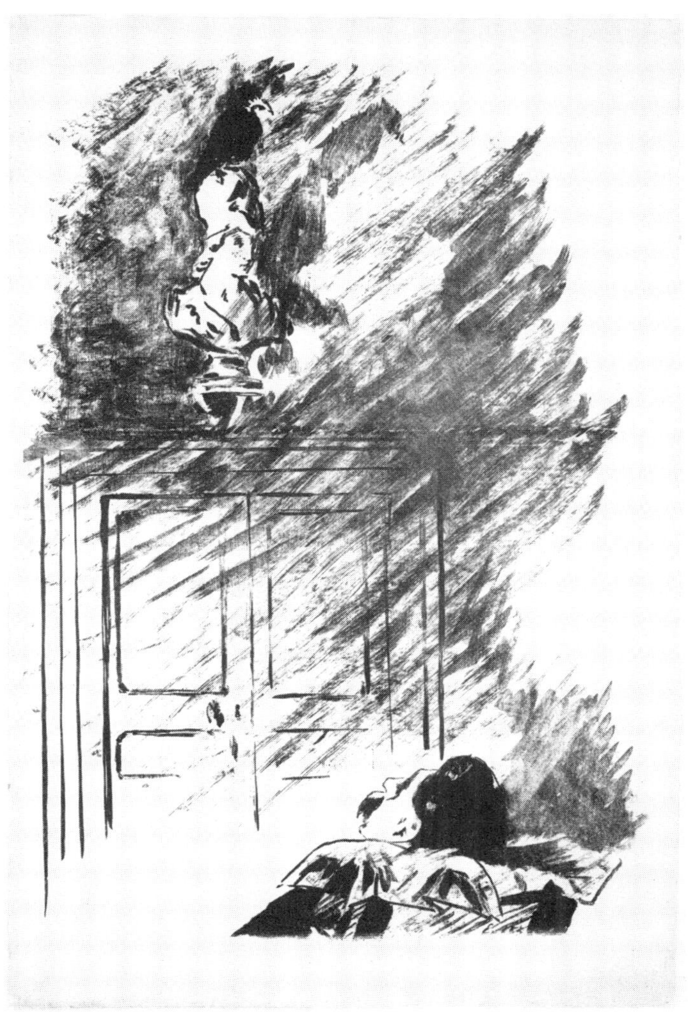

Figure 2.5. Édouard Manet, *Le Corbeau* (1875), third image (*Le Buste*), transfer lithograph, British Museum, London. Lesclide named the third image, prosaically, after the Bust of Pallas.

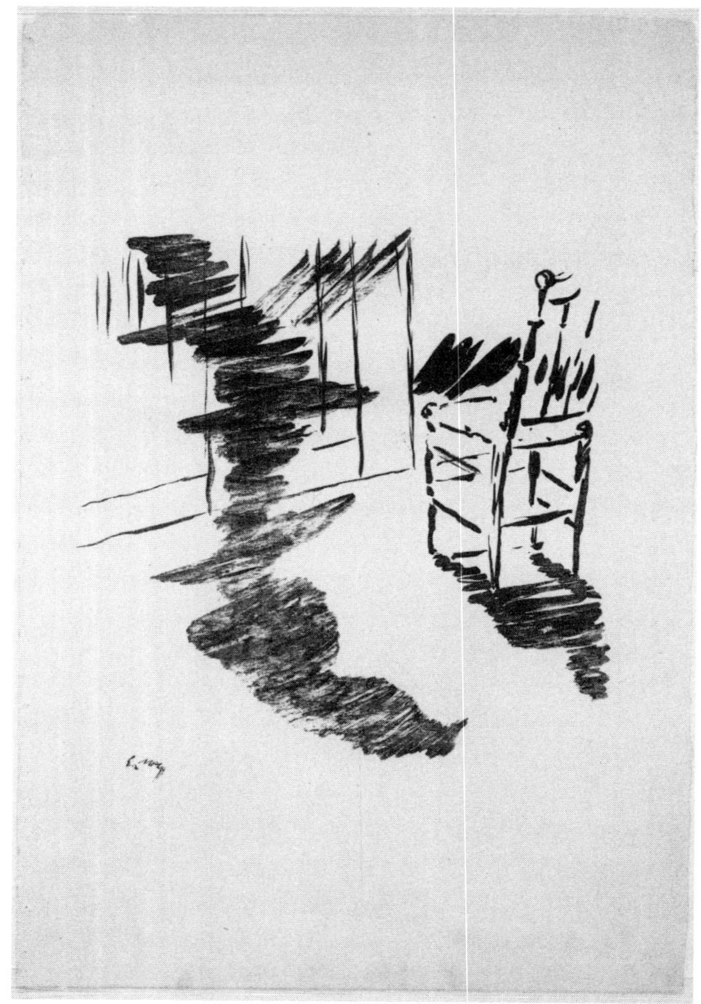

Figure 2.6. Édouard Manet, *Le Corbeau* (1875), fourth image (*L'Ombre*), transfer lithograph, British Museum, London. A critic using the pseudonym Gygès in *Paris-Journal*, in the most widely read notice of the book (reprinted by Ernest Hoschedé in the *Chronique des Arts et de la Curiosité*), found that "through the play of summary silhouettes and violent shadows, Manet has transposed from one art to another the sense of nightmare and hallucinations that Edgar Poe has so powerfully realized in his work." "The Shadow" provides the best evidence.

translation with such force that a few decades ago a debate raged whether these images were not in fact made by gillotage, wherein a photographic process is used to create a zinc relief plate which is in turn inked and printed.[12] A comparison with the few posthumous gillotages actually made after Manet makes it hard to believe he would have accepted this shoddy commercial expedient. But that the misunderstanding could erupt in the late twentieth century is a testament to the enduring shock value of Manet's technique, his "rapid, spirituel, running pencil," as Arsène Houssaye put it in the *New York Tribune* in his Frenchified English.[13]

A symmetrical misunderstanding manifested itself in the (by and large favorable) press of 1875. Reviewers spoke almost invariably of Manet's illustrations as *eaux-fortes*, etchings, an error that we may take as attesting an impression of artisanal finesse on Manet's part.[14] The "etching-myth" must have gratified Manet, insofar as the painter agonized about the fine points of these prints, especially that of the flying bird, producing four states and destroying a hundred valuable sheets of China paper in his attempt to balance interior and exterior lighting.[15]

Why go to such pains, and endanger the venture by publishing delays and the costly destruction of printed sheets, for hardly perceptible differences in tone? Manet's obsessive procedure resembles the epic etched sequences made by his friend Félix Bracquemond, who taught Edgar Degas and others in his circle the technique. Bracquemond was a versatile artist, particularly in a technical capacity; he sometimes attempted scenes of city life, even of the siege of Paris, but his heart was in animal studies, which he could make comically overblown, like the *Storm Cloud* (*La Nuée d'orage*) threatening a flock of geese in eleven distinct states, lit like a Crucifixion by Rembrandt. He printed a bombastic "Raven" of his own in 1854, fresh on the heels of Baudelaire's translation of the poem, at a time when Manet was a student in Couture's studio (Figure 2.7).[16] Echoes of his compact, pyramidal monster can be found in Manet, particularly the thrusting beak of *The Crossing*; more striking perhaps are the multiple echoes of the bird's shadow wrapped around the base of the gallows. But

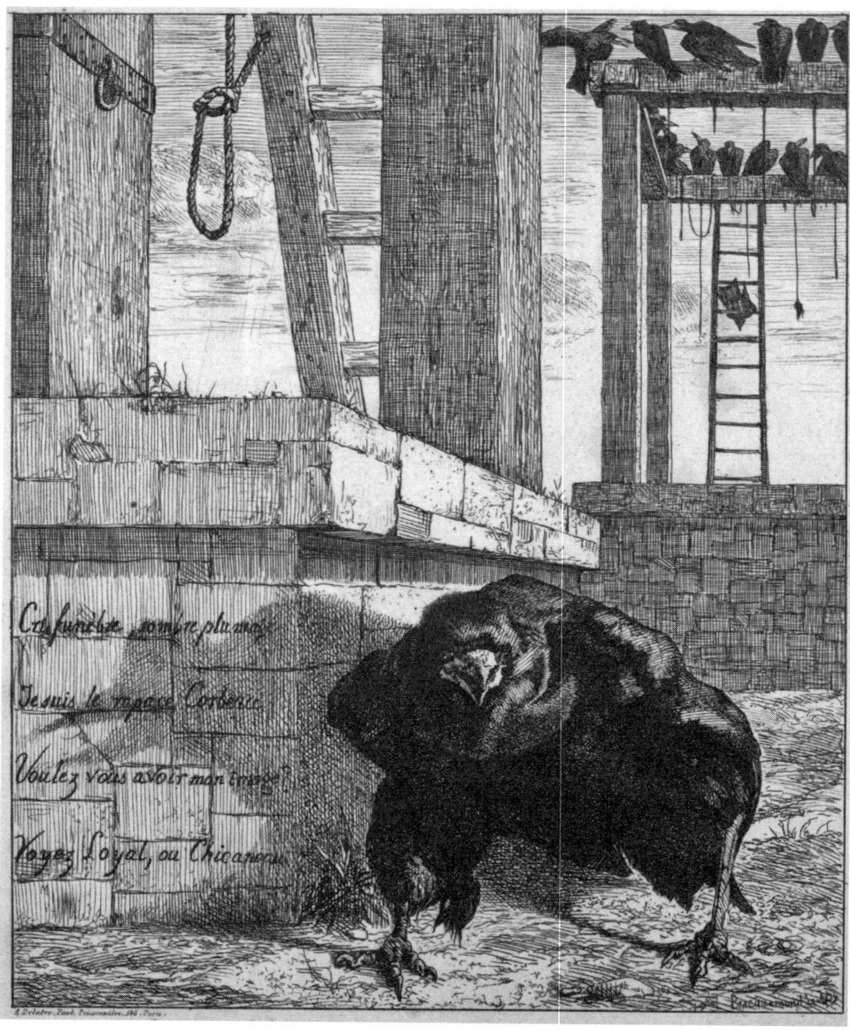

Figure 2.7. Félix Bracquemond, *The Raven* (1854), etching, published in *L'Artiste*, British Museum, London. This earlier raven is also accompanied by a poem, a piece of moralistic doggerel of Bracquemond's own design.

over and beyond iconographic analogies, what the two artists shared was an ideal of seeking out subtle perceptual effects that had come to seem particular to oil painting, but that printing also allowed in an unpredictable push and pull of revision, proof, and further revision. The theoretical justification for such tinkering lay to hand: like Poe in "The Philosophy of Composition," the painter-etchers (*peintres-graveurs*) were after a consistent mood, and that is why they calibrated light and dark, figure and atmosphere, rather than bother about the depiction of precise times of day, as the impressionists would later.

Of the two friends, Bracquemond was by far the more experienced printer, but Manet too had etched since near the beginning of his career; a portfolio of eight prints was published by Alfred Cadart in 1862.[17] Art historians who have noted the close working relation between the two artists have tended to minimize it in the service of Manet's originality.[18] Bracquemond was well aware of it: an *ex libris* he fashioned for Manet from a well-known 1871 photograph, bearing Auguste Poulet-Malassis's witticism *Manet et manebit*, "He endures and will endure," attests to this high esteem of his friend's talent (Figure 2.8).[19] We should pay attention to how Bracquemond abetted it: we can do this best by examining an etching aided by Bracquemond, the third state of the very free etching, perhaps intended for but not included in Cadart's portfolio, after Manet's first Salon submission, *The Absinthe Drinker* of 1859 (Figure 2.9). In earlier states, Manet's etching is a lucid drawn paraphrase of the iconoclastic oil painting: a vigorous quilt of crosshatching conveys the shabbiness of the drinker and his environs, but neither the painting's murk nor its displacement of viewer and viewed.[20] The third state amplifies the painting's atmosphere in the direction of Grand Guignol and *Les Mystères de Paris*. From the tenebrous shadow thrown by the drinker to the shock of light that turns his face into a mask, to say nothing of the sheer dirtiness of wall and street and bottle, the print celebrates visual difficulty. The grungy reality of the subject is inextricable from the grungy reality of the artist's gestures and marks, and it seems at times to be at war with them, as if expressing oneself and one's subject were mutually implying and yet contradictory actions.

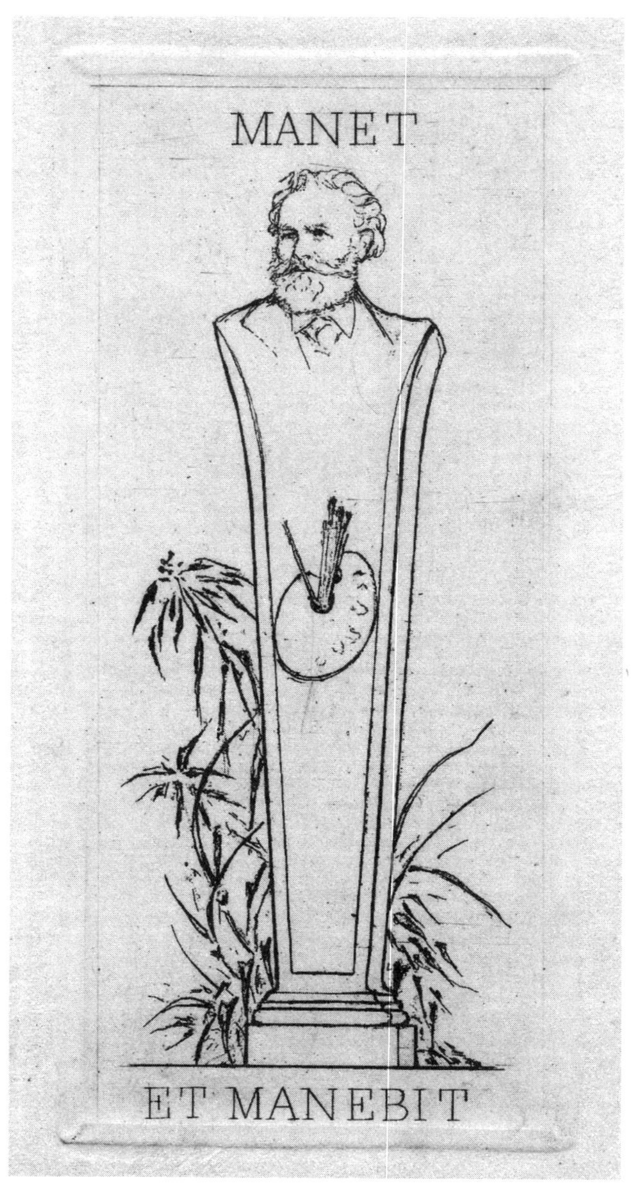

Figure 2.8. Félix Bracquemond, *ex libris* for Manet (1870s?), proof etching, British Museum, London. Henri Béraldi (*Les Graveurs du XIXé siècle*, vol. 3, *Bracquemond*, nos. 508–10) notes that the phallic palette and brushes, "reminding one vividly of the god of the gardens," are found only in the proof stage—an eloquent tribute to the etcher's art.

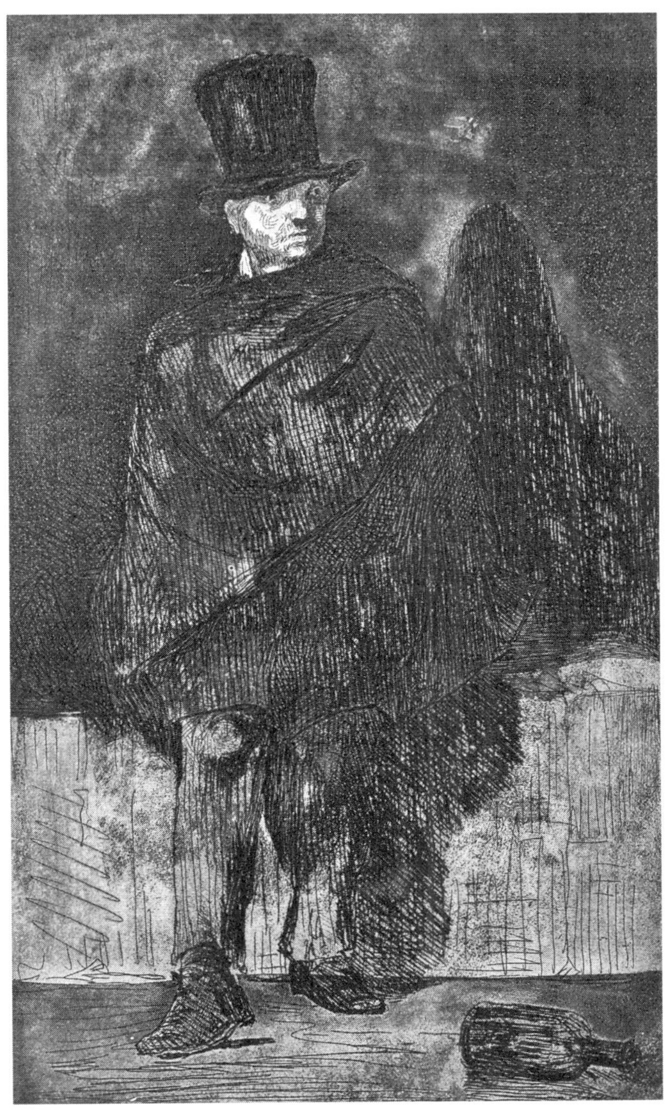

Figure 2.9. Édouard Manet, *The Absinthe Drinker* (1867 or 1874), etching and aquatint, third state, Art Institute of Chicago. The shock of light on the drinker's face, hardly noticeable in the first state, makes for nearly Picasso-like fragmentation.

Bracquemond's role in this etching, as in that of *Olympia*, is certainly subsidiary: to help Manet master the gamut of light, dark, and suggestive mark-making in printing as he had always mastered it in oil. The two artists' relative merits should not blind us to a shared project of making the print a medium of sensitive perceptual distinctions.

The Palpability of Privacy

To grasp how well Manet prepared for his visionary suite of illustrations to *Le Corbeau*, one other collaboration must be mentioned. In 1874, in his first project for Lesclide, Manet contributed a running commentary of etchings to Charles Cros's narrative poem "The River." The rather plodding realist couplets are broken felicitously by Manet's vistas: most striking is the conclusion, which sandwiches a pedestrian parting joke from Cros (warning a *perverse* antiquary not to meddle with his *verse*) between a windswept seascape and a birdwatcher's close-up of a swallow skimming the water, its fleeting shadow suggested by fine pencil-like marks (Figure 2.10). It is as if Manet jogs along, leisurely outpacing the poem while pressing the reader close to the surface of the titular water. If Manet never again interwove his images so playfully with a text, that is probably because the texts got more complicated (Poe's and Mallarmé's), as did the images Manet furnished for them. From this five-finger exercise, Manet retained a sense of drama, the impeccable timing with which his prints play counterpoint to the English poem (they face every page of Poe's original, their blank versos facing Mallarmé's French). A jolt results, requiring the reader to become a viewer, halting Poe's breathless words with their exclamation marks and imperious dashes to reflect, as does the baffled narrator, on the mysterious visitation of the bird with its one word.

What is the joint effect of Mallarmé's translation and Manet's lithographs? Mallarmé wished to translate the text as literally as possible to "preserve something of the original song" of Poe. Some critics took him to task for being so literal as to become "more American than Poe."[21] If critical taste has caught up to Mallarmé's "punctilious" style of translation, there is good cause to think that he intended

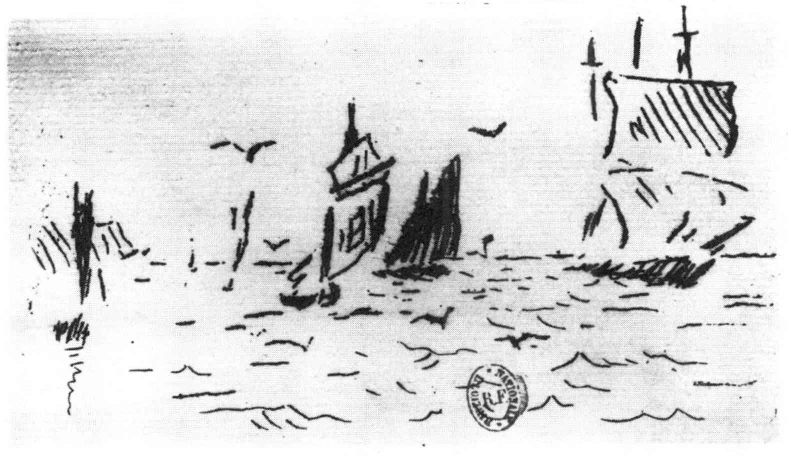

Qu'on se lise entre soi ce chant tranquille et fier,
dans les moments de fièvre et dans les jours d'épreuve;
qu'on endorme son cœur aux murmures du Fleuve.

Va, chanson! Mais que nul antiquaire pervers
n'ose jamais changer rien à tes deux cents vers!

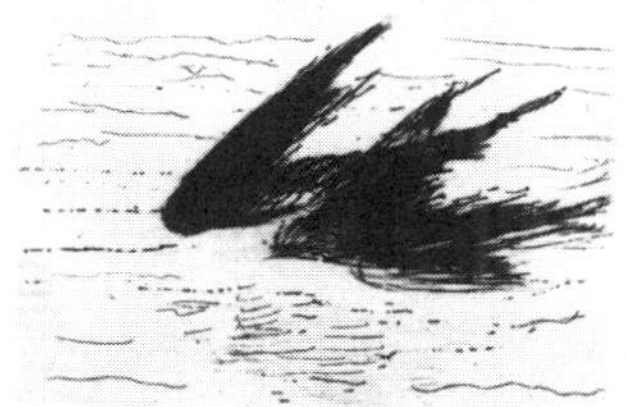

Figure 2.10. Édouard Manet, etchings to Charles Cros, *Le Fleuve* (Paris: Librairie de l'Eau-Forte, 1874), p. 15. The two views not only contrast two kinds of views (panorama and bird-watcher's glass), but two kinds of line: the flowing pen and the nervously scratching pencil (refined to the point of imperceptibility in the shadow of the bird on the water).

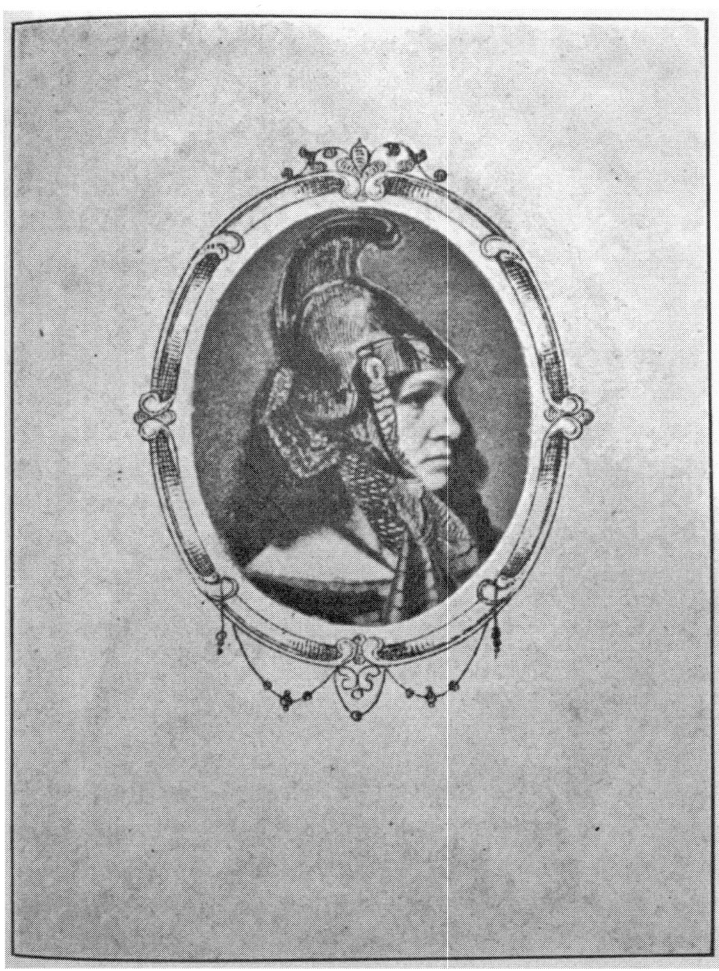

Figure 2.11. *Mrs. Whitman as Pallas*, retouched photograph by William Coleman first published in Caroline Ticknor, *Poe's Helen* (New York: Scribner, 1916), p. 200. The Pallas helmet is evidently hand-drawn (collaged) onto Whitman's hair; a letter cited by Ticknor suggests Whitman modified the photograph herself: "Thinking that fine feathers make fine birds I have costumed two of my photographs…"

a more self-effacing performance; by printing his highly rhythmic translation in blocks of italicized prose, he intended to stay out of Poe's and Manet's way. In turn, the ragged, vignette-like frames of the lithographs prepare us for flights of visual poetry.[22] The poet was both proud of Manet's prints and conscious of their overwhelming effects.[23] Consider Sarah Helen Whitman (Figure 2.11), Poe's onetime fiancée and biographer, with whom Mallarmé exchanged several gallant letters and portrait photographs in 1876 and 1877.[24] A draft of one of Whitman's letters praised the second and third lithographs, both of which "illustrate the walls of my boudoir," and are "wonderfully unique and impressive." But Whitman confessed that "as for the one where we see of the Raven only what purports to be 'his shadow on the floor,' it is so far out of the reach of my appreciation that I hardly know where to class it! *Entre nous* I should like to do with it what the Greeks did with their honoured dead, i.e., *cremate* it."[25]

This is not the kind of invective Manet suffered at the hands of no less a personage than Dante Gabriel Rossetti, who fumed about "a huge folio of lithographed sketches from the Raven, by a French idiot named Manet, who certainly must be the greatest and most conceited ass who ever lived."[26] Whitman is not engaged in scatology, but rather taken aback by the intensity of her negative reaction to this most cursive of the illustrations, which even a century later may appear "almost indecipherable in the density of its real and abstract references."[27] Just such a blend of admiration and alarm informed the most searching critical response to the book, by Richard Hengist Horne, an English poet who in his youth had been reviewed by Poe:

> The Artist has taken the hint of getting rid of the body altogether, by showing only the empty chair, with its equivocal shadows half suggesting some mortal remains, and the long, bedeviled sort of shadow of the Raven blackening the floor. By what kind of light, and where the light comes from, is a question for artists to settle. It is grandly grim and self-contained.[28]

Grandly grim and self-contained is a good motto for the book. But this chair and its equivocal shadows remain a raw nerve, throwing

into relief its aesthetic challenge — which, as we will see, is a philosophical one as well.

Whitman herself, though new to Manet's work as to Mallarmé's,
was no simple Poe hagiographer. As a writer, her volume on *Edgar
Poe and His Critics* (1860) inaugurated serious criticism of Poe's art and
philosophy, refuting the scandalous imputations of his first editor
and malicious biographer Griswold.[29] At the same time, she inhabited
his works to the extent of posing for a photo in the garb of Pallas, as
if she belonged on the poem's mantelpiece with the bird on her helm.
Her joke about the burning, by the same token, is less a philistine act
of violence than a sympathetic enactment of what the image does to
Poe's language. Whitman is attuned to the tenuousness of figuration
in this image (it depicts "of the Raven only what purports to be 'his
shadow on the floor'") and even more to her own failure to intellectually order it ("I hardly know where to class it!"). It is as if, in
speaking of her secret desire to burn the image, she put her finger on
Manet's own iconoclasm, the way his vivid image combines raven and
bust and "shadow on the floor," vibrating, threatening to blow clear
of its moorings like greasy smoke.

In his tactful reply, Mallarmé welcomed Whitman's appreciation
for Manet's work, "so intense and modern at once," and confessed to
sharing her discomfort with the last print. "The shadow of the bird
in the last does not displease me, being mobile and accurate; but I
like less the presence of the chair, and understand how you found
the whole too summary. Manet belongs completely to the contemporary artistic movement; and as far as painting is concerned, he
is its chief."[30] We may ignore the manipulative appeal to authority.
The idea that to the heroic modern artist all things are permitted,
including being careless or "summary," has too often been used to
explain the triumph of modern art, when it is no more than part of
the problem: *why* is this permitted or even salutary? More revealing
is Mallarmé's own admitted dislike of the chair. To understand it,
we have to gauge how closely Manet stuck to the poem. Beside such
witticisms as the distinctly Parisian skyline in *The Crossing*, there are
no conspicuous anachronisms or deviations from Poe. The occasion

for each print in specific verses is easy to spot: *The Desk* (see Figure 2.3) illustrates the opening couplet: "Once upon a midnight dreary, while I pondered, weak and weary, / Over many a quaint and curious volume of forgotten lore." *The Crossing* (see Figure 2.4) accompanies "Open here I flung the shutter, when, with many a flirt and flutter, / In there stepped a stately Raven of the saintly days of yore."[31] *The Bust* (see Figure 2.5) is an obvious enough rendering of "Perched upon a bust of Pallas just above my chamber door," while *The Shadow* (see Figure 2.6) illustrates "And the lamplight o'er him streaming throws his shadow on the floor."[32] The rest of the closing triplet illuminates this final image: "And my soul from out that shadow that lies floating on the floor / Shall be lifted — nevermore!"[33] The empty chair evinces a dramatist's eye for consistency, since the penultimate stanza begins with the hero jumping out of it: "'Be that word our sign of parting, bird or fiend!' I shrieked, upstarting."[34] Inspection of the verses is useful, whatever we think of the poetry, for it reveals a red thread in the illustrations that is *not* immediately visible: the passages contain the auctorial first-person pronoun, "I" (pondered, flung, shrieked), modulating finally into the possessive pronoun "my" (soul). This emphasis on the first person may seem to fail for the third (Pallas) picture, but that is because I provided the customary citation, which is wrong; the image itself occurs not when the bird first alights on the bust, but a page later, when the narrator seats himself before the bird: "But the Raven still beguiling all my sad soul into smiling, / Straight I wheeled a cushioned seat in front of bird and bust and door." This comfortable posture lends itself to philosophizing on the nature of meaning: "Then, upon the velvet sinking, I betook myself to linking / Fancy unto fancy, thinking what this ominous bird of yore — / What this grim, ungainly, ghastly, gaunt and ominous bird of yore / Meant in croaking 'Nevermore.'"

This, I hope, clears up the mystery of Mallarmé's discomfort with Manet's chair: as striking and distinctive a design as the simple wooden model might be, it is surely not the overstuffed affair of the poem, into which the narrator "sinks" (Figure 2.12). Has Manet sacrificed fidelity to elegance? The answer matters, since the apparently

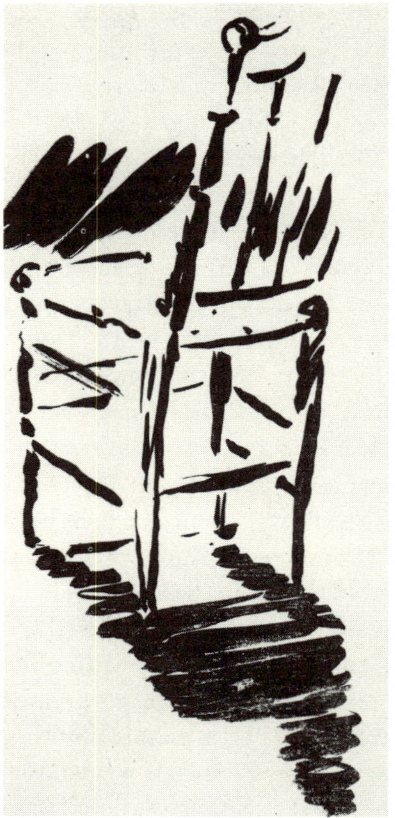

trivial encounter between narrator and chair is at the heart of the poem, and contributes significantly to the more dramatic encounter between man and bird. After his linguistic musings, the narrator is caught by surprise, amid his garrulous thoughts, by a sensation filled with emotional import: "This and more I sat divining, with my head at ease reclining / On the cushion's velvet lining that the lamplight gloated o'er, / But whose velvet violet lining with the lamplight gloating o'er, / *She* shall press, ah, nevermore!" The italicized pronoun *she*, standing for the absent Lenore, who has been named by the despondent narrator several times, makes its first and sole appearance here,

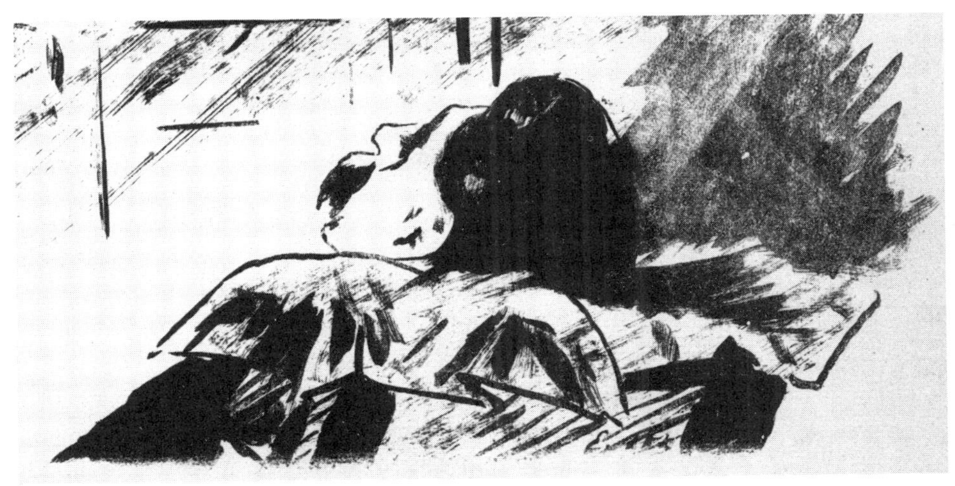

Figure 2.13. Édouard Manet, detail of
The Bust (Fig. 2.5), with stuffed armchair.

called forth by the memory of her body's pressure on the violet velvet
cushion: the detail, with its forceful alliteration, is of such moment
that it would be silly to leave it out for the sake of a Spartan taste in
furniture.[35] But *did* Manet leave it out? The verse speaks of "my head
at ease reclining," and in *The Bust* we have that upturned head, resting on a cushion propped on the backrest. We don't see the chair very
well; only the frame is indicated in black strokes that seem thicker
than the filigree silhouette in *The Shadow,* but then, the chair is closer
to the viewer here. And it can't be quite Poe's stuffed model either,
because the light-colored cushion is a separate object, its slack corners flopping over the edges of the backrest (Figure 2.13).[36]

Why does Manet diverge from Poe, feeling Lenore's weight on a
pillow instead of a stuffed chair? As with the window thrown open to
admit a swooping raven in *The Crossing,* rather than the stately walking bird of the poem (akin to Bracquemond's), it might be that Manet
has striven here to make something visible which, in a too-literal

rendering, would have evaded visual thought: there a bird *entering* the chamber, here a head *pressing* a pillow, and finally a shadow seen through the bars of the chair's backrest. A succession of stuffed chairs wouldn't have done the job.

If this account of Manet's stage-management of Poe's drama is right, it places a premium on direct perception on our part — a faculty that matters enormously in the poem, for without its intervention the narrator would not have been drawn from his cheerful speculations on the bird to melancholy memories of the lost Lenore. The role assigned to the cushion by Poe is no less than that of a medium between the living and the dead. Its yielding softness, the suggestion of another body pressing into it — the feel of that body against the narrator's, or his imagination of what it feels like to *be* that body — all this makes the experience of the chair a séance of sundered souls. And this is handled, as Poe insists in "The Philosophy of Composition," without super-natural or metaphorical machinery; it is nothing but the pressure of *his* body in the chair, which by a trick of language and of light ("On the cushion's velvet lining that the lamplight gloated o'er, / But whose velvet violet lining with the lamplight gloating o'er"), is declined into the lost, unknown pleasure of *her* body, Lenore's, in the very same chair (which "*She* shall press, ah, nevermore"). Readers may be reminded of Proust's madeleine. The procedure in Poe is just as idiosyncratic and concrete. Through an exterior sensation, an inner train of associations opens the door to something like a private language, a set of sensations with peculiar meaning for their bearer.

Pronouns and Private Language

We have arrived at the real nub of the difficulty, not just of read-ers and viewers of *Le Corbeau*, but also for the symbolist art which followed and traced its descent from this book. The crisis of sense diagnosed by Mallarmé in his prolix and ever-mutating essay *Crise de vers* is made tangible in Manet's flickering chair with its fluctuating shadows, and in the poet's invocation of violet velvet cushions and the pressing they will no longer receive. If this diagnosis appears to foist on Poe a symbolist sensibility a half century and an ocean removed, that is, I think, because we have become less attentive readers than Baudelaire, Manet, and Mallarmé were. They might have known, from the 1850 *Works of the Late Edgar Allan Poe*, a curious meditation first published in *Graham's Magazine* a month before "Philosophy of Composition" under the catchall title "Marginalia," and which also quotes and elaborates on "The Raven."[37] In this text, Poe avows, as Mallarmé or Frege might a half century later, that he does not believe "that any thought, properly so called, is out of the reach of language."[38] Yet he was drawn to a class of experiences on the border between waking and sleep "which are *not* thoughts" and seem to elude the writer. Dogged investigator that he was, Poe trained himself to fall into these 'psychal' states, as he called them, at will, and to wake himself in order to "embody them in words." Though he did not yet have the definitive results he hoped for, Poe hazarded a prognosis:

> I am not to be understood as supposing that the fancies, or psychal impressions, to which I allude, are confined to my individual self — are not, in a word, common to all mankind — for on this point it is quite impossible that I should form an opinion — but nothing can be more certain than that even a partial record of the impressions would startle the universal intellect of mankind, by the *supremeness of the novelty* of the material employed.... In a word — should I ever write a paper on this topic, the world will be compelled to acknowledge that, at last, I have done an original thing.[39]

What Poe promises in this text, and introduces in his account in "The Raven" of the experience in the armchair, is nothing less than the prototype of what the Austrian philosopher Ludwig Wittgenstein was to decree an absurdity — private language, a means of expression to which one person alone has access. The classic scenario through which Wittgenstein was to introduce this idea — the sentence "No one can have THIS pain," said while beating one's breast — sounds silly, but there must be more to it.[40] Surely we feel that whatever "*this*" may be, "this pain" touches on *something*, whether we then go on to say that it is a unique and proprietary thing or something like what others have experienced, or, more fastidiously still, like Poe, that "on this point it is quite impossible that I should form an opinion." Every person's idiolect or personal speech, to the extent that it contains such phrases, is potentially a private language.

Though baptized only in 1953 with the posthumous publication of Wittgenstein's *Philosophical Investigations*, the idea of private language first emerged in the late nineteenth century, notably in a work of philosophy published in 1884, a mere year after Manet's death.[41] Its roots lie in romantic literature, as the next chapter will show, but its philosophical inventor, Frege, who in old age was Wittgenstein's mentor, struggled throughout the fin de siècle (starting around 1880) to articulate his ideas on the nature of logic and its relation to mind and privacy. He finally published them at the end of the First World War in an essay on "Thoughts," a copy of which reached Wittgenstein while the latter was a prisoner of war in Cassino, Italy. In it, the case for private language and that for public comprehension are inter-twined. "Everyone," notes Frege,

> is presented to himself in a particular and primitive way, in which he is pre-sented to no-one else. So when Dr. Lauben thinks that he has been wounded, he will probably take as a basis this primitive way in which he is presented to himself. And only Dr. Lauben can grasp thoughts determined in this way. But now he may want to communicate with others. He cannot communicate a thought which he alone can grasp. Therefore, if he now says "I have been wounded," he must use the "I" in a sense which can be grasped by others, per-haps in the sense of "he who is speaking to you at this moment."[42]

So far, so good — as in Poe and Wittgenstein, the nature of meaning for Frege is such that by definition public intelligibility is secured. Yet Frege expresses a fundamental doubt: "Is it at all the same thought which first that man expresses and now this one?" Here the threat of private language looms in what we might call its "private sense" variant: it is not that the word "I" is new or exotic, but that its *use* in a private way threatens intelligibility of the sentences containing it. We avoid this in communication by substituting a public sense of "I" for the private sense that comes naturally to each speaker, but Frege ends the passage asking whether the resulting public thought is the same as the earlier, allegedly private thought. If not, private senses might lurk behind public utterances, masked by homonyms that make their presence undetectable.

Is Frege's worry in this passage about the threat of private language one we should share? Philosophers impressed by Wittgenstein ridicule the very suggestion of attaching a private sense to words like "I."[43] More recently, sympathetic attempts have been made to credit Frege with the discovery of singular thoughts that only one person may have, perhaps even only on particular occasions, because their expression depends essentially on context: a private language if ever there was one.[44] But both this peculiar notion of public thoughts that only I may express and the behaviorist assertions of the publicity of all sense miss what is most fascinating in nineteenth-century explorations of private language. What makes Frege's — and Poe's — idea more valuable than Wittgenstein's radicalization of it is that it takes seriously the need for objective communication between thinkers, *and* also the subjective aspect of thought that these thinkers may struggle, and often fail, to share with or at least make manifest to one another. Private language, in its "private senses" interpretation, stands for the bare possibility of unverifiable subjective divergence: different uses of the same sentence or sign or artifact, which agree as to facts but differ in their subjective import. Poe's lamplight and shadow fall on uses of the "I" submerged in such an introspective sense; we wrest meaning from it, but not all its meaning.

The candidate for private senses in "The Raven" is obvious: the insistent "I" of the narrator, though it might seem beyond the reach of visual art, is identified in Manet's mustached figure not just with Poe but with the voluminously mustached Mallarmé (Figure 2.14).[45] Given the modesty of the translator, it seems churlish to attribute this to self-aggrandizement on his part, or to a misguided tribute on Manet's, suggesting that his friend had supplanted Poe. What Manet hints at is rather an imaginative transfer akin to that of Whitman photographed as Pallas, the "becoming I" of the translator inhabiting his source text, the seeping of a private sense appropriate to Mallarmé's French text into Poe's poem; the painter cannot, of course, more than hint at it, and it is done most fully in the image of

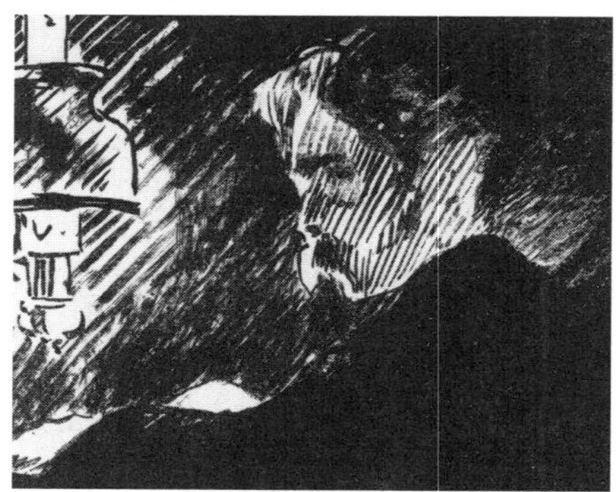

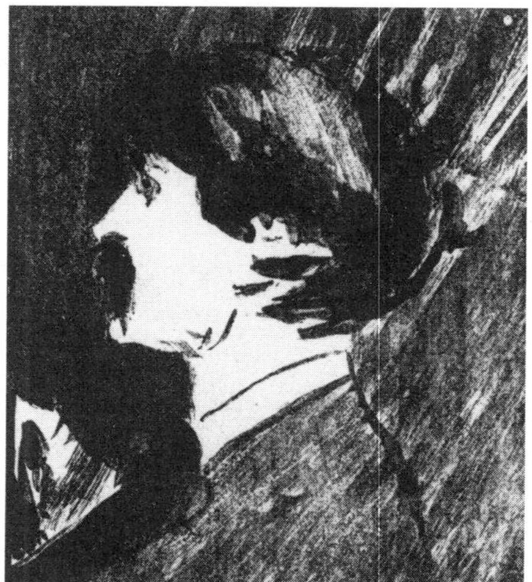

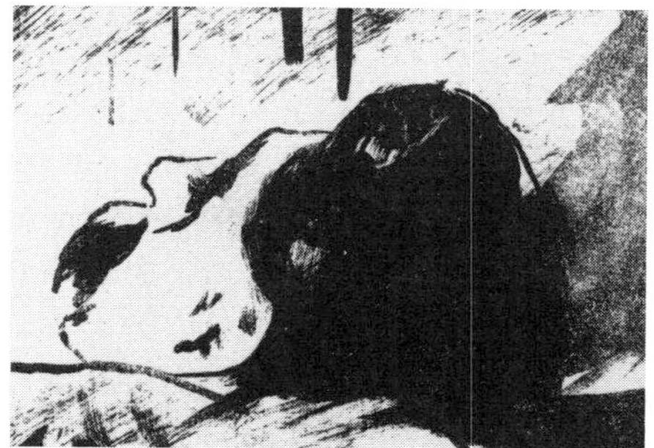

Figure 2.14. Édouard Manet, details of
The Desk, *The Crossing*, and *The Bust*.
Is it the same protagonist? And is it Poe—
or Mallarmé?

the empty chair. Here the disappearance of the mustached figure, replaced by the meandering, nearly abstract shadow of raven and statue, together with the drastic vignette format of the scene, which leaves out most of the door below the bust and bird, and even pointedly cuts off the upper right corner of the chair (and its shadow), suggests that we ourselves, or rather, that *I myself*, have been put in the place of the narrator, left staring at the puddle of shadow on the floor, in a posture of despair from which I "shall be lifted—nevermore."

Such a poetic animation of the first-person perspective, flitting from protagonist to author to reader, fits well with what Poe says about the denouement of the poem. In "The Philosophy of Composition," we are told that the narrator's crescendo of self-laceration by means of the bird (who reliably answers "Nevermore" when queried whether he may see his lover again in heaven) is meant to give rise to a precise, replicable state of mind. "This revolution of thought, or fancy, on the lover's part, is intended to induce a similar one on the part of the reader—to bring the mind into a proper frame for the *dénouement*."[46] He affirms of his poem what some commentators have said of Manet's illustrations: that until that denouement "every thing is within the limits of the accountable—of the real."[47] In the penultimate stanza, the hero shouts, "Take thy beak from out my heart," the poem's first metaphorical expression, according to Poe. This is meant to lend emblematic value to the raven, that is to say, the kind of logical generality that would allow it to apply to a heterogeneous family of feelings, the narrator's as well as our own.

The emblematic raven, wherein Poe finally presents a symbol of *"Mournful and Never-ending Remembrance,"* was the *first* thing readers of Manet and Mallarmé saw, in the "portrait head" of the raven staring forth from the book's covers (Figure 2.15).[48] This degree of attention to Poe's understanding of the poem is perhaps exceptional to Manet and Mallarmé. But the conception of the bird as emblematic in the context of the final verse with its oppressive shadow is in fact typical of the poem's illustrators. The baroque wood-engravings of Gustave Doré in 1884 don't omit it, but they may be merely following Manet, whose posthumous exhibition that year finally cemented his

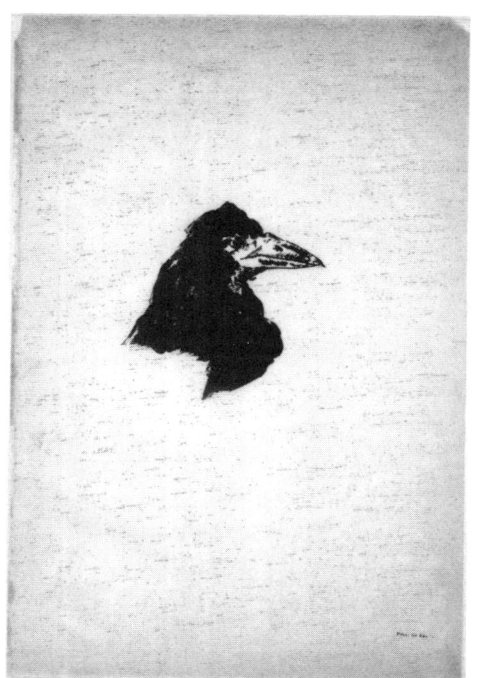

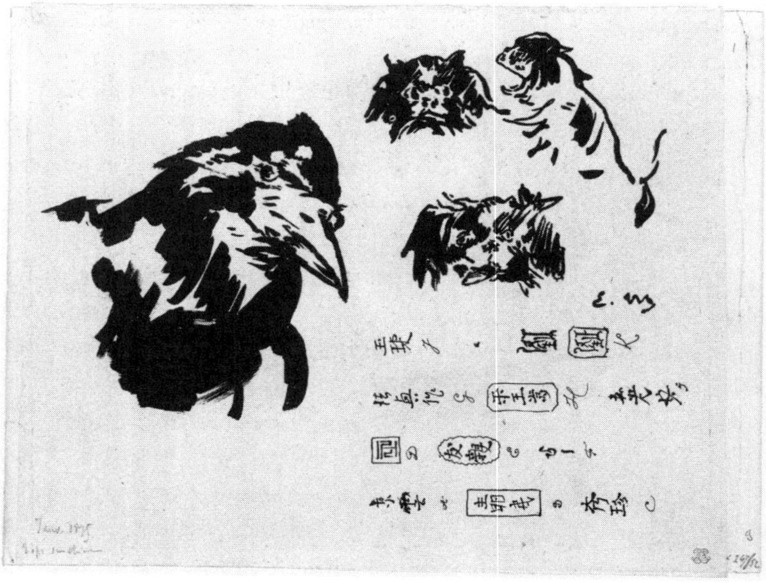

Figure 2.15. Édouard Manet, simili-parchment cover of *Le Corbeau* (1875), and first sketch of raven (with dogs and Japanese printmakers' signatures), lithographs, Art Institute of Chicago and British Museum, London. Of all the changes, the raven's resolute stare in the final version is most significant. Although the Japanese woodcut signatures and sketches of the Pekingese dog have little to do with the raven, there is a disquieting suggestion of a raven's beak emerging from the lower dog's head, just by its right ear.

classic status. More striking are the comparatively modest engravings of a German translation published in Philadelphia in 1869, that is to say, half a decade before Manet set to work: here the shadow projected from the bust ("by what light?" as Horne rightly inquired) appears not in one but in two versions, as if to drive the point home, illuminating the poem's conclusion in English and in German (Figure 2.16).

The melodramatic quality of such efforts to visualize the terminal melancholy of the narrator (by equating him with the bust; as in Manet, the figure of the man is omitted) brings home the symphonic subtlety of Manet's solution, with its winding, arabesque shadow melding bird and bust, and the fleeting shadows on the seat of the chair that suggest man, woman, *and* bird. It is as if not only the narrator's emotions, but his thoughts and powers of perceptual discrimination, have been plunged into the gloomy negation of the "Nevermore."

A final confirmation that such a subjective habitation of the linguistically prepared role is at work in "The Raven" illustrations is provided by what may turn out to be the solution to the chair riddle. Another look at the first print, the lamp-lit study, reveals *two* chairs: the familiar stuffed armchair, with its distinctive square frame, upon which the protagonist sits, and a simple wooden chair upon which two outdoor items — a top hat and a cane — are casually thrown (Figure 2.17). It has been suggested that these items are a lighthearted allusion to Manet himself, come to pay Mallarmé a visit.[49] But granted that this may be the etiology of the hat and cane, to say that they "only mean" Manet, in the midst of Poe's text, with its cloistered protagonist startled awake by the tapping at his chamber door (""'Tis some visitor,' I muttered") is to fail to engage the spirit of the poem. The anxiety of the narrator, with its grumbling repetitions (""'Tis some visitor entreating entrance at my chamber door — / Some late visitor entreating entrance at my chamber door; — / This it is and nothing more") gives the poem momentum, just as it gives the incongruous hat and cane their place in the interior drama. It is this feeling alone that differentiates the two chairs across the prints: we may plausibly

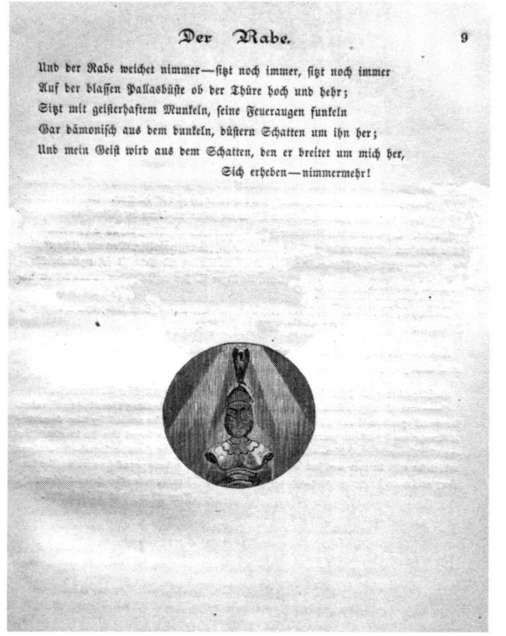

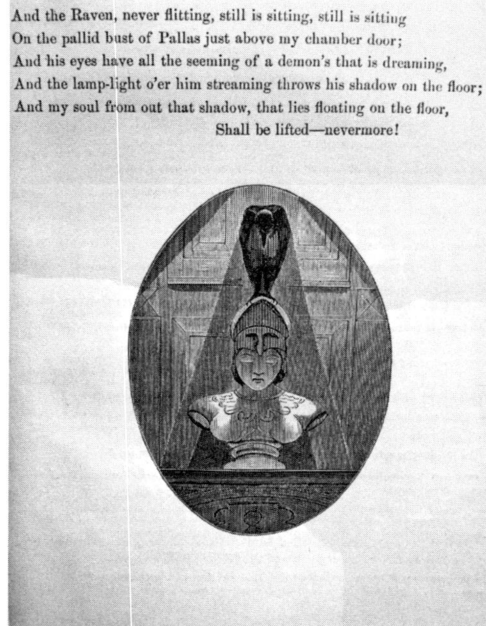

Figure 2.16. David Scattergood, wood engravings to *Der Rabe: Ein Gedicht von Edgar Allan Poe*, trans. Carl Theodor Eben (Philadelphia: Barclay & Co, 1869), pp. 9, 29. Note the conspicuous expansion of the image in the English text; not just the mantel, but also the closed eyes of Pallas first become perceptible here. It is as if the illustrator wanted to underline Eben's modest claim that, the poem not being translatable, the real experience belongs to Poe.

see in the cushioned seat of the third print the overstuffed, wood-framed armchair on which the protagonist sits at his desk in the first print, but the wooden chair of the fourth is certainly that occupied by the hat and cane. The substitution from *Bust* to *Shadow* stands for the game of imaginative musical chairs enacted by this elaborate composite artwork, Poe's poem with its French translation by Mallarmé and its visual translation by Manet.

Imaginative self-insertion of this kind is what separates a sympathetic reading of "The Raven" from the indifference and ridicule that so often met it, and of which Rossetti gave such a resounding sample.[50] Divergences in aesthetic judgment are the stock in trade of modern art, with critics splitting violently over what they often acknowledge as the same overt achievements. This is just the kind of disagreement that the existence of private senses would have us

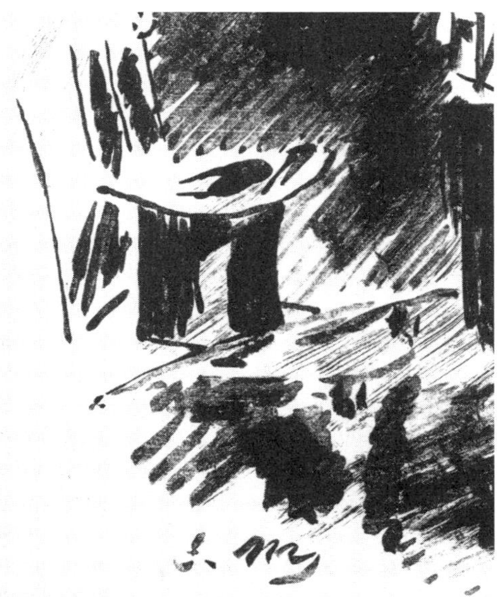
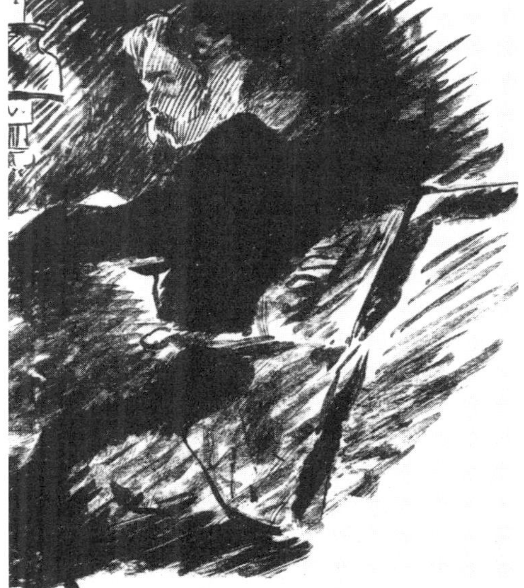

Figure 2.17. Édouard Manet, *The Desk*,
details of hard chair (with hat and cane)
and stuffed armchair.

expect. Frege himself conceived such unverifiable, but suspected, divergence in two subjects' access to a public thought in aesthetic terms. He imagined two beings, one of whom grasps the theorems and axioms of plane geometry exclusively in terms of points, while the other does so in terms of lines. Because of the principle of duality, the two would agree in all judgments of truth and falsity; their radically alien experiences would manifest themselves as "divergences in judgments of aesthetic value."[51] This, Frege stressed, would give the participants in the dispute no point of entry to one another's experience, but only a hint that such a difference obtains.

Learning by Playing

At this point in the argument the danger arises that if I am right, I cannot know for certain, since I have no access to the private senses of Manet and Mallarmé, let alone Poe, and am only able to point to indices that might have other explanations. I accept this reservation as a consequence of the existence of private senses. I do not regard this difficulty as decisive against the position, not only because it brings to light the anxieties that humanistic interpreters of artworks have so often registered, but because regardless of what field of research one is embarked upon, the point is to get at the truth of the matter, not to make one's task easier by ignoring difficulties.

To check that we have not gone astray, that the crisis of sense we have identified in "The Raven" is that announced by Mallarmé over a decade later, it is worth shifting from our inspection of the book to a more detached historical perspective. Like such distinguished predecessors as Goya's *Caprichos*, the illustrated *Corbeau* flopped for complex reasons, only some of which are related to its challenging content.[52] Further collaborations of this scope became impracticable for the two friends.[53] True, the publisher Derenne took on their next project, Mallarmé's *Afternoon of a Faun* (*L'Après-midi d'un faune*), printed the very next year, 1876. Yet, though it dwarfs the "Raven" translation as a poetic achievement, Mallarmé's *Faune* is a more modest undertaking as far as book design goes. Which is not to say that it is a step backward: the visionary density of the *Corbeau* lithographs and the subtle textual counterpoint of the *Fleuve* etchings are combined here in a disarmingly classical manner. But they have been yoked to a decorative imperative: there is a still life of wild plants for an *ex libris*, a frontispiece of the titular faun dreaming of his lost nymphs, an opening vignette of the bathing nymphs, and a closing ornament (*cul-de-lampe*) in the form of a cluster of grapes. The book's front matter proudly names these traditional printer's devices, calling the poem an *églogue avec frontispice, fleurons & cul-de-lampe*.[54] Curiously, Manet is not mentioned on this title page in connection with these illustrations; the credit is given in the poet's dedication. As for the four items listed, the "fleurons" must refer to the little vegetal still life at the bottom of the proof sheet, whose heavy leaves resemble the heart-shaped foliage of the traditional fleuron, namely ❦. But these fleurons don't punctuate the text. Freed from their traditional function, they occupy a separately printed *ex libris*, with its copy number to be completed by hand; unless perhaps the nymphs, themselves hiding among the vegetation, are meant to serve as fleurons (Figure 2.18).[55]

The vigorous element of visual projection from the poet's text has been retained in this little genre scene of the nymphs scrubbing themselves as vigorously as any bather by Degas. To balance this earthiness, Manet has attended to the dry bureaucratic language in which Mallarmé's faun opens his daydream — "These nymphs I would perpetuate." Manet does not plunge them into the atmospheric chiaroscuro of *Le Corbeau* but gives their wet hair and vibrating silhouettes the same restless graphic line as the reeds, bent by the wind. To the faun's dreamy question, "Did I love a dream?" (*Aimai-je un rêve?*), Manet answers, in effect, "Yes, one made of the letters and punctuation of this poem." As Léon Rosenthal puts it, not without admiration, the nymphs "sont très écrites."[56] Only the faun, in his own spacious setting on the frontispiece among hillocks and scratchy scrubs, is endowed with dabs of pink wash suggesting, just suggesting, bodily presence (Figure 2.19). The dilute red watercolor applied to the body of the faun, but also to sky, ground, and the blank page that defines them, confirms the painter's touch without letting go

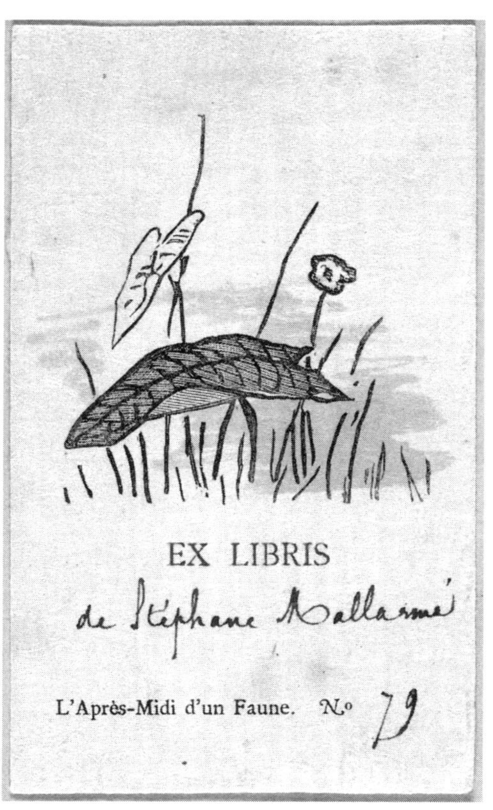

Figure 2.18. *Ex libris* and nymphs from Mallarmé's copy of *L'Après-midi d'un faune*, Bibliothèque Nationale, Paris. Which are the fleurons?

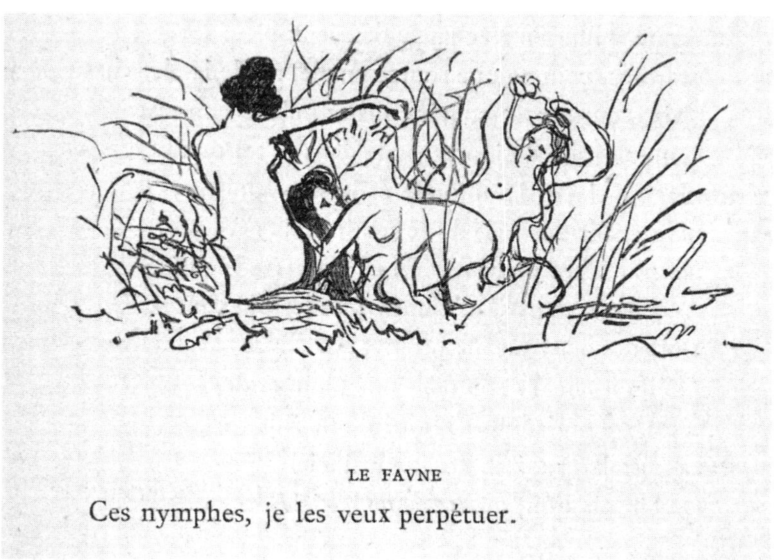

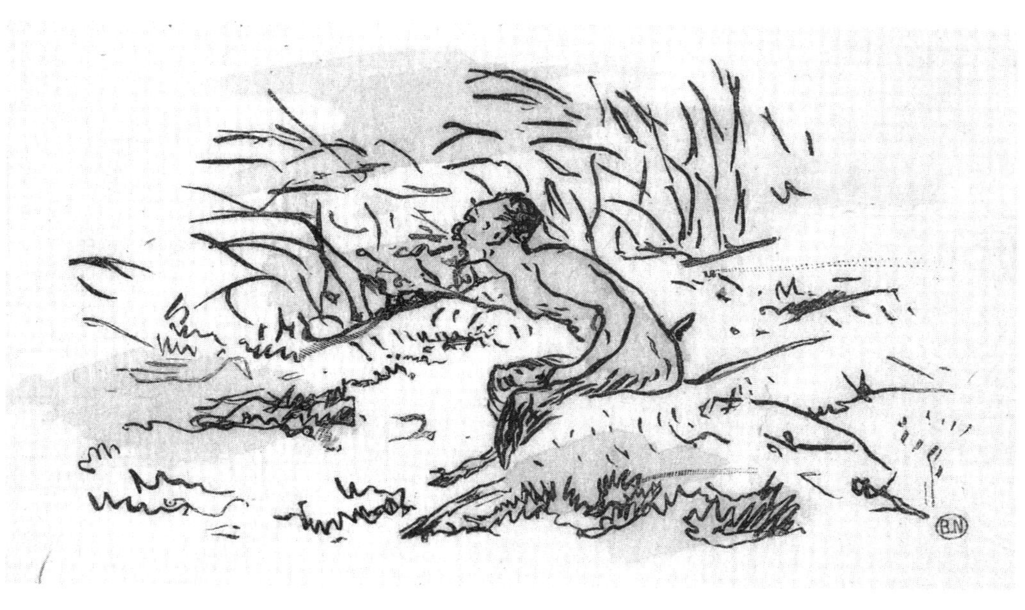

Figure 2.19. Édouard Manet, frontispiece to
L'Après-midi d'un faune, woodcut with watercolor, Mallarmé's copy, Bibliothèque Nationale,
Paris. In a letter, Mallarmé called this a
"curious illustration: melding within a very-
true modern sentiment both the Japanese
and the Antique" (*Correspondance*, vol. II,
p. 119). The absent-minded pink hand-coloring
adds to the windswept, dreamy ambiance.

of the unreliable, shifting tangibility of a daydream. The *Faune* as a whole, and its image in particular, is poised between the immersive subjectivity of *Le Corbeau* and a knowing, textual aesthetic.

The sophisticated swirl of romantic sensuousness and modernist self-doubt of the *Faun* is a fit preamble to Mallarmé's 1897 tour de force, *A Throw of Dice Will Never Abolish Chance (Un coup de dés jamais n'abolira le hasard)*. First printed in the remarkable multilingual journal *Cosmopolis* with many typographical compromises, the poem's revolutionary singular placement of each phrase on the page consummated a lifetime of book design by Mallarmé (Figure 2.20).[57] Mallarmé wanted *Un coup* to appear in an illustrated edition as sumptuous as *Le Corbeau*, with three lithographs by Odilon Redon (see Figure P.5), financed by art dealer Ambroise Vollard and printed by the venerable firm Didot.[58] This edition got as far as proofs, which Mallarmé edited again and again with calligraphic gusto, but was scuttled by Mallarmé's death in 1898. It was printed finally in 1914, three weeks before the outbreak of the First World War, by Mallarmé's son-in-law Edmond Bonniot — without Redon's illustrations, whose definitive placement in the book must remain a matter of conjecture.

The reception and interpretation of *Un coup* is even more intricate than its bibliography, but for our purposes only its art-historical reputation matters, and this is due largely to Rosalind Krauss's 1989 lecture on cubism, "The Motivation of the Sign."[59] Krauss suggested that rather than achieving an ineffable union of sign and meaning, as intended, Mallarmé's last poem anticipated Picasso and the linguist Ferdinand de Saussure in showing that meaning is not internal to the mind, as the private linguist thinks, but emerges, even in the most hermetic art, as a public play of differences between arbitrary signs, like that between the words *cat* and *bat*.[60]

This kind of reading has been derided for its faith that "that nice man Saussure in Geneva had got it right," as T. J. Clark has acidly put it.[61] But here I want to draw attention to another aspect of Krauss's argument. What makes *Un coup* relevant to cubism, surely, as much as its breakup of stanzaic structure, is its pictorial qualities. These

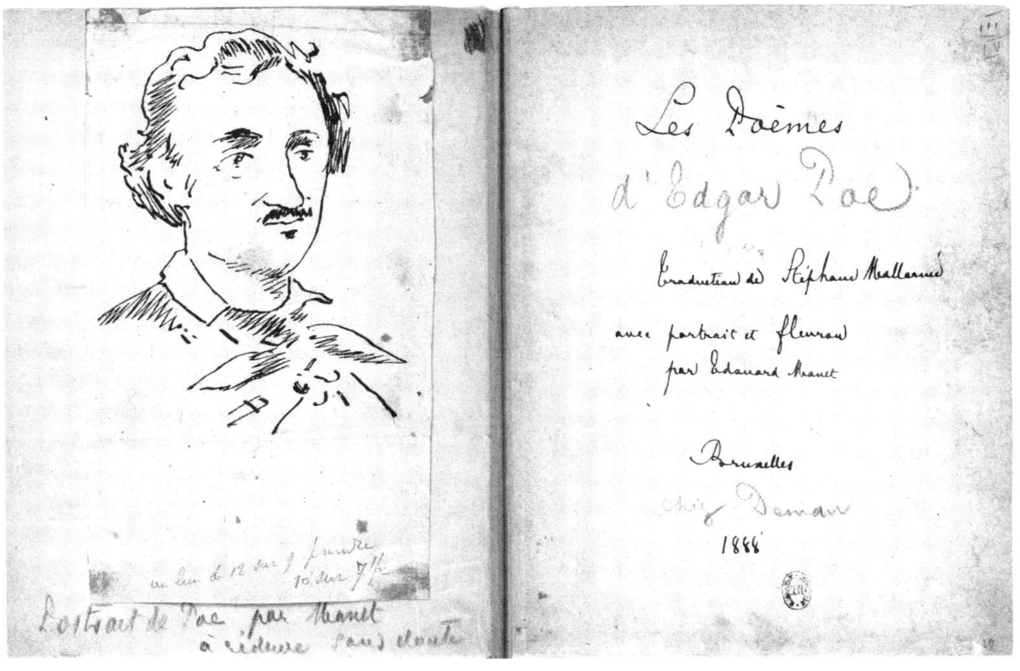

Figure 2.20. Stéphane Mallarmé's maquette for *Les Poèmes d'Edgar Poe* (Brussels: Deman, 1888), Bibliothèque Nationale, Paris. Note the etched portrait of Poe by Manet used as a frontispiece, usually regarded as an early etching of the 1860s.

are implicit in the vigorous graphic gestures of Mallarmé's edits; the poet himself in his correspondence referred to these pages sometimes as drawings, sometimes as prints.[62] And as his son-in-law Bonniot pointed out, Mallarmé's two-page layouts did away with any distinction between recto and verso: each page opening is a poetic unit, but also in effect a landscape (Figures 2.21 and 2.22).[63]

Given the expansive imagistic combinations of ink and paper that result — from single openings to the book-spanning *throw of dice* — what right have we to speak of pictures becoming linguistic and arbitrary rather than language becoming pictorial and subjective? In recent years, editors have taken issue even with Bonniot's loyal printing, for while he toiled to preserve Mallarmé's spacing, he substituted a blocky Garamond type for the delicate Didot of the proofs.[64] We might thus go further than the poem's first editor and say that symbolism reveals pictorial, subjective notes in even the most impersonal and arbitrary of the artist's tools: the printer's type.

The stakes of this debate are not quite as recondite and academic as Mallarmé scholarship sometimes appears. The issue of private language is important, and art has a say in it, precisely because of the vast potential to *misunderstand* one another that plays such a central role in our social, political, and everyday lives, and on which modern art has put a premium.[65] The attraction of reviving the idea of private language, or rather languages, to explain divergence in understanding lies partly in the possibility of seeing how our limitations are born of our strength as subjectively thinking, conscious creatures. The challenge, as yet unmet by theory, but boldly explored by Mallarmé or Manet or Poe, is to comprehend how we have shared knowledge of the subjective aspects of our mental life, as they flicker in one mind at a time.

Here we make contact again with Poe in "The Raven." For as he first pulls up his chair, the hero is an "arbitrary sign" theorist of language. The raven's first "Nevermore" is just a word: idly the host had asked his guest for its name. The man marvels that it can talk: "Though its answer little meaning — little relevancy bore." But he doesn't stop reflecting on the cause of this particular word choice: "'Doubtless,'" said I, "'what it utters is its only stock and

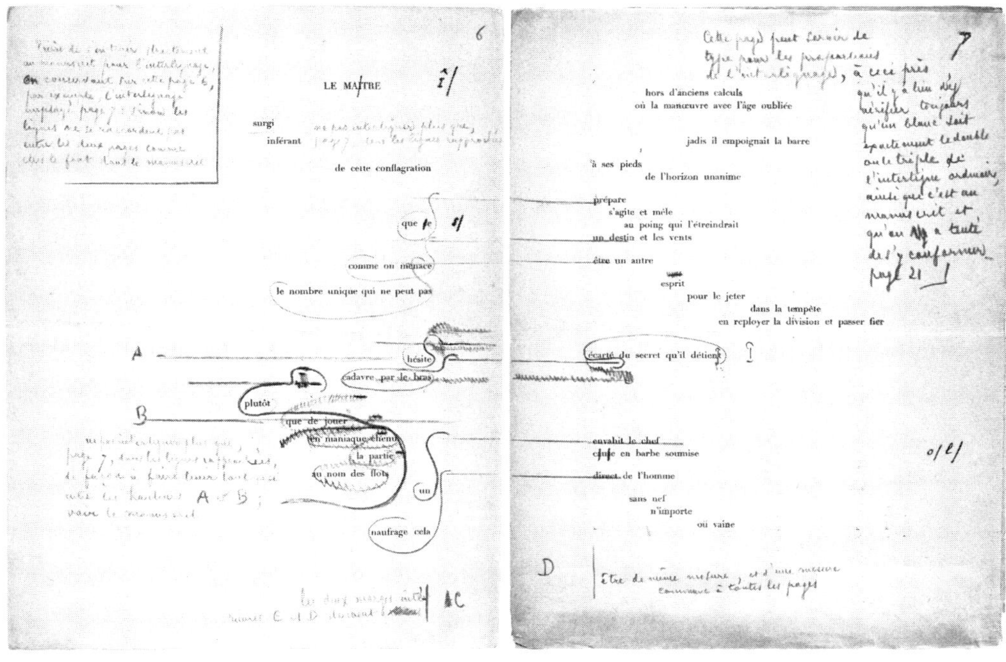

Figure 2.21. Stéphane Mallarmé, *Un coup de dés jamais n'abolira le hasard*, printer's sheets corrected by the author (1897), Bibliothèque Nationale, Paris. The first pair of consecutive sheets demonstrates in Mallarmé's calligraphic hand the unity of each opening.

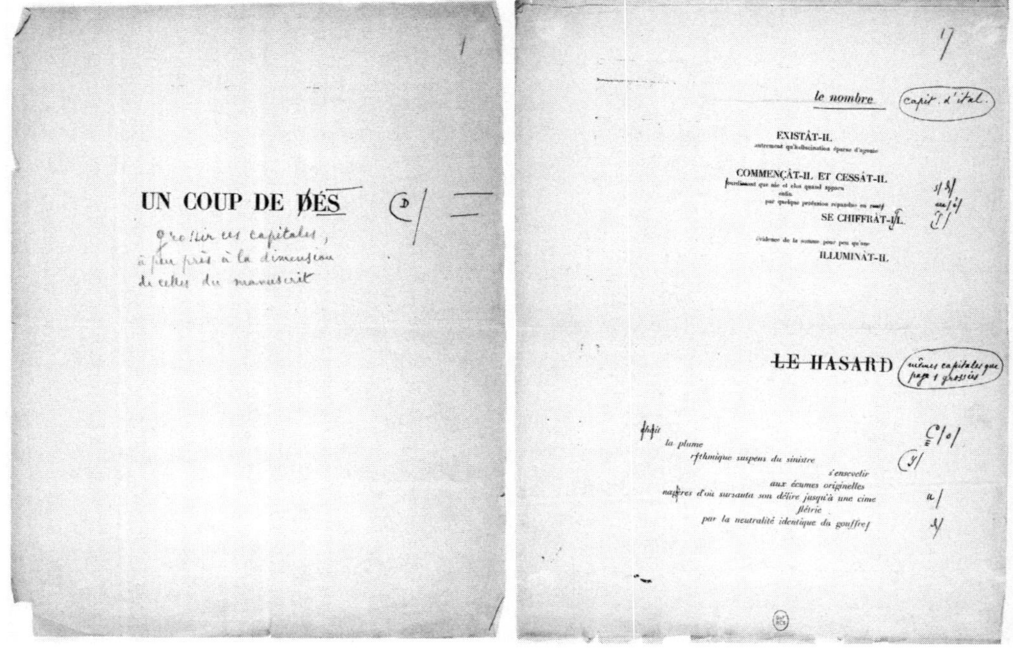

Figure 2.22. The two sheets here form the beginning and end of the title phrase of Stéphane Mallarmé's *Un coup de dés jamais n'abolira le hasard*. The poet's insistence that the font match "Un coup de dés" to "le hasard" is not meant simply to preserve the intelligibility of a sentence but also of a distributed multipage image.

store / Caught from some unhappy master, whom unmerciful Disaster / Followed fast and followed faster." The hero is soon disabused of this hypothesis, as the bird's one word proves an apt riposte to all his queries. So we get the true moral of the arbitrariness of words and images: they are not empty counters we associate with meaning by public fiat. Rather they are tangible things in the world on which we hang experiences, which often resemble one another sufficiently to allow us to grasp the same thoughts, and in the limited case allow us to sense divergences in how we see the world by our very inability to share them.

Mallarmé's last work shows that we cannot take the arbitrariness of signs to guarantee the publicity of all language. For as Frege drily put it, "One cannot forbid anyone from taking any arbitrarily produced object or process as a sign for something else."[66] Why can't that something else be private, like the experience called "THIS pain" by Wittgenstein's interlocutor? To see how such apparently solipsistic means of expression may rejoin the stream of communication, it would be best to start at the opposite end of the spectrum from *Un coup*, setting aside for the moment visionary graphic arrangements of poetic language to consider how language is used in everyday subjective experiences. Astonishingly, Mallarmé was involved in such a project too. In his day job as English teacher he designed a funny little textbook *cum* board game called Recreational English, or Box for Learning English by Playing Alone (*L'Anglais Récréatif, ou Boîte pour apprendre l'Anglais en jouant et seul*).

The "box," to which the Musée Mallarmé recently devoted an exhibition, is filled with engaging watercolor drawings by the poet in the service of quite pedestrian language exercises: a clock whose hands you can turn, a tongue you can pull out to learn the difference between "this" and "thick" (Figure 2.23), and perhaps most imaginatively, a butterfly anchored to a potted flower, which can alight on words like "over," "under," "toward," and "away" (Figure 2.24). For all the ingenuity Mallarmé displays in making language thus tangible, this combination of self-help and picture book could hardly function without a French key. Even with the key, a Mallarmé scholar notes wistfully that

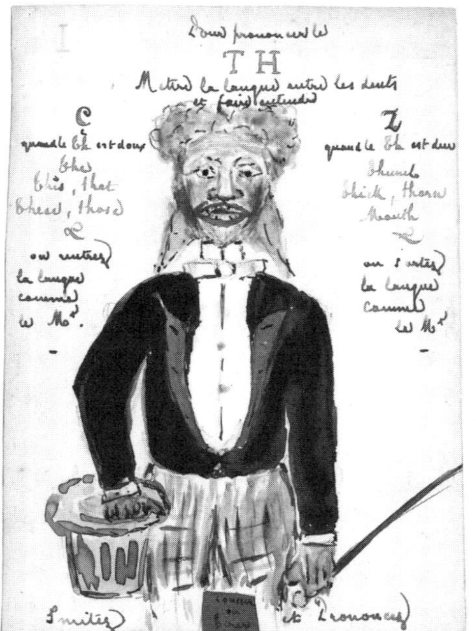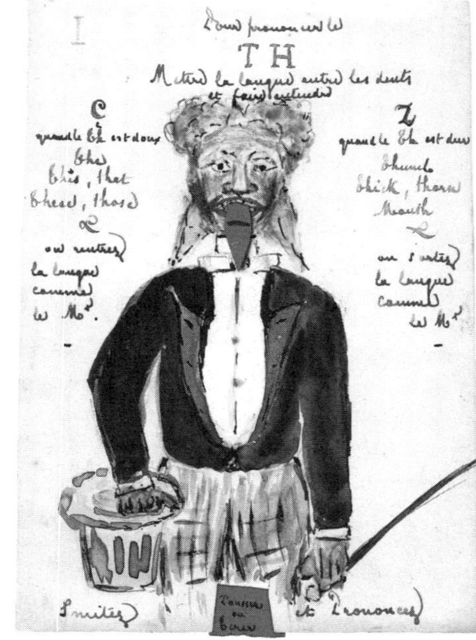

Figure 2.23. Stéphane Mallarmé, pages from *L'Anglais Récréatif*, paper, cardboard, metal, string, watercolor, Bibliothèque littéraire Jacques Doucet, Paris. The wild Gene Simmons-worthy tongue, movable through the tab at bottom, does not quite make for linguistic accuracy in pronouncing the two English "TH" sounds.

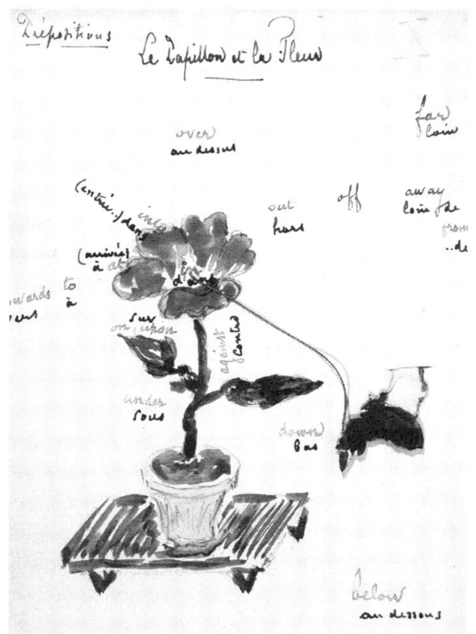
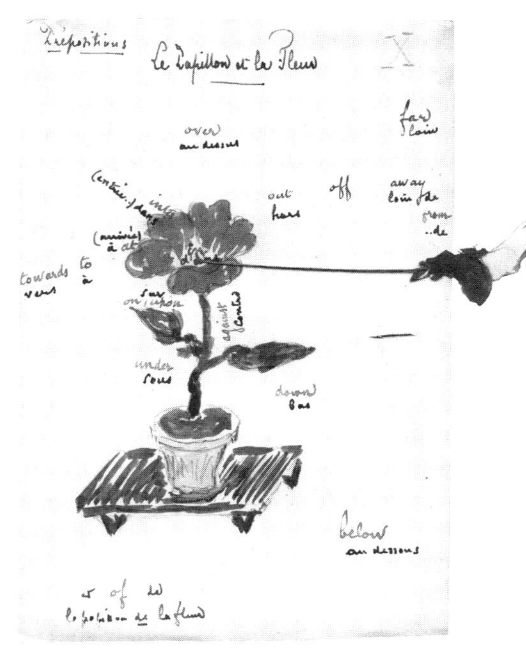

Figure 2.24. Two configurations of "The Flower or the Butterfly," Stéphane Mallarmé, *L'Anglais Récréatif*. By accident or design, the flower-butterfly, though it can reach every preposition, can also stretch off the page, into the void of real space.

the tongue is "a drawing destined to mislead anyone trying to use it."[67] One has the feeling this would be the case no matter *how good* an English teacher (and applied artist) Mallarmé had been. Insofar, Wittgenstein's critique of private language as a model of language *learning* has a point: the Promethean achievement of the young Augustine, stealing language from the hearsay of his elders as if it were a jealously guarded secret, may be too heroic an account of learning to speak to serve for classroom instruction (one might as well read Shakespeare aloud and hope for the best).[68] And so the "box for learning English by playing alone," though delightful, is no substitute for a teacher.

Be that as it may, Mallarmé was also in possession of an insight about language learning that Wittgenstein overlooked.[69] Far from being a purely social practice mysteriously passed to the individual from those already initiated (a view that raises the uncomfortable question of how language arose in the first place), language is both a shared possession and embodied in the conscious use of words by individuals. There is no substitute for the "aha!" moment encountered, say, in moving Mallarmé's butterfly "over" the flower, or in pasting color samples from a clothing catalogue into a notebook, as my English teacher did years ago in Romania. The experience of binding words to particular subjective experiences anchors one's grasp of the word as no rote learning can. The result is a social game with arbitrary counters, but also a subjective play in which words are associated with bodily memories. Private experience becomes the basis of public learning. This insight is at the center of the sometimes abstruse efforts of many symbolist artists and theorists.

Here the theorist must rejoin the historian, asking: is such a basis for public knowledge in private experience in fact *possible*? The theorist we followed in making a compelling case for the conceptual possibility of private language, Frege, did not directly answer his question whether the thoughts containing private and public senses of "I" are one and the same.[70] What he argued instead is that every *thought*, which is a sense with a particular kind of completeness, necessary to its being true or false, must be publicly graspable. That seems to settle the question as far as he is concerned, since it was thoughts and the pursuit of true ones by science that most interested him. But as he did not go back to deny the possibility of a private sense of "I," this leaves mysterious how "I am in pain" and "you are in pain" express the same thought, in case I self-identified myself in "that special way available only to myself," for which psychologists and empirically minded philosophers are nowadays finding much evidence.[71]

No wonder people think it more cautious to deny private senses entirely. Cautious or not, what we want is to get the matter right: and according to the critics of private language, we can no more coherently deny than affirm their existence. That leaves no reason to think that some perfectly clear "public" thoughts (the only kind Frege, unlike some of his modern interpreters, would admit) are not grasped with the aid of private component senses available only to single individuals and varying subtly between them. I have presented "The Raven" as an investigation by Mallarmé, Manet, *and* Poe into the intelligibility and aesthetic force that can be imparted in an effort to publicly articulate private experience. If it's intelligible, you might object, it is public. But if it *works* aesthetically, I reply, subjective resources are in play that cannot be shared, and the resulting combination of private and public resources may merit the name "private language," despite the public thoughts it consists in. What emerges is not *a* private language in the singular, the shared lingua franca of thought, but private *languages*, or better yet *private language* as a mass term, the possibility of complex aesthetic experience articulated by each person.

How the nineteenth-century discovery of private senses, and their circumscription in language and images (and sound, of course), issued in the art and theory of symbolism, the next chapter will try to show. In taking the full measure of this collision of symbolist experimentation and the romantic theory of subjective ineffability still alive in Poe, it will be necessary to backtrack to the early and mid-nineteenth century, engaging the realist and impressionist radicalization of romantic theory. This will allow us, in the final two chapters, to confront on their own terms fin-de-siècle efforts in art and theoretical science to marshal subjective experience to a practice of rational communication through new symbolic languages and images.

CHAPTER THREE

Where Do We Come From?

Symbolism's Psychologistic Roots

In the last chapter, Manet's and Mallarmé's efforts to come to grips with Edgar Allan Poe's poem "The Raven" transformed symbolism's troubled fascination with psychology into the philosophical problem of subjective thought, with all its personal and worldly implications. That chapter aimed to show that the problem is real and interesting, and not just an illusion caused by confused thinking — what Wittgensteinians, behaviorists, and cognitive scientists dismissed as the muddle of private language. We have seen some oblique ways to conduct a shared conversation about private experience, and that this possibility strengthens our conviction that there *is* such a thing to discourse about; but many questions, both historical and theoretical, were left open. Above all, in the symbolist return of "The Raven," language may seem to play an oversized role in cementing the possibility of private senses, one that may impede the extension of such practices to other pictures — and other nonlinguistic forms of art. In particular, the pivotal role of the first-person pronoun "I" in Poe's text made possible an appeal to the Fregean notion of private means of access. But pronouns, it would seem, have no direct visual parallel. I do not think this is quite right — self-consciousness about the very fact of looking is central to much art, as it is to the wordless, yet dramatically first-person, view of the raven's shadow in *The Chair*. But again that might seem like an accident of illustrating a particular, idiosyncratic text full of the first person. We can better judge the significance of private senses in art by shifting attention from drawing to color, which has so often been taken to exceed the reach of conceptual thought.

Van Gogh Paints a Photograph

In September 1888, Vincent van Gogh requested from his sister Willemien a carte-de-visite photograph of his mother that she had mentioned in an earlier letter.[1] On receiving it, he was at first glad to see their mother in good health, but soon enough he grew "impatient" with the monochrome print (Figure 3.1; see color plates). He painted an oil copy of the photograph in "the same ashy coloration," adding in a letter to his brother Theo that he has "gone to terrible trouble to find the combination of ashy tones with grey pink."[2] In October he described the palette as "ashy, on a green background, and her clothes carmine."[3] We can only assume that he had in mind the final state of a painting now in the Norton Simon Museum in Pasadena, California, whose color scheme evolved according to a rationale he revealed in another letter to his brother just one day earlier: "I can't look at the colorless photograph, and I'm trying to do one with harmonious color, as I see her in my memory."[4] A century of hyperbolic myth-making, culminating recently in a wildly romantic exhibition, *Van Gogh's Bedrooms* (Art Institute of Chicago, 2016), may have schooled us to expect anything but sobriety from Van Gogh's letters: which is too bad, because he is one of the most precise and thoughtful writers among major artists, whether he is discussing philosophy, aesthetics, the mind, or the concrete business of artmaking. Alas, his account of painting his mother's picture does not seem to have satisfied recent viewers. Wall labels in the Norton Simon are apologetic about the "pallid, unnatural green," calling the mother "somewhat pale and sickly in tone."[5] Van Gogh might have responded that instead of faking flesh tones he could not see, he applied cobalt blue and yellow in response to the narrow tonal range of the photograph, imbuing it with "a sense of warmth"—a sense of warmth he attributes to the image of his mother in his memory. Of course, the phrase "a sense of" is the sticking point:

in painting her thus, the result is no longer a private experience provoked by certain colors, but an object visible to others, who have no access to Van Gogh's memory any more than he had access to his sitter. And here an odd fact crops up: even the physical painting itself is not so firmly shared as all that, for the photograph I shot myself differs considerably from the cooler blues and warmer yellows of extant reproductions. And the colors of my own photographs diverged when I took more than one, though certainly the image *looked* the same as I moved around it (Figure 3.2; see color plates).

Having seen this painting repeatedly over the past three decades, most recently on taking these photographs shortly before writing this, I vouch that the colors in the image at left are truer to the picture than those at right, or in any of the printed and digital reproductions known to me. But perhaps that is only a local truth. Color is, as much as pain, a paradigm case of the private. Gottlob Frege, in trying to answer his own question whether thought is always and necessarily public, perceived a limit in color, and explored it most sensitively:

> My companion and I are convinced that we both see the same field; but each of us has a particular sense-impression of green. I notice a strawberry among the green strawberry leaves. My companion does not notice it, he is colorblind. The color-impression, which he receives from the strawberry, is not noticeably different from the one he receives from the leaf. Now does my companion see the green leaf as red, or does he see the red berry as green, or does he see both as of one color with which I am not acquainted at all? These are unanswerable, indeed really nonsensical, questions. For when the word "red" does not state a property of things but is supposed to characterize sense-impressions belonging to my consciousness, it is only applicable within the sphere of my consciousness.[6]

What is nonsensical is not the possibility of seeing differently *as such*, but only the question of this private difference put into public language; of course, I can imagine the field all green or all red. In doing so I can name my imaginings, marking them as private senses. I cannot ask which of these senses are my colorblind friend's, because for me, the two states are distinguishable, and we know, by his own report, that he has no two such distinguishable imaginings. So I can hardly ask him which of *mine* he possesses. The more general point is that even with someone who sees "as I do," it is only correspondences in public language, backed by analogous behavior (finding the strawberry), that reassures me of our like-mindedness. To expect more from the argument would be to deny the subjectivity of vision, if not to deny that we see at all.[7] So we may go on and argue about Mallarmé's tethered butterfly or Van Gogh's mother without the dogmatic assumption that there is no difference at all in what we see. Without being Van Gogh, we may take his picture as his public account of a private experience.

The twentieth-century habit of considering private language as an obstacle to insight in both art and science thus has to be overcome if we are to make sense of what lies beyond it: how objectivity is possible, and how it makes possible not only the communication of knowledge and the coordination of practical action but also what knowledge we have of our diverse subjective resources, how they bind and separate us. But the distrust of subjectivity is no purely twentieth-century phenomenon. We have seen already in the preface that symbolism in art and theoretical science emerged in reaction to a putatively scientific treatment of human and animal cognition in the latter half of the nineteenth century, one that put great emphasis on the mind but insisted on treating it mechanistically. This materialistic psychologism manifested itself also in a distinctive assumption of symbolist theory, especially as applied to painting: a resolute aesthetic formalism, the idea that "a painting — before it is a battlehorse, a nude woman, or some anecdote — is essentially a flat surface covered with colors assembled in a certain order."[8] Maurice Denis's first axiom of his pseudonymously published 1890 manifesto *Definition of Neo-Traditionism* is a prototype of twentieth-century formalist art criticism from Paul-Henry Kahnweiler to Clement Greenberg.[9] But what might take later formalists aback is the psychological rationale the symbolists gave for formalism. Thus Denis's friend Paul Sérusier, in his *ABC de la peinture*, a kind of symbolist catechism: "Nature is the totality of objects our senses reveal to us."[10] As with Frege's optician, who is forced to concede that optic nerves and light waves are finally nothing but sense impressions in the mind of the experimentalist, we are left with "nothing but" canvas and pigment, which supply the sensations available to the artist to combine. The world, being purely material, is a result of our psychology. Denis in his *Définition* agrees, if hesitantly: "[Nature] probably means: the totality of optical sensations."[11]

Monochrome and Arbitrary Signs

How do we get from a flat surface covered with paint to the image imbued with meaning? Or is it the other way around? A picture will help. Consider the fifth plate of Alphonse Allais's 1897 *April-First Album*, which contains "1) a spiritual preface by the author; 2) *seven magnificent plates* engraved in different colors in copper; 3) a second preface, almost as spiritual as the first; and finally, a funerary march specially composed for the funeral of a deaf great man."[12] The album, in keeping with its April Fools' Day title, is of course harmless fun. Yet the "Tomato Harvest by Apoplectic Cardinals on the Shore of the Red Sea (*Effect of Aurora Borealis*)" (Figure 3.3.a; see color plates) carries in its subtitle a sharp parodic barb aimed at impressionism. The familiar French parenthetic *effet*, first applied by Anne-Louis Girodet a century earlier to a painting he nearly ruined by excessive admixture of olive oil (*Endymion: Effet de lune*), becomes familiar, and finally respectable, in works like Caillebotte's *Vue de toits (Effet de neige)*, shown at the fourth Impressionist Exhibition in 1879.[13] Allais, a comedian, first exhibited a picture under his own name in Jules Lévy's charitable Exposition des Arts Incohérents in 1884, "affixing to a wall a sheet of Bristol paper absolutely blank," as critic Félix Fénéon put it, under the title *First Communion of Anemic Young Women in Snowy Weather* (Figure 3.3.c; see color plates).[14]

The comic modus operandi militates against attributing to this monochrome any serious critical or aesthetic intent. The given title, an unlikely mishmash of elements, human and landscape, conform-ing to the tonal key of the picture, is used to ridicule both avant-garde complacency and pompous academic narrative. Allais has a sharp eye (and ear) for the awkward.[15] But there is more to this comedy than meets the eye: first, as Allais himself claims, he is not the inventor of the monochrome (or "monochroïdal painting," as he calls it), but gives credit to the dubiously titled *Combat of Blacks in a Cave*, which poet Paul Bilhaud showed in the first Arts Incohérents (1882), and which is now known only from the first plate of Allais's *Album*.[16] Allais too had followers, notably Émile Cohl, who in 1884 exhibited another black monochrome, and as late as 1910 presented, in the film *Le peintre neo-impressioniste*, hand-colored monochromes as the work of his hero.[17] If the monochrome was nothing but a joke, it is remarkable that it did not wear thin sooner. Allais, as if recalling Manet's fussing with Holland and China paper in printing the "Le Corbeau," went so far as to publish a second white monochrome in the *Album* (Figure 3.3.b; see color plates).

There is a method to this madness, lighthearted or otherwise. Allais in his preface declares the painter not to be the "ridiculous artisan who needs thousands of different colors to express his petty conceptions." Who is to say that, being printed in jest, these works do not work precisely as the symbolists claimed all paintings do? The Nabis also strove to "repress" traditional painterly virtuosity, an element with no apparent place in their ontology of the world as a pre-perceptual field *cum* canvas and pigment.[18] In the void, a kind of subjectivity would find its place: "The primary function of the monochrome, in other words, is to give *pleasure* (by affording conceptual play), which is only sanctioned spuriously by appeal to any thoughts it entails," as a distinguished historian of postimpressionism reads Allais.[19] The monochromes, in their simplicity of means, may well be the most radical application of Denis's and Sérusier's strictures that all painting is pigment on a support (printer's ink on paper, in this case). We should not hurry to dismiss the whimsical monochrome as mere silliness, even if it is that, too. For the same status has to be accorded to Kazimir Malevich's 1915 *Black Square*: a 2015 radiographic study of it reveals Bilhaud's motto "Negroes battling at night," probably taken from Allais.[20] The generative status of the monochrome — both in setting very minimal restraints on meaning and in serving as a consistent visual field to which meaning may be attributed — was already acknowledged in Allais's titles and prefatory remarks. Indeed, the humorist zeroed in on a sore point for those avant-garde artists who wished to combine coloristic sensitivity with the right to dictate meaning. Take the only apparently opposite case of James McNeill Whistler, who lent his compositions titles emphasizing their formal properties:

> Why should I not call my works symphonies, arrangements, harmonies and nocturnes? I know that many good people, whose sense of humour is not very capacious, think my nomenclature funny and myself eccentric. Yes, eccentric is the best adjective they find for me. I admit that it is easier to laugh at a man than to appreciate him. . . . My picture of a "Harmony in Grey and Gold" is an illustration of my meaning — a snow scene with a single black figure and a lighted tavern. Now that to me is a harmony of colour only. I care nothing for the past, present, or future of the black figure, placed there because the black was wanted at that spot. . . . They say, "Why not call it 'Trotty Veck,' and sell it for a round harmony of golden guineas?" I reply simply that I will do nothing of the kind. Not even the genius of Dickens should be invoked to lend an adventitious aid to art of another kind from his.[21]

The *Nocturne in Grey and Gold*, also called *Chelsea Snow* (Figure 3.4), is a lucid, if subtly atmospheric, cityscape; there is no danger of mistaking what is up or down, snow and sky, man or lamp. Whistler is merely asking us to regard the picture as if such distinctions did not matter. But the striving after such psychic *effect* is not too much different from, if the reverse of, Allais's asking us to see apoplectic cardinals in a red field. At stake in both kinds of pictures, and in the laughter they provoke, is the breakdown of what Stephen Eisenman calls the social contract between artist and viewer, giving or at least implicitly suggesting standards by which the viewer may make sense of the artist, and, in the other direction, the kind of reception the artist may expect from a viewer.[22] Sheridan Ford, in reprinting Whistler's manifesto, sternly called it "The Picture Defined," subtitling it "Pictorial Art Independent of Local Interest."[23] But what is left when *all* the "local interests" of painter and spectators have been swept

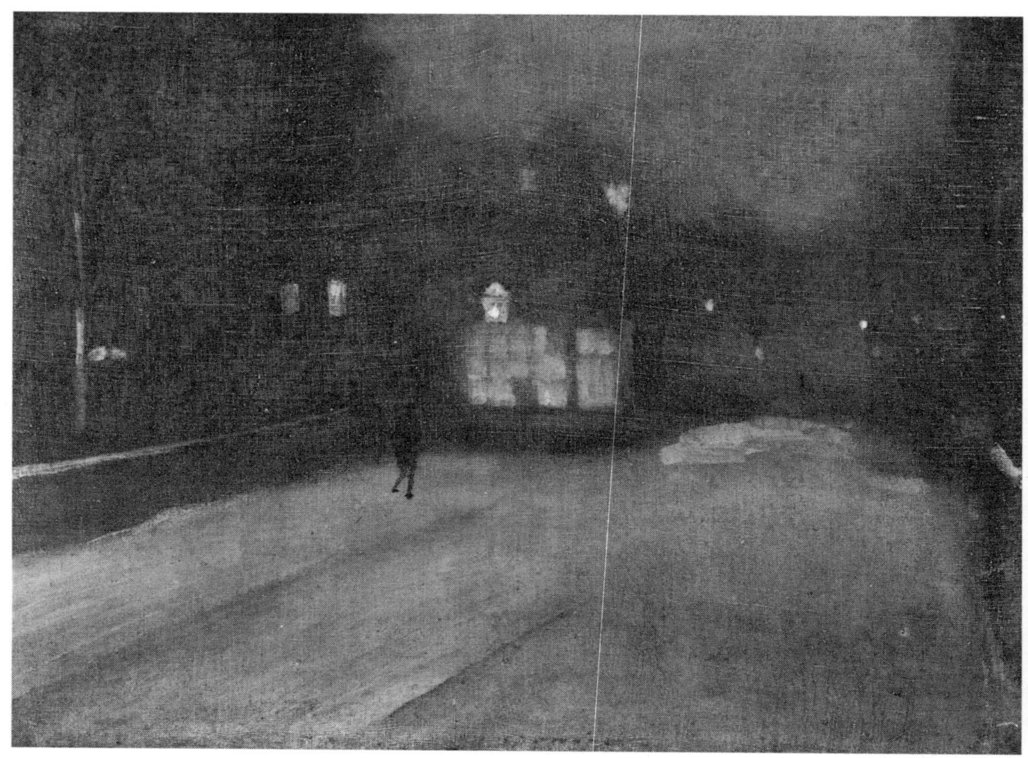

Figure 3.4. J. A. M. Whistler, *Nocturne in Grey and Gold: Chelsea Snow* (1876), Fogg Art Museum, Harvard University. Whistler is as usual ironically undermining himself with the suggested title "Trotty Veck" (the hero of Charles Dickens's 1844 serial *The Bells*) — the gaslight and shivering figure in the snow are very Dickensian, once he mentions it.

away? Only the visible physical object and, lacking such objectively shared structures as narrative and subject matter, wishful ascriptions or formalist descriptions of the object arrived at in accordance with psychological laws of action and reaction, consistent with, but by no means necessitated by the thoughts of the artist. If the relation of content to object were more stable, Whistler would not have had to swat away the imposition of parody titles like *The Yacht Race, a Symphony in B sharp*. A symphony is in B-sharp major if that is its key, in a way that Whistler's works are not "self-explanatorily" harmonies in one color or another.[24] The nineteenth-century painter is in much the same position as the nineteenth-century "formalist" mathematician, who assigns to empty figures an arbitrary role in games of counting:

> Let *a, b, c*... be some experienced or mental objects or relations of objects; then one can think of *a* and *b* as in some way purely conceptually and formally linked to one another, and view as the result of the linkage a new object or a new relation *c*, which, because it can appear in all further deductions in the place of the two members *a, b*, insofar as they are linked, may be called *equal* or *identical to* (German *gleich*) the linkage.[25]

The formalist, whether in science or art, claims to institute, or discover, rules by which mere visual figures may be fruitfully manipulated, as in a game. But this Humpty Dumpty practice is dependent on more familiar practices of meaningful sign use. As Frege pointed out, the formalist rules of sign use, like $a + b = b + a$, depend for their plausibility on implicit knowledge of their content, in this case the commutativity of addition. "Indeed no one would be tempted to see a rule of play in ◁ § ⟫ ⋈ ⟫ § ◁ ," though it is obtained merely by substituting unfamiliar symbols into the equation expressing the commutative law as printed above.[26] Analogously, the supposedly arbitrary generation of pictorial themes by Allais and other "incoherents" depends on standard practices of pictorial reading that are loosened, made contingent, but not at all rejected by the more atmospheric works of Whistler and the impressionists. The danger, Allais seems to be suggesting in the same comic tone affected by

Frege, is that truly arbitrary signs may stand for anything. The free subjective interpretation of mere figures, then, is a kind of dead end of the psychological understanding of pictures as sense impressions to be ordered by a conscious mind. They represent a logical, if not a historical, end of painting. To see how the psychologically astute art of the nineteenth century got there, and got around it, we must backtrack and start anew.

Romantic Ineffability

Difficulties concerning the purely formal image, though widespread in a milieu like symbolism that emphasized a dualism between mental image and its material vehicle, were by no means new in the history of art. Baudelaire, reckoning Eugène Delacroix's achievement shortly after the artist's death in 1863, asked why he was such a *suggestive* painter, calling to mind both novel thoughts and recollections.[27] In answering, Baudelaire anticipated his friend Whistler:

> A well-drawn figure fills you with a pleasure entirely alien to the subject. Voluptuous or terrible, this figure owes its charm to nothing but the arabesque that it cuts in space. The limbs of a martyr being skinned, the body of an enraptured nymph, if competently drawn, comprise a kind of pleasure into whose components the subject does not enter at all; if it is otherwise for you, I shall be forced to think you a torturer or a libertine.[28]

Baudelaire's point is not that Delacroix produced "arabesques and nothing but," much less that he trafficked in uninflected invitations to tell tales à la Allais, but that his works' formal properties alone were equal to the task of producing the feelings and thoughts their narratives demanded: a shortcut from content to the desired effect, bypassing the visible painterly structure, would presuppose the kind of *parti pris* the critic sardonically disavows as immoral. Again, as with Denis three decades later, it is *psychological depth* that demands this physical flatness:

> But, in the end, sir, you will doubtless ask, what is this mysterious *je ne sais quoi* that Delacroix, for the glory of our century, has translated better than any other? It is the indivisible, the impalpable, the dream, the nerves, the

soul; and he did it — observe him well, sir — with no means but contour and color; he did it better than any other; with the perfection of a consummate painter, with the rigor of a subtle writer, with the eloquence of a passionate musician.[29]

Beneath the hyperbole, Baudelaire in fact ascribes a subtly literary function to Delacroix's purely optical means: that of plumbing the "dream, the nerves, the *soul*." Here, as elsewhere, Baudelaire was improvising on a venerable tradition. The early nineteenth century, with its cult of genius, was sensitive to the unavailability of other persons' interiority. E. T. A. Hoffmann faced up to it in his 1816 story "Der Sandmann," interrupting an epistolary exchange to complain volubly about his inability to put experience into words:

> Have you, gentle reader, ever experienced anything that totally possessed your heart, your thoughts, and your senses to the exclusion of all else? Everything seethed and roiled within you; heated blood surged through your veins and inflamed your cheeks.... And wishing to describe the picture in your mind with all its vivid colors, the light and the shade, you struggled vainly to find words.[30]

Hoffmann, being a romantic, thought it the artist's business to attempt the impossible. Not surprisingly, he reached for a visual metaphor, fantasizing that, had he been an "audacious painter," a dashed-off sketch might have succeeded where more deliberate means failed.[31] Art historians have not always taken seriously such romantic efforts, though of course they have recognized their pertinence to romantic art. The trouble isn't that the efforts, and especially the painterly metaphors that often clothe them, are naïve: it is that, if they were fulfilled, the interpreter would be left without a rational task, unless it is that of intuitive guru articulating the obvious.

This risk is palpable in a work that perhaps more than any other signals the transition from romanticism to realism, from an excess of inarticulate subjectivity to an excess of worldly texture. The picture comes fairly early in Gustave Courbet's series of self-portraits of the 1840s, and it is said to represent, somewhat dubiously, a man driven mad by fear (Figure 3.5; see color plates). The picture has the loud vapidity typical of the young Courbet: the man foppish in a striped jacket, cape, and slippers meant to evoke the Renaissance but recalling rather cheap provincial theater. Courbet may have wished to catch him "in the act of leaping from the edge of a cliff," but he is certainly *not* "springing directly toward the beholder," as Michael Fried hopefully suggests, for his left knee is planted on the turf, an awkward position from which to tumble to one's doom.[32]

When Fried more cagily ventures that "the image as a whole perhaps suggests that he has been driven to his insane deed by contemplating the abyss before him," then transposes this to a familiar concern with the "vertiginous gulf between sitter and beholder — and ultimately . . . between painting and beholder," we may wonder where the artist's subjectivity ends and that of the interpreter begins.[33] To push the question away, citing "intuition" or the "ideal spectator" is not just incurious: it is one trap of psychologism, relying on myths of psychological uniformity to bridge a gap — which cannot be bridged, for it is logical, not empirical — between private and public sense.[34] Can we do better, aware of the difficulty of what the painting, without quite being able to make it clear, is trying to accomplish?

What we see in Courbet's painting is exceptionally revealing for the difficulties incurred by a follower of Hoffmann and a contemporary of Delacroix. The mental disorder of the hero is splashed upon the landscape: The precipice meant to swallow him, if there is one, is effaced by the dull green underpainting of the grass, which also makes an incursion into the man's left knee. Below it, the brown, white, and desultory specks of darker green, far from revealing the strata of an unfinished painting, are dramatic tricks through which the painter has sought to embody the figure's madness — as if he went mad himself while painting (himself). It is the heart's sketch: Hoffmann taken at his word. But of course the effect is inadequate and confusing: we stand nearby, hovering over the void, and stare the man earnestly in the face, taking in the details of his costume and the far-off landscape, while below, nature or at least vision has gone haywire. The odd separation of the man from his dissolute way of seeing was to haunt the nineteenth-century art that took subjectivity seriously: from the romantics, whose heir Courbet is, through realism and the invention of photography, to impressionism, color photography, and beyond.

The Doubling Problem

The evidence of Baudelaire, Hoffmann, and Courbet suggests that psychologism arose not just from the progress of science, but from an admirable romantic ambition to reveal minds in all their disorienting subjective richness. If the rise of experimental science made psychological explanation appealing in domains as diverse as politics and physics, the revolution in optical technologies, culminating in the invention and popularization of photography, held the promise of making visible the very act of seeing.[35] Late romanticism, and the realism, impressionism, and symbolism that built on it, may have learned from the camera "their detachment from the material with which they worked," issuing in the "icy aestheticism" noted by T. J. Clark.[36] But photography also made possible a new ideal: the look and *feel* of subjectivity made objective, printed out into the world for public inspection. It is from this point of view that I approach photography, and the painting that claimed to go beyond it in the objective transcription of subjective interiority.[37] If the failure of that project was as conspicuous as Courbet's passionate self-derangement, it carried within it the first, tentative efforts to articulate a balance between the subjective and objective dimensions of human thought through picturing.

In the remainder of this chapter, pictorial realism, and the impressionism and symbolism that wrest from it the claim of psychic accuracy, will be treated in terms of a *philosophical realism* affirming the existence of a world prior to representation. Such a project seems to run counter to the dominant understanding of late nineteenth-century art as turning aside from the midcentury realist's *vrai* in search of a subjective *vraisemblable*, if not a world of unmoored feeling and fantasy. But it does not in fact, if reality includes subjectivity, and furthermore, if the latter is the key to our grasp of the former. The subjective turn in nineteenth-century art, well beyond painting, really did put the problem in this nearly Kantian manner.[38] Thomas Hardy, in the 1895 preface to the book version of a serial novel, excuses the story's bleakness by claiming that "*Jude the Obscure* is simply an endeavour to give shape and coherence to a series of seemings, or personal impressions, the question of their consistency or their discordance, of their permanence or their transitoriness, being regarded as not of the first moment."[39] Hardy's noun "seemings" alone presages the new attitude toward subjectivity among reputed realists. The complementary nature of research into the subjective and the objective — in short of an attempt to delineate the boundaries of the real, be it private and mental or public and physical — was also central to nascent art history, which saw in it a long-term development, if not the discipline's own raison d'être. Thus Alois Riegl, summing up the trajectory of Western art in his 1902 review of a book by the Salpêtrière doctor Paul Richer:

> [Art] is rooted in our ambivalent approach to nature, that is to things, which art has to simulate. On one the hand we still live to some extent in the traditional view that things exist wholly independent of us and thus confront us, the observing subjects, as objects. One finds this view in its purest naïveté in ancient Oriental art; the ancient Greeks already knew that the actual appearance of things always carries with it a subjective moment, belonging not to the object, but to the subject, who is its carrier. . . . The relation between the objective and the subjective in the appearance of things ruled with its periodic shifts the whole development of the visual arts until today; the development was in general always bent on the increase of the subjective moment, that is, it sought nature less and less in an independent Objective and more and more in a Subjective determined by the observer.[40]

Riegl was not out to decry the growing subjectivity of art, as Richer did, advocating a return to Greek standards of anatomical precision. The art historian expressed only modest scruples that we may "not be ripe" for what contemporary artists (he cites the symbolist Jan Toorop) count as nature.[41] Yet the view he attributes already to the ancient Greeks — that *any* appearance of an object contains a subjective moment — struck with a vengeance in the art and science of his time.

For if every object, as we perceive it, already carries with it a subjective component, the realist dream of seeing the object as it really is (even to a subject) recedes with every leap forward in verisimilitude. The invention of light-fixing in photography, and even of color-fixing in color photography, to say nothing of the faithful observation of social type in realism and momentary plays of light in impressionism, only raised fresh theoretical doubts about the achievability of the task of mimesis. Symbolism has often seemed a skeptical or exhausted reaction against this kind of naturalism. There is something to that, as we saw in reading Aurier. But there is also more: symbolism tacitly accepted the modern ideal of representation summed up by Riegl as ever more faithful depiction of the object by means of bringing more of its subjective import into view. Must traveling this path result in the kinds of unintelligibility we have been exploring in Courbet, Allais, or Whistler? It is my conviction that even symbolist critics of mimesis, like Aurier, did not reject the realist project merely out of prejudice, but because they found the demands of subjective realism impossible to meet, even in principle. The ever more credible mimesis of experience ended in apparent alienation. Why this should be so will emerge in the course of this chapter.

Riegl's "subjective moment" came to dominate criticism with the rise of photography. I will not rehash that often-told story but will only register its little-noted impact on a formerly central aspect of artmaking: the practice of perspective. In 1836, the obscure English miniature painter Arthur Parsey, previously noted for a manual of ivory painting and a particularly hapless attempt to square the circle, published a treatise called *Perspective Rectified*. Despite the immodest title, the book was only an introduction for artists and amateurs unschooled in geometry; it showed how to construct various solids with and without the use of vanishing points. In the closing remarks, Parsey did bring up the convergence of objects perpendicular to the direction of vision, for "to be critically correct, perpendiculars should only be drawn so, when the eye is central."[42] But in his practical instruction he brushed off this technicality as one among many:

> If we insist on rigid and microscopic nicety, no line is purely horizontal, any more than any two perpendiculars are parallel; for, as by the laws of gravitation, every thing tends towards the centre of the earth, lines must in a trifling degree diverge; and as we inhabit a spherical body, the tops of equal perpendiculars will describe a circle; but it would be difficult to make it evident to the senses; it would be a puzzling calculation to compute the area of a segment contained within a horizontal line of one hundred yards long. As we draw perpendicular lines parallel, let us be content to draw horizontals straight…[43]

As if to underscore his conviction that such subtleties may be ignored in art, the book's last diagram is an impressive foldout of a staircase, devoid of any unsightly recession (Figure 3.6).

Four years later, Parsey was less temperate. The second edition of 1840 bore an excited title, *The Science of Vision, or Natural Perspective!, containing The Natural Language of the Eye*, and devoted itself to substantiating the convergence of vertical lines brushed aside as a quibble in the first edition. Parsey went so far as to call his frontispiece (Figure 3.7), reproducing a drawing exhibited at the Royal Academy in 1837, "the first picture drawn with optical accuracy." As English mathematician Augustus De Morgan later observed in his *Budget of Paradoxes*: "Of course the building looked very Egyptian, with its sloping sides."[44] In his own time, Parsey's results were attacked with more vehemence than insight: in the *Westminster Review*, George Henry Lewes wrote that since Parsey's point is so simple and self-evident, and yet *no one* had adopted it before, something *must* be amiss with it, suggesting finally (and more plausibly) that convergence is often unnoticeable.[45] A more substantial critique came from the young John Ruskin, writing as Kata Phusin ("according to nature," in Greek) in the *Architectural Magazine* of 1838, to which Parsey had contributed.[46] Ruskin granted Parsey the obvious fact, denied by most other critics, that in looking up, we see the sides of a building converging: "Let him go to the bottom of the monument, stand 12 yards from its base, and look up; and then let him talk about the nonconvergence of perpendiculars, if he can."[47] Indeed, Ruskin and Parsey agreed on principles, but not on how to draw. For, Ruskin argues, we behold

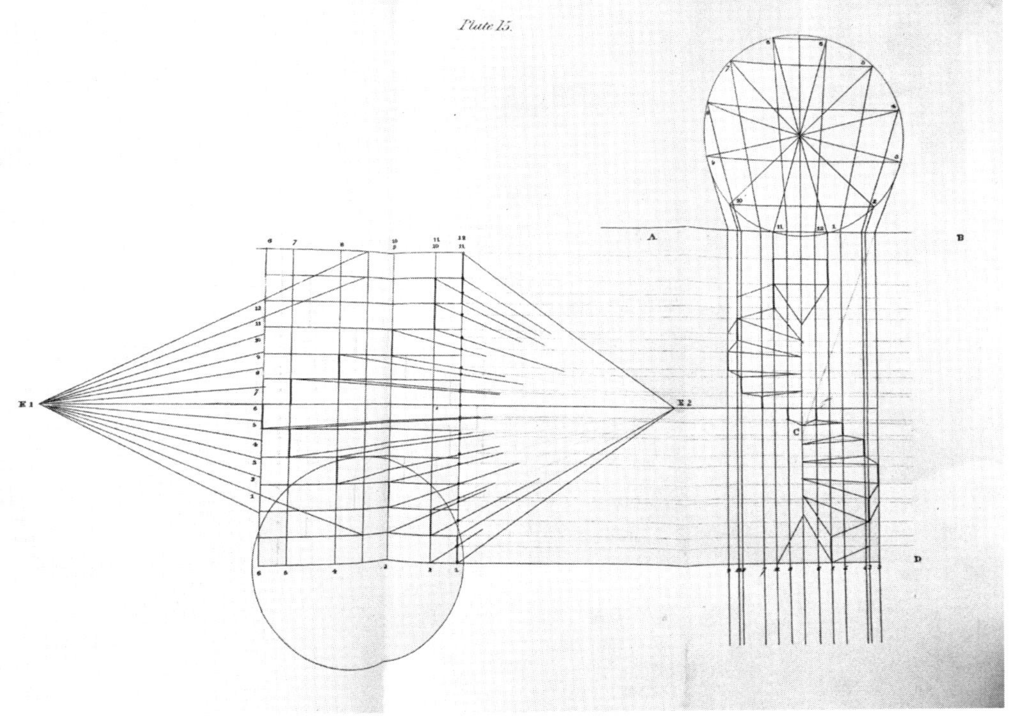

Figure 3.6. Arthur Parsey, *Perspective Rectified; or, The Principles and Application Demonstrated* (1836), n.p.

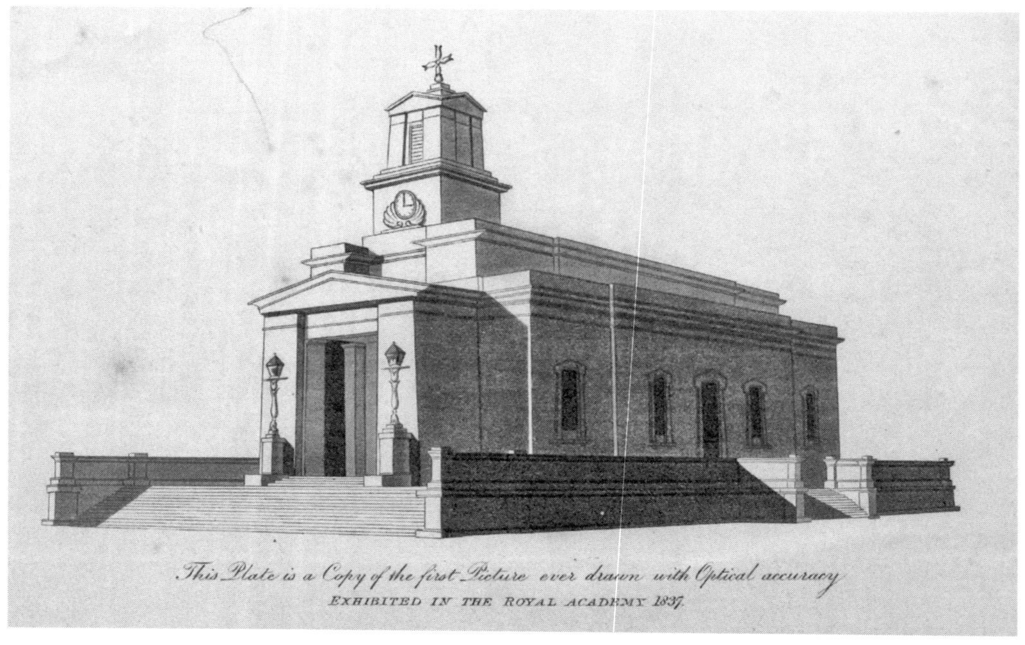

Figure 3.7. Arthur Parsey, *The Science of Vision, or Natural Perspective!* (1840), frontispiece. The text (pp. 81–82) concedes that only a very near approach would give such "extreme" results: in effect, a giant eye positioned at the railing.

pictures in perspective as well. If the picture is conceived as a pane of glass between the motif and the viewer, the light from the object and that from the picture plane will converge equally; so on the picture the lines should be parallel, as they are in the building itself, and we will get the right amount of convergence just from the recession of the pictorial support.[48] At stake is the question whether painters should represent "pencils of light" connecting viewers with the scene depicted and thus liable to the same perspectival distortions (Ruskin), or as self-sufficient windows on the mind, depictions of how objects appear to a viewer *not* standing before a picture, and thus not liable to further distortion (Parsey). Neither view can show things "as they are" rather than "as they seem": for both views make provisions, albeit differently, for the position of the eye and the movement of the head.[49] Ruskin goes so far as to try to show, with the help of a sketch (Figure 3.8), how a vertical vanishing point below one's feet and one above one's head add up to "points of sight on a vertical horizon."

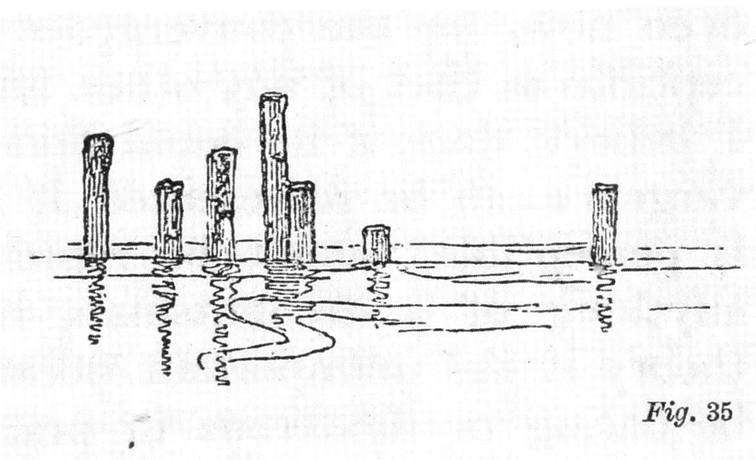

Figure 3.8. Kata Phusin (i.e., John Ruskin), Figure 35 in "Remarks on the Convergence of Perpendiculars," *The Architectural Magazine* 5.48 (1838), p. 96.

To see the posts converge upward, we would have to turn our head until water and reflections vanish; to see them converge below the surface, we would have to look down until we no longer saw the sky. Ruskin concludes triumphantly: "From all this, it appears that perpendiculars only appear to converge under peculiar circumstances, which can never be represented in a drawing."[50]

The two disputants enjoyed a solid understanding of perspective, but their understanding of pictures was less clear, and far from stable. Parsey himself was radicalized between the timid first edition of 1836 and the second edition in 1840 by the invention of photography, which made such an impression that he inserted the "daguerrèotype" in the work's rambling subtitle.[51] Not that photographers were unanimous in celebrating the "proofs" of subjective distortion that Parsey felt vindicated his perspectives. Thomas Sutton, editor of *Photographic Note* and Clark Maxwell's collaborator on the first color photograph, a technical virtuoso among the early photographers, inveighed against vertical convergence in his three-volume novel about the life and crimes of photographers, *Unconventional*:

> Distortion is a thing which nobody seems to care two straws about. The public have been taught to regard it as inseparable from photographs, and a part and parcel of the process. The other day I was shown in London a set of views of the Crystal Palace, taken for a leading Firm, in which the lines of the architecture were tumbling about in all directions, because the blockhead who took the views had "cocked" his camera, as they call it.[52]

One such blockhead was Ruskin himself in more advanced years. Among his exquisite Venetian watercolors, there is one in the Ashmolean Museum showing the southeast corner of the Doge's Palace with the Bridge of Sighs just visible (Figure 3.9). The drawing's execution in a boat may excuse the pitch of the orthogonals of the Palace: the roofline dips to the left, as does the vertical of the pilastered corner of the building. Yet on glancing at the building across the canal, which pitches the other way, there can be no doubt that Ruskin observes the convergence of verticals à la Parsey — whether out of theoretical repentance or, more likely, from

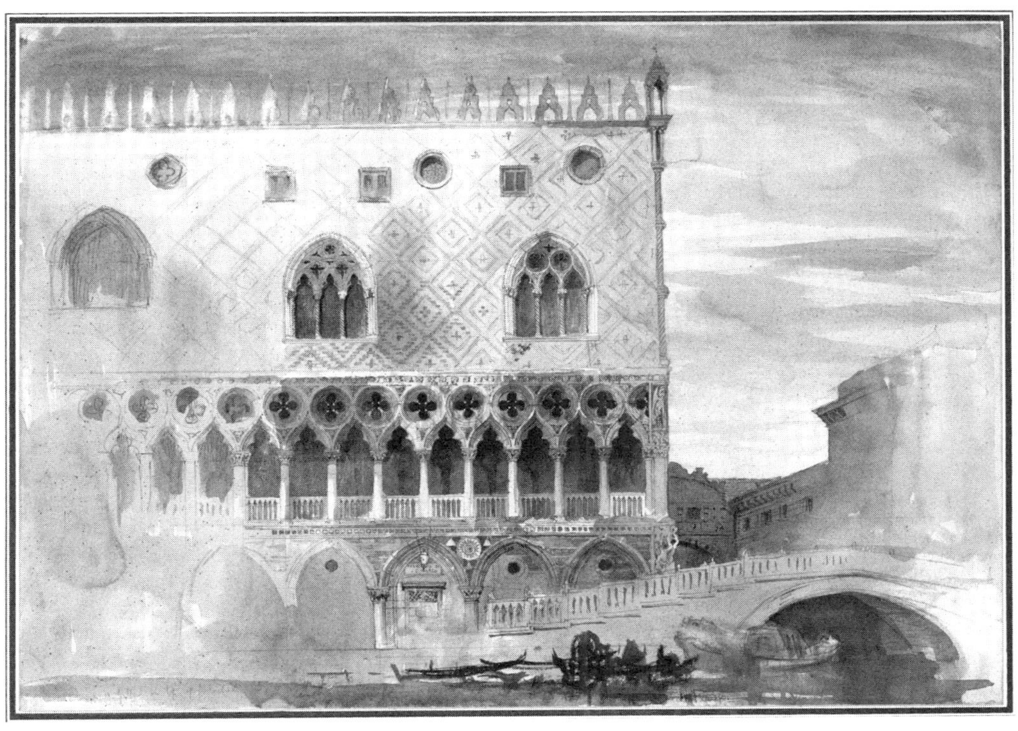

Figure 3.9. John Ruskin, study of the Ducal Palace, Venice (1865), watercolor, Ashmolean Museum, Oxford University. Extant daguerreotypes formerly owned by Ruskin of this stretch of the Canale Grande are too distant to manifest recession.

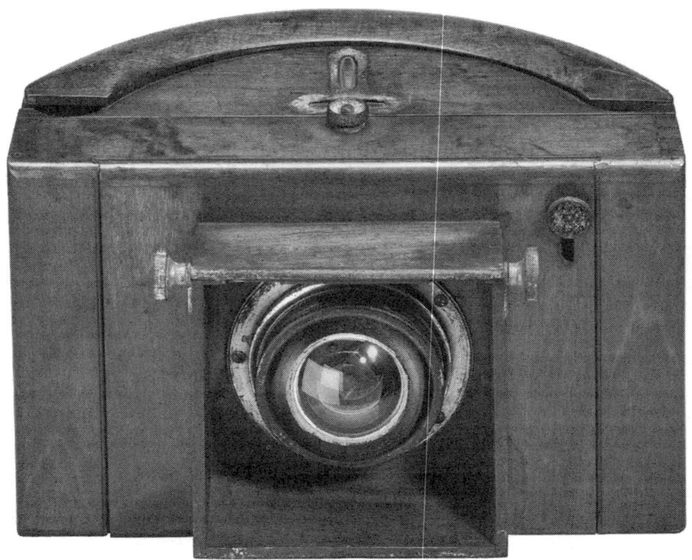

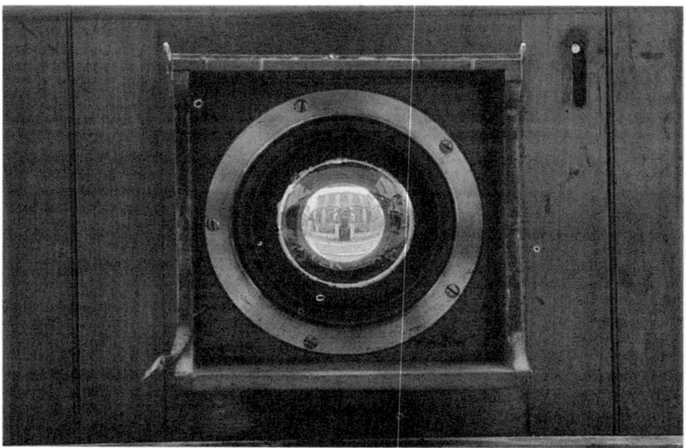

Figure 3.10. Thomas Sutton's panoramic camera (c. 1861–63), front and back. Museum of the History of Science, Oxford University.

having forgotten theory and merely gotten carried away craning his neck to draw.[53]

Ruskin may have set aside youthful scruples in carrying out painstaking observations of the stones of Venice three decades later, but it is worth noting that Sutton too, though he devised a triple-lens system designed to banish distortions, experimented with some wildly revisionist technologies.[54] His panoramic camera employed as a lens "a thick glass globe filled with water," or "to speak more properly . . . two hemispherical glass shells screwed together with a small stop between them."[55] The resulting view commanded an angle of 120°, but it did not, due to the focus achieved with the oblique rays, print onto a flat plate, requiring instead a cylindrical paper to compensate for the distortion introduced by the spherical lens. This proved no match commercially for the simple roving-lens "pantascopic" camera introduced shortly afterward. But more interesting to us is the analogy between Sutton's water-filled, "irised" lens and the human eye (Figure 3.10). Sutton may have intended to produce perspectively correct panoramas, but the anamorphic view through his camera, with its eerie simulation of the viewer's own visual apparatus, betrays the same "ambivalent approach to nature" Riegl attributed to modern painters.

The problem of correlating pictorial artifacts with the experience of vision that exercised Parsey, Ruskin, and Sutton is certainly a matter of technology and physical theory — but only in part. It comprises also a philosophical question, namely: What part of subjective seeing do we expect to see reproduced in an artwork? Are we to regard the camera, the drawing, or canvas as a slice of the physical pencil of light reaching us?[56] As standing in for a perceiving human being? Or as a phantasmagoric eye behind the eye, exposing the powers and limitations of my own vision or mind (I almost wrote "our")? The problematic doubling of the world implicit in "cocked" perspective and indeed in any picture claiming optic fidelity threatens to turn triple: besides reality and the artwork, there is the image we experience, standing in uneasy relation to both external facts. As De Morgan remarked of Parsey: "I am inclined to think it is commonly supposed that the artist's picture is the representation which comes before the mind: this is not true; we might as well say the same of the object itself."[57] What exactly is a philosophically realist artist after then? Writing a year after Sutton's death, in defense of what would come to be called the second Impressionist Exhibition, Edmond Duranty formulated a novel realist analogy between camera, eye, and world in the service of *la nouvelle peinture.* "Suppose," he imagined,

> that at a given moment one could take the color photograph of an interior, one would have a perfect accord, a typical and true expression, all things participating in one sentiment; let one wait, a cloud veil the daylight, and just then pull another print: one would obtain a result analogous to the first. It's up to observation to supply these instant means of execution that we do not possess and to conserve intact the memory of aspects they would have rendered. But if one were now to take some details of the first print and join them to some details of the second to form a painting! Alas, homogeneity, accord, truth of impression, all will have disappeared, replaced by a false, inexpressive note.[58]

What is the moral of this cautionary tale? Color photography was science fiction for most readers in 1876. But the pastiche Duranty despises was business as usual: "just what painters do every day who do not deign to observe." The argument against academicism should not blind us to the breathtaking oddness of Duranty's analogy between painting and photography, which historians have dismissed as scientist. Yet the critic hardly looked to the camera as a savior: he disliked the stiff poses of studio photography.[59] The consistency of the color photo is necessary but not sufficient to the aesthetic effect he sought in painting. More was needed, a kind of generality, "the *real expression* of all the facts of a certain type (*genre*)" which Duranty, like others, finds in Gustave Courbet's *Burial at Ornans* but *not* in the romantic kitsch of Vigneron's *Convoi du pauvre* (Figure 3.11), wherein a single dog follows a hearse: "In our epoch, a poor man's convoy is not *generally* followed by a single dog. This remark is less puerile than it seems."[60]

Puerile or not, Duranty first published his thoughts on the relation of the particular to the general two decades before *La nouvelle peinture,*

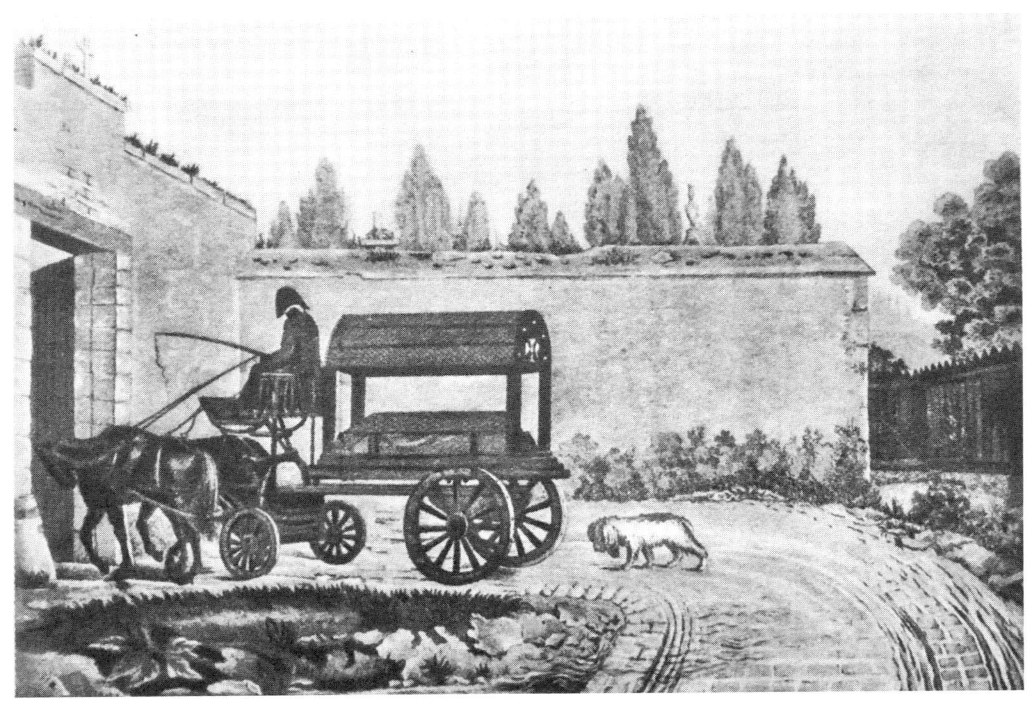

Figure 3.11. Jean-Pierre Jazet after Pierre Roch Vigneron, *Le Convoi du pauvre* (1820), aquatint, London. A rich mythology grew up around this print: Beethoven saw in it Mozart's funeral; Balzac paid sardonic tribute to it in his study of bureaucrats, *Les Employés* (1844); and poet Tristan Corbière named his 1873 tribute to Courbet after it.

in his short-lived journal-manifesto *Réalisme*, in November 1856. That same issue's "Confession of a forty-five-year-old painter" asks painters to "treat man as the landscapists treat trees, houses, grasses."[61] Such artworks, which Duranty self-consciously calls documents, share a teleology if not an ontology with photographs: they secure generality by the truthful specificity of what is shown. This is guaranteed, as it would be in Duranty's 1876 fantasy of the color photo, less by optical fidelity to the scene depicted than by the internal consistency of its light and atmosphere.[62] That is because, to follow Duranty's thought, the general validity of a symbol is not due to it being as impersonal and lacking in attributes as the artist can make it, but to its representing a plausible case, one we may believe is actually instantiated, that is, in showing a "real" thing falling under a particular concept.[63]

To see how Duranty's ideas work in practice, we may return to Caillebotte's *View through a Balcony* (see Figure 1.1).[64] This canvas has been celebrated for "a new self-conscious removal from nature through concentration on the surface of the picture," a removal which remains incomplete, for as Kirk Varnedoe and Rodolphe Rapetti have noted, the acute triangle of empty space above the railing at top right, which positions the viewer to the left, impairs any full identification of the grille with the surface of the canvas.[65] Yet one must concede the justice of Varnedoe's general impression: the grille does *not* take up the whole canvas, but it *seems* to, just as Caillebotte, through this stunning choice of standpoint, kneeling and pressing his face to the balcony grille, as it were, does *not* present us with a pure visual sensation, but *seems* to. The crisp focus and resulting legibility of the grille, and of the horse and coach visible behind it on the street, do not match any photographic technique then available (it might be done digitally by superimposing shots of varying focus) — but again the visual impartiality of photography *seems* to dictate the striking juxtaposition. Moreover, the eventful silhouette of the metalwork gives us a coordinate system for adjudging, in Duranty's way, the rich particularity and hence the general validity of the scene: were the pedestrian or coach viewed a moment later, their bodies would impinge elsewhere on the iron pattern.

Duranty's criterion of comparability to an instantaneous color photograph certainly favors Caillebotte's disjunctive approach to unifying a pictorial space perceptually; it is as if the more incoherent the scene, the more believable it is as the product of a moment of sight. Yet how valid is Duranty's criterion? Though the critic was critical of the commercial photographic practice of his time, does his argument, followed to its logical conclusion, not reduce the artwork to an instantaneous color photograph, taken as a kind of ideal transcript of nature at an instant?

The early history of color photography would certainly have complicated this ideal. In 1861, to accompany lectures by James Clerk Maxwell on perception of color by sensitivity to three pure tones (by what were soon to be distinguished as the cone cells of the retina), Thomas Sutton projected something like the first permanent color photograph, by first exposing three monochrome collodion glass negatives with light passed through green, red, and blue filters, then projecting light through the negatives and color filters onto one screen (Figure 3.12; see color plates).[66] Thus it was possible to photographically produce a color image without color negatives. Maxwell had built machines allowing the breaking and reunification of the spectrum since 1852; the machines kept getting smaller and more portable, so he could take them with him in taking measurements from colorblind observers (an important source of data). But he noticed a psychological, perhaps even an aesthetic, obstacle to the enterprise: "It is difficult at first to get the observer to believe that the compound light can ever be so adjusted as to appear to his eyes identical with the white light in contact with it."[67] Hence the need for a color photograph to *show* the impossible — the fusion of monochrome light of differing hues to render the appearance of ordinary color in white light. So Sutton produced the optical mixture through overlapping projected light. In fact the ghostly picture of the tartan ribbon only resembles its subject because Sutton's emulsion, though insensitive to red, was sensitive to ultraviolet, which leaked through the filter.[68] It is really in the eye, and not in any other object, that reality was captured in its real polychromy.

Whether or not Duranty knew of such experiments, his analogy was not meant to capture an ideal aesthetic object, much less a real one, but an ideal subjective experience. Painting should be like a color photo not in its body but in its *effect*, directing us to an instantaneously observed, colored reality, rather than resembling a pallid paper print. Though this disposes of the naïve scientism that first marred our understanding of his analogy, it poses a new and more difficult problem. The color photograph that subjective vision ought to resemble is itself, unlike consciousness, an object in space, liable to change aspect with time of day and atmospheric conditions. To put the matter simply: one cannot look at the overcast sky in glaring sunlight. It does not suffice to prevent conflict between moments *within* a depiction, as Duranty asks — conflict also arises between the conditions depicted and the conditions under which we see the picture. This demand must be met or sidestepped if possible, since the modern artist is in no position to dictate how an artwork is seen and by whom.

Brutal Realism

My argument in this chapter, thus far, has been by and large critical. Increased technological ingenuity in fixing aspects of reality in pictures (perspectivally and chromatically) only served to open up new chasms between the art object and subjective perception, even as the former came to simulate traits of the latter. An analysis of optics, or of light, could not by itself bridge this divide, because pictures are not perceptions but are themselves perceived, thus multiplying whatever effects of the real world they were supposed to have successfully incorporated. Would a constructive approach work where imitation failed? Perhaps Sutton's projected image can point the way: what if careful aping of subjectivity in its particulars gave way to mechanical means to elicit effects associated with the subjective?

The question was addressed in the latter half of the nineteenth century above all by the more experimental printmakers. Of the four etchings exhibited by Bracquemond in the Fourth Impressionist Exhibition, *Au Jardin d'acclimatation* (Figure 3.13; see color plates) distinguishes itself through its technical process.[69] The print was executed by the "procédé de Debucourt," invented nearly two centuries earlier but never adopted due to its difficulty: it required separate design for each color, to be printed from four separate plates (three colors and black).[70] One impulse came from the Japanese woodcuts Bracquemond had championed since cofounding the Société du Jing-Lar in the 1860s, a fraternity of aesthetes who met to wear kimonos, eat with chopsticks, and manifest their shared enthusiasm for Japanese art and, perhaps more surprisingly, democracy.[71] This context may explain the interest in color printing — but what does it have to do with the pictorial conquest of subjectivity? Can we say with Jean-Paul Bouillon that "the mere act of dividing up the image into its fundamental colors . . . in some ways resembles the Impressionist technique"? Or must we conclude more cautiously, as Bouillon does of another print, that "it suffices to prove the fragility of the stylistic classification as well as the lack of coherence in the movement"?[72]

Without any special pleading, I think Bracquemond and printmaking in general bear on the crisis of impressionism, its ancestral ties to realism, and its mutation into symbolism by suggesting a way out of the doubling problem of Duranty and artists dedicated to the mimesis of subjective perception. Etching offered, if not fully realized, a constructive means of perceptual mimesis, by putting an image together out of clearly synthetic elements that do not pass for the contents of consciousness. For a start, we should note that the perceptual analogy is a tenuous one: the color separation of Seurat, or halftone printing, blending in the eye to produce tonal intensity, is a far cry from the flat colors of the print, interacting only in the orange of the birds (a combination of the yellow and red plates), and running the gamut from flat saturated color to grainy, dull gray textures of foliage and chain-link fence, an effect no doubt aided by Bracquemond's virtuosic application of aquatint.[73] Henri Béraldi, the chronicler of the print revival, notes that Bracquemond designed porcelain services by cutting the figures of animals out of his prints.[74] The juxtaposition of forms, conspicuous in the single-color trials, results in angular networks of visual stimuli, descriptive yet resolutely strange (Figure 3.14; see color plates).[75]

The angularity and abstraction that might seem to be mere artifacts of the production process are sharpened into a strange standoff between women and pheasants in the color print. Bouillon is right to detect "a certain moralizing spirit . . . in the confrontation" between woman and bird, but it does not really issue into a satire on female vanity.[76] Marie Bracquemond, the artist's spouse and an impressionist painter, who exhibited with the group for the first time in 1879, supports her head and a glove casually in her palm, sunk in thought: her eyes do not settle on any bird.[77] Her sister, standing behind her in white, surveys a broader view than we are granted. The identity of the women must have been obvious, given that one of Bracquemond's other exhibited works, *Une terrace de Sèvres*, shows Marie Bracquemond, dressed in black, painting a model whose features and dress match those of her companion in *Au Jardin d'acclimatation* (Figure 3.15). If the women in the *Jardin* are preoccupied, it is the birds who engage the humans perceptually, turning their heads, as birds do, to cock laterally placed eyes at the women. What is moral here is the leveling spirit: humans, too, are perceiving animals. Félix Bracquemond, called "the Michelangelo of Ducks" by one of his first American collectors, Walter Carter of Brooklyn, returned incessantly to the theme of perceiving animals, with or without human presence.[78] The shared feature of manifold studies of animals moving, feeding, or otherwise reacting to their surroundings is that they are shown as sentient agents responding to the world. Of course, the conventional, if virtuosic rendering of texture and atmosphere in these prints is far indeed from advancing a claim of subjective comprehensiveness—the world as we see it, or as the animals see it. They are just illustrations of animals seeing (sniffing, biting, flapping)— nothing more.

A print published by Bracquemond in 1882, and often reprinted (notably by *La Revue de l'Art Ancienne et Moderne* in 1900, as Bracquemond was being lionized for the Exposition Universelle), and titled *Surprised Ducks* or *The Surprise* (Figure 3.16), suggests a more ambitious

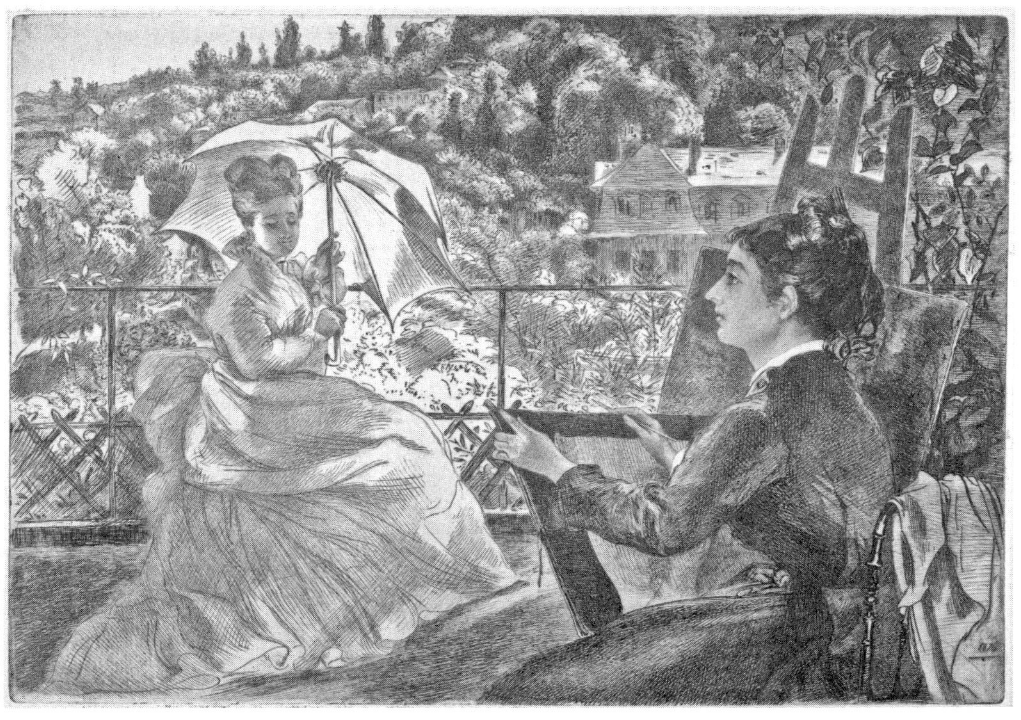

Figure 3.15. Félix Bracquemond, *La Terrasse de la Villa Brancas* (1876), etching and drypoint with aquatint with some surface tone, printed on oriental paper, seventh state, British Museum, London. Note that Marie Bracquemond, a student of Ingres's, is drawing outdoors, though a canvas and easel indicate that she also paints *au plein air*.

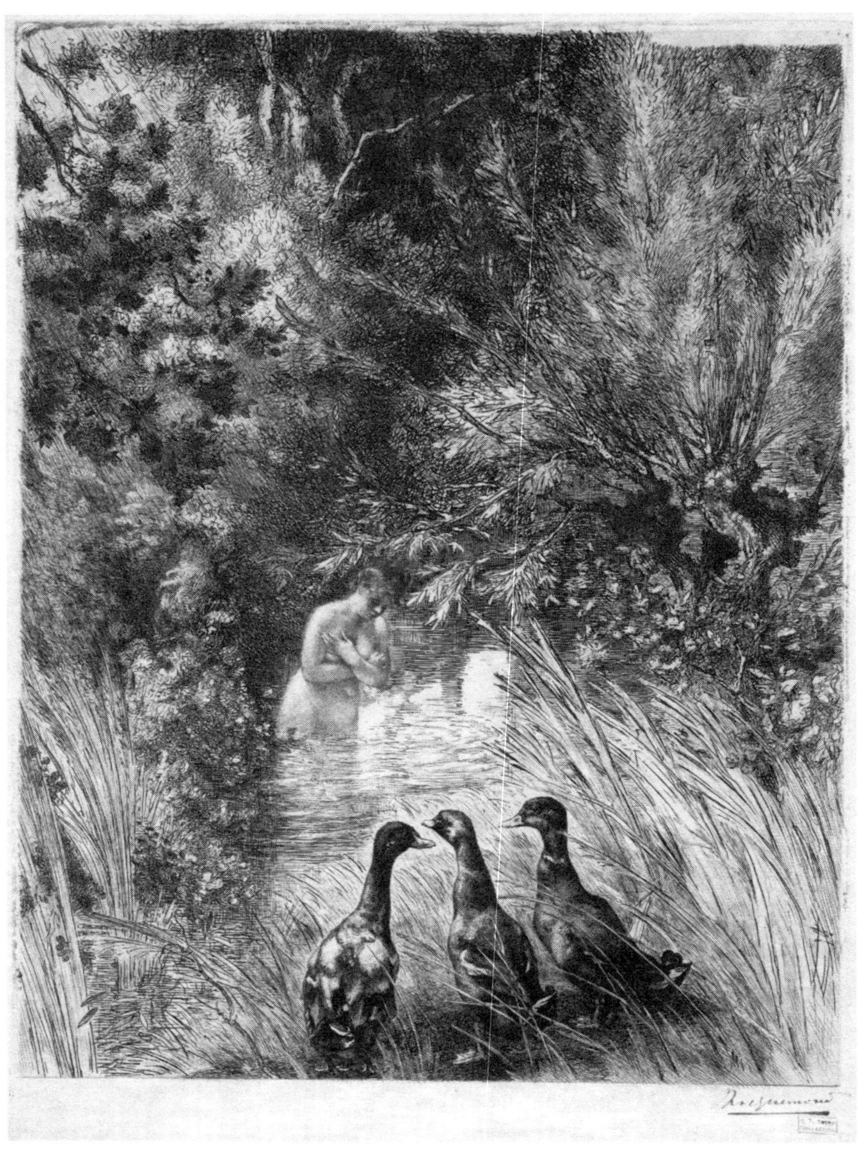

Figure 3.16. Félix Bracquemond, *Canards surpris* (*La Surprise*) (1882), etching, ninth state, British Museum, London. The abstract title captures what is both funny and radical in this print—that we don't know who is the object and who the subject of vision.

project.[79] The briefer title is more elegant in its suggestion of a state shared by the nude bather and ducks; the longer title rightly specifies that it is the visual position of the ducks that we inhabit. The fullness of the nude bather, recalling the later nudes of Auguste Renoir, and the bent covering gesture, just slightly recalling the background figure in Manet's *Déjeneur sur l'herbe*, are programmatic. However modest, this is a modernist bathing scene, one in which perception itself has been put on view. The extraordinary thicket of trees and reeds, and the bravura effects in the water, give way to hard, clear silhouettes of the birds (who are perceivers and perceived), as statuesque as the nude is delicately stippled. The effect is less one of sudden appearance from the water than of sudden awareness of another living thing.

Is there any broad aesthetic import to such an animal and print-centered impressionism? Yes, for two theses about perception were widely shared by psychologists and artists working after 1870.[80] The first was a firm distinction between sensation and perception, that is between physiological stimuli received by a living thing and the subjective, conscious representation of that stimulus in the mind. Taine, in his grand 1870 summation of empiricist psychology, *De L'Intelligence*, put the matter thus: "In effect, by its bare presence, a sensation, notably a tactile or visual sensation, engenders an interior phantom that seems exterior object." Of course sensation, in itself a physical fact, could give rise to a "percept" that is misleading: dream, hypnosis, and phantom limbs suggest that sensation stands no surety for an object outside the mind. This did not trouble Taine, who was more interested in the mind than in knowledge of the world: "Once the sensation is present, the rest follows; the prologue ushers in the drama."[81] It is but a small step from this to Taine's notorious formulation that "instead of calling hallucination a false internal perception, we must call exterior perception a *true hallucination*."[82] Though it has precursors in Bishop Berkeley and Taine's hero David Hume, *De L'Intelligence* is the immediate source of twentieth-century sense-datum theory, according to which both genuine and erroneous perception share a common denominator in the subject's experience.

The second thesis is an art-theoretical corollary of the first: the objectlike quality of paint strokes or other pictorial components (e.g., etched points or lines) is inversely correlated to its ability to contribute to conventional illusion. As Bracquemond put it in his 1885 treatise *Du Dessin et de la couleur*: "The more the touch (*tache*) assumes importance in itself, the more modelling disappears."[83] This is no criticism; Bracquemond distrusted Jean-Léon Gérôme and the *lécheurs* (lickers, thus nicknamed for their "licked-smooth" canvases), insisting that the academic discipline of blending (*blaireauter*) "should in no way be confused with the methodical distribution of values."[84] Rather, the goal was to achieve dynamic equilibrium between the substantiality of sensations and the object they constitute in perception.

The sense-datum theory of perception with the "touch" theory of art add up to a simple account of impressionism: in playing up the autonomy of the elements of picturing (be they patches of oil on canvas or ink on paper), it offers to vision a scene closer to sensation than to perception. Compare the orthodoxy, passed from Castagnary to Lionello Venturi and John Rewald, that impressionists do not depict objects but their subjective impression on the artist. This view falls prey to the doubling problem, for a subjective impression presented in a physical artwork becomes the object of another subjective impression for the spectator. Bracquemond tried to evade this by presenting subjects engaged in a kind of animal sensation, which the prints only partly order — whether with the professed indifference of the *Jardin d'acclimatation*, or the visual humor of *La Surprise*. What pure sensation might have looked like to Bracquemond can be seen in an 1885 aquatint of the park at Saint Cloud, as close to "abstraction" as the century produced (Figure 3.17). The print is a proof, as indicated by the lack of a lower edge. The rudiments of a landscape motif are provided in the grotto-like center and just-recognizable sloping tree trunks at top right; or one may imagine a tree-lined *allée*, the branches denuded in winter.[85] But the lightning-like stabs of black, the livid texture of rock or foliage, executed in the evenness characteristic of aquatint, contribute little to the traditional *veduta*.[86] The residue of recognizable objects serves only to secure the

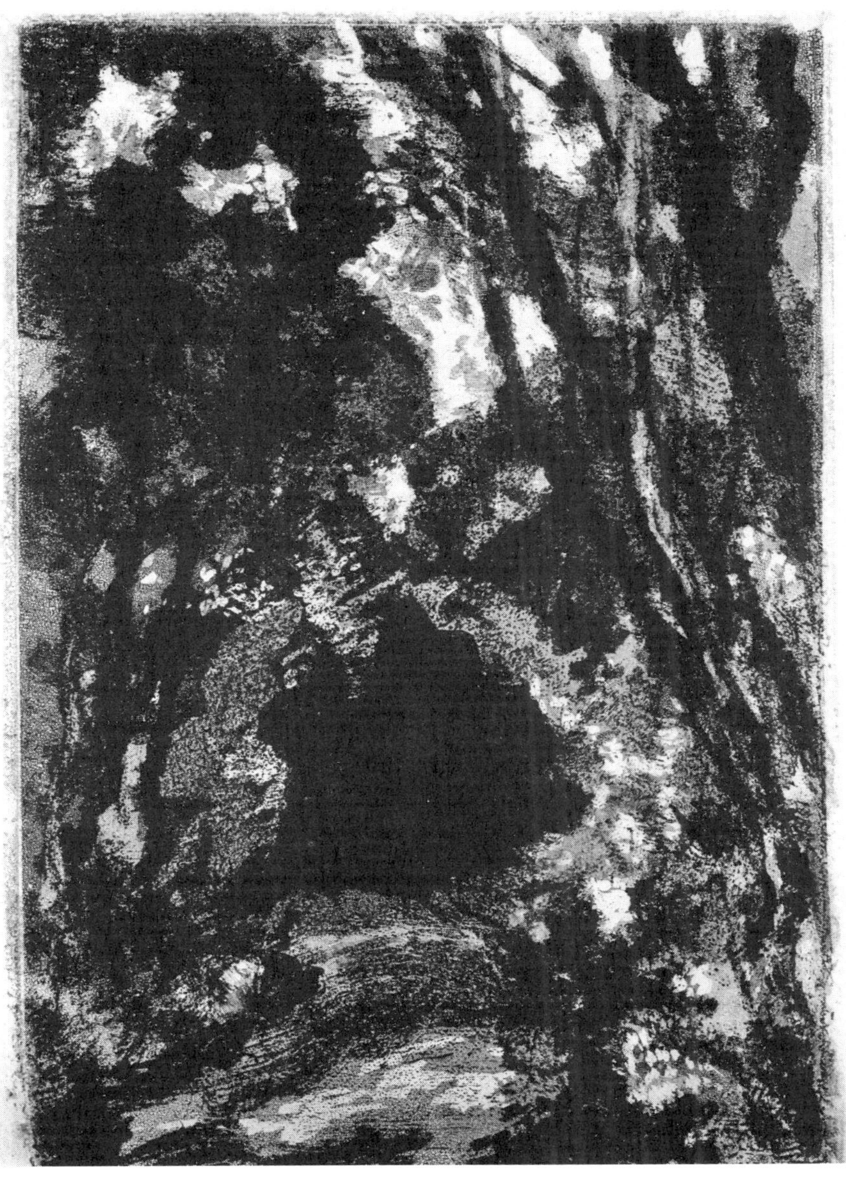

Figure 3.17. Félix Bracquemond, *A Clearing in the Parc de Saint Cloud* (1885), aquatint, British Museum, London. Whether one sees an anecdotal "clearing" in the central void is less salient than the hypnotic rhythm of alternating light and dark.

patches of tone as perceptual raw material — individual sensations, arranged one by one by the printing process, rather than broken up as in Caillebotte's *View*, whose decorative grillwork divided an urban landscape into its parts.

Whatever its motivation, the 1885 *Saint Cloud* is atypical of what Bracquemond showed with the impressionists. Though held in high esteem by Degas and certain critics (Burty and Duranty), he was, with the other anecdotal realists in Degas's circle, a target of ridicule for the colorist Auguste Renoir.[87] And Gustave Caillebotte, the group's aesthetic gendarme, though brought into the fold by Degas in 1879, turned immediately against "the crowd he drags along with him."[88] Félix Bracquemond did not exhibit with the group after 1880, though Marie persisted until the bitter end in 1886. The painters of modern life qua optical illusionism won the day; the print revival fell out of avant-garde favor before being reclaimed by symbolism in the 1890s. But its significance to the project of making subjectivity visible in art is hard to overrate. In 1891, on the heels of symbolism's first public triumph and loss (Aurier's article on Van Gogh and the latter's death), Mary Cassatt started printing a set of multiple-plate etchings in a process akin to Bracquemond's, but combining aquatint with drypoint rather than etching.[89] They are celebrated for their blending of Japanese perspective and domestic observation, but Cassatt's ongoing experiments, with their uniform areas of color and ragged, collage-like contours, go beyond Japonisme into a kind of meditative perceptual symbolism.

The unusual standpoint of the earlier print (Figure 3.18; see color plates), outside the boat, in fact hovering over the pond, so that the yellow reflection in the water (presumably the woman's blouse) bleeds off the plate, is pure Bracquemond. The steely, cautious eyes of the ducks are as much involved in the transaction as are the patient, quiet faces of the humans. If here Cassatt still pursued very subtle effects of perceptual commensuration by inking the plate directly, *By the Pond* dispenses with such fussiness (Figure 3.19; see color plates). It achieves instead a monumental repetition in the yellows of the woman's jacket, broken up into three angular regions, and exploding in the blond Apollo curls of the child she holds up, whose gaze seems to survey a counterpart of the oddly slipping, elliptical view of trees, sky, and lake around him. The scratchy drypoint silhouettes hardly seem capable of delimiting the color fields, which seep and bleed into one another as if in anticipation of Mark Rothko. Against this indeterminacy, the stretches of opaque ink, modulated only by modeling in the aquatint application, as if in simulation of a tinted photograph, struggle to hold on to their conceptual identities: tree, grass, water, mother's shirt. The effect is poignantly as if the young subject of the print were perceiving the landscape. The uncanny, unstable balance is hard to imagine holding under other circumstances than those so carefully chosen by Cassatt.

The problem whose twists and turns we have charted may be summarized thus: a project in force since romanticism of expressing in art the contents of experience throws up the baffling suggestion that consciousness, or its visual aspect at least, is doubled — a phenomenon made visible, and verbalized, in the debate over Parsey's claim that perspective should manifest vertical recession. Photography and plein air painting threw up new versions of the doubling problem, which painters — and the apparently more modest *peintres-graveurs* — tried to sidestep by focusing on an imagery of disjointed sensation, rendered in a technique aspiring to approach consciousness from the side of nonconscious, bare phenomenal qualities, sensations rather than elaborated perceptions. This "ground-up" approach was often allegorized by the printmakers in images of animals perceiving. Joined to more literary themes, it may be the only coherent principle that unites such precursors and allies of symbolism as Puvis de Chavannes, Gustave Moreau, the Italian Macchiaioli group, and Jean-Jacques Henner. Their diverging approaches to paint handling — flat and neutral (Puvis, the Italians) or translucent and ethereal (Moreau, Henner) — are brought together by the way they fuse apparently isolated groups of sensations into simplified perceptual wholes. Of course, not all were impressed. Caillebotte, in paintings like the bracing *Calf's Head and Ox Tongue* (Figure 3.20), seems to deny emphatically the possibility of uniting mere sensation,

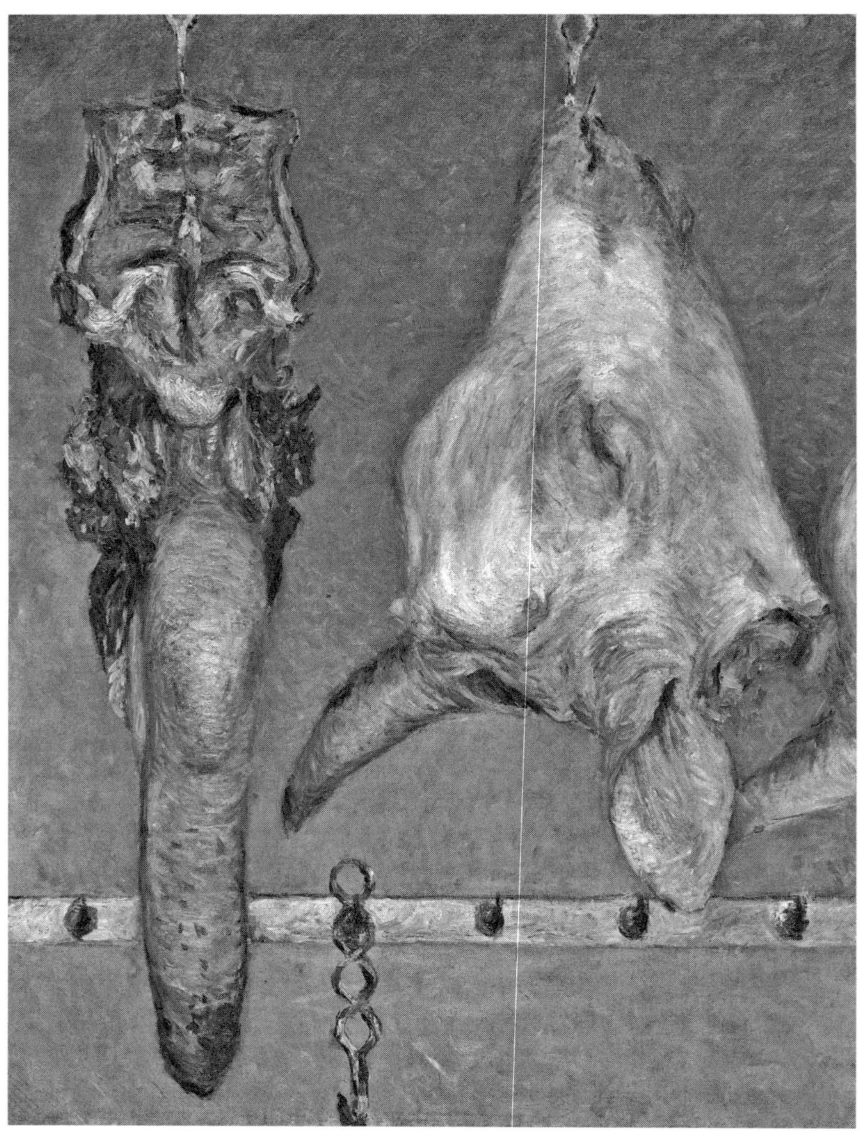

Figure 3.20. Gustave Caillebotte, *Calf's Head and Ox Tongue* (c. 1882), oil on canvas, Art Institute of Chicago. One of four meat still lifes, all of which remained in the artist's family collection until this one was sold in 1999.

such as an animal tongue or brain might have, and the subjective mastery of the impressionist, to whom modern life reveals even its bloodiest corners.

The conflict between an aesthetic of bare sensations and one of conscious reflection on subjectivity is most acute in criticism sympathetic to the new painting but unfamiliar with its effects, such as that around the British pre-Raphaelites. It is hard to find criticism of Manet so courteous and destructive as that of William Michael Rossetti, brother of Christina and Dante Gabriel Rossetti, and correspondent of Mallarmé.[90] In April 1876, in response to an exhibition of French painting at the Deschamps Gallery in London, Rossetti wrote:

> From Manet we receive *Les Canotiers*; the strong, coarse, ungainly, capable picture, which made so much noise in Paris last year — the work of a leader who may perhaps some day be absolutely a master; but that day threatens to be one when the blind shall lead the blind, the perversely-aiming painter shall lead the perversely-appetent purchaser and public, and both shall fall into the ditch. Aesthetic realism and brutal realism are two different things: the former admirable, the latter unendurable.[91]

The Academy, which ran this review, was no reactionary paper; the immediately following article is a Paris letter by Burty concerning Duranty's *Nouvelle peinture*.[92] What Rossetti had in mind in decrying "brutal realism" emerges best in a review of the French Salon in the next issue, signed by Emilia Francis Strong Pattison, a rapier-pen reviewer and philosophical positivist then beginning to make her mark with articles on the art of the French Renaissance.[93] Pattison begins by considering two pictures refused by the Salon jury, which Manet chose to exhibit in his own atelier. The pictures are *Les canotiers*, which Rossetti had attacked, and *The Laundry*, which Burty had defended. Pattison soon moves from specific works to general principles:

> In so far as all painting is a language — a set of symbols in which the artist expresses a certain number of facts *selected* by him, while he rejects others which do not suit his purpose — the so-called Realism of M. Manet and of the

class to which he belongs might be called Idealism — Idealism which has its own mannerisms, its own types of convention, its own fitly corresponding methods of procedure. The kind of facts for which the so-called Impressionist class seek are of a very different order from those looked for by another class, but both sets of facts have an equally real existence, and the same principle of selection comes into play in handling either the one or the other.[94]

Pattison, in her reduction of painting to a language of symbols, as convention-bound in its vanguard as in its academic dispensation, anticipates by a decade and a half the symbolist art criticism of Aurier.[95] More, her capacious conception of the selection and rejection of facts allows her to judge impressionism by its own standard of truth to appearances, while emphasizing its "idealist" emphasis on subjective perception. And the rhetoric of art as a language of symbols is not left idle, but drafted into a criticism of Manet's portrait of Desboutins, which Pattison compares with the man himself:

> Throughout we recognise in the treatment a distinct *parti pris*. The fatigue, the melancholy of the habitual expression are passed over, while the signs of will and character are strongly accentuated. The searching look becomes, as a friendly critic has it, "d'une fixité saisissante"; it is, indeed, a stare, the tremulous nostrils are widely dilated, the curves of the mouth are hard set. In the handling we get a directness of attack which is all but brutal, the forms are indicated with unhesitating frankness, there is no caress in the touch, no delicate and incessant drawing: all is blocked in squarely in violent broken tones, which only find their place at some yards' distance, and so left — left without that grace of added finish which is, indeed, but the grace of added fact.[96]

Setting aside Pattison's preference for literary narrative — she reserves highest praise for an artist, Puvis de Chavannes, as summary in his paint handling as Manet — there is some justice in thinking of Manet's lack of finish as itself an "added fact." Though Caillebotte would likely have been outraged, the model of seeing from the ground up he shared with Manet — and to some extent with Bracquemond — did not have built into it a rationale for when to halt the process of analyzing perception into constituent sensations. Duranty, in his dubious analogy, might have said that the process is over when the picture resembles a color photograph. Artists committed to simulating subjectivity from the ground up have no such alibi.

CHAPTER FOUR

What Are We?

A Symbolist Picture Theory

The second chapter of Taine's *On Intelligence*, dedicated to images, begins by recounting a personal experience. The scene is "yesterday evening about five . . . on the quay by the Arsenal, watching in front of me, across the Seine, the sky reddened by the setting sun." A footnote gives yesterday's exact date as November 24, 1867. Here is what happened:

> Fleecy clouds rose in the form of a half dome, and bent over the trees of the Jardin des Plantes. The whole of this vault seemed encrusted with scales of copper; countless indentations, some almost burning, some nearly black, extended, in rows of strange metallic luster, up to the highest part of the sky, while, all below, a long bronze-coloured band, extending along the horizon, was streaked and cut by a black fringe of branches. . . . In half an hour, all this had died out; there was but one patch of clear sky behind the Pantheon; reddish-coloured smoke was wreathing about in the dying purple of the evening, and the vague colours intermingled. A blue vapour hid the arches of the bridges and the edges of the roofs. The apse of the cathedral stood alone, looking with its pinnacles and jointed buttresses, in size and shape like an empty crab-shell. Things prominent and coloured but a moment ago, were now like mere sketches on a dull paper.[1]

The text might be called the primal scene of impressionism. Indeed, it goes beyond the cursive visual discriminations of that group, describing a sky discretely divided into "scales of copper," such as no artist attempted before the second generation of neoimpressionist painters (Figure 4.1; see color plates). This is all the more remarkable in that Taine's taste in art was frankly conservative.[2] In any case, whether pioneering or just out of time, there is also a sense in which Taine's landscape is *not* that of a painter. The reader of this book will have become accustomed to ask: is Taine's written account of this sunset a public sense, granting us imaginative access to his perceptions? The case seems almost the opposite of an etching like Charles Meryon's *Ministère de la marine*, with its Bosch-like vision afflicting the skies over Paris (Figure 4.2). That artist's associative thinking begat a concrete phantasmagoria, spinning the dry iconography of a navy building into a riot of archaic Greek charioteers and flying fish, whereas in Taine a straightforward syntax with pretensions to great perceptual discrimination (the scales of copper) is apt to provoke divergent visual impressions in its readers. We can argue about the text, compare it with images like Cross's; but, as Frege noted of his colorblind companion, we cannot compare our experiences with Taine's.

A generalized version of the doubling problem confronts us in Taine. The problem concerns any person who paints, draw, writes, or photographs something other than "the object itself," namely an experience of it. No wonder art was reconceived not on the model of a mirror doubling reality, as perspective had been since the Renaissance, but as a language whose ground level (sensation, usually taken to be a physical process) plays a syntactic role, while its combinations (perception, taken as a psychological process in a conscious mind) are expressive: "painting as a language," as the English critic Emilia Pattison put it with reference to Manet. But here, too, we are on the outside looking in, wondering whether the sensible signs of another's conscious experience (be they visual marks, sounds, or words) can serve as a vehicle for our own.

The attempt to solve this problem is at the core of the transition from impressionism to symbolism. In his article on "The Impressionists and Édouard Manet," which exists only in an English translation

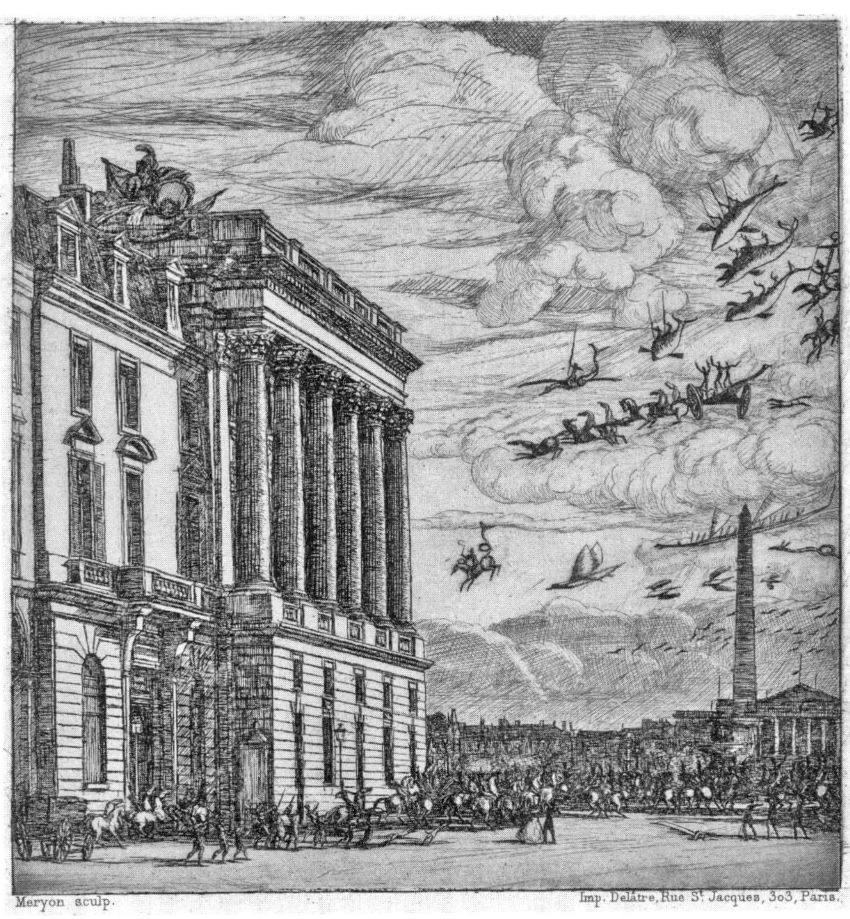

Figure 4.2. Charles Meryon, *Ministère de la marine (Fictions & Voeux)* (1865), etching, British Museum, London. An incredible (but allegorical—because of the navy) sky overlays an accurate rendition of the Place de la Concorde, with its obelisk.

he had published in a London monthly in 1876, Mallarmé faced the problem squarely, quite an achievement considering that his main concern was to puff his friend's art for a foreign audience unfamiliar with it. Mallarmé announced boldly that plein-air painting, with its appeal to everyone's experience, was a political art, corresponding to the increased access to democratic participation in post–Second Empire republican France.[3] But this requires a technique with claims to generality: "as no artist has on his palette a transparent and neutral colour answering to open air, the desired effect can only be obtained by lightness or heaviness of touch, or by regulation of tone."[4] Mallarmé, like some of his painter contemporaries, is dismissive of linear perspective — "that utterly and artificially classic science which makes our eyes the dupes of a civilized education."[5] Discarding this, Manet makes space legible through an "absolutely new science" of framing or "cutting down the pictures," as well as through the cursive brushstroke, which makes one think objects "are only seen in passing." Alas, the rendering of light effects in the new arbitrarily framed views, which Mallarmé hopes will prove intelligible to all, in fact puts artistic convention, and social distinction, back between canvas and viewer.[6] Mallarmé sees the difficulty: "But will not this atmosphere — which an artifice of the painter extends over the whole of the object painted — vanish, when the completely finished work is as a repainted picture [that is to say, the *tableau* is complete]?"[7] His reply, recalling Duranty: "from the first conception of the work, the space intended to contain the atmosphere has been indicated, so that when this is filled by the represented air, it is as unchangeable as the other parts of the picture."[8] It is then something conceptual, a kind of framing of space, that makes possible the mimesis of atmospheric givens. This in turn complicates that definition of painting we found applied from romanticism to the end of symbolism, according to which its subject matter is introspective rather than outwardly descriptive. "They are impressionists in the sense that they render not the landscape, but the sensation produced by the landscape," as Jules Antoine Castagnary put it in 1874, echoing Delacroix's commandment to the young artist: "Everything is a subject; the subject is yourself; it is your impressions, your emotions before nature."[9] As Castagnary worried, the danger in this injunction is that artists painting their own emotions rather than nature might turn out to be solipsists painting nothing at all.

Pictures without the Mind

The deep root of the doubling problem lies in the conviction that there is more to what we and other intelligent beings think and feel than we can share directly with one another. Even Wittgenstein concedes to the "solipsist" that "what you have primarily discovered is a new way of looking at things. As if you had invented a new way of painting; or, again, a new metre, or a new kind of song."[10] Could this new way of looking not be embodied directly, meeting the doubling problem head on, as it were, to show the world as it looks to the picturing subject? What would that require?

Complaints about deviation from linear perspective were routine for critics of Manet and the impressionists, but Caillebotte's *Déjeuner* is perhaps the first painting to explicitly abandon the premise of an instantaneous view comprising the whole space, in favor of a mobile view of the plate and silverware before the viewer-participant, and of the more distant prospect of the artist's somewhat aloof bourgeois family (Figure 4.3). This uncompromising picture had few imitators, at least in the realm of high art: by the early twentieth century, enterprising illustrators like the American Winsor McCay did not hesitate to shock the world with plunging perspectives understood to represent the first-person point of view of their fictional characters (Figure 4.4).[11]

The patient's-, or corpse's-eye view (through the glass window in the coffin) is an object of phantasmagoric fun in McCay's 1905 comic, a nightmare the spectator wakes up from to be greeted by the homier confines of a third-person view of the dreamer's bed. There were radical subjectivists willing to go even farther in placing the subject visibly in space, with no suggestion of the fantastic; though none more so than a deceptively casual drawing, or diagram, in a book on the border of physics, philosophy, and psychology: Ernst Mach's

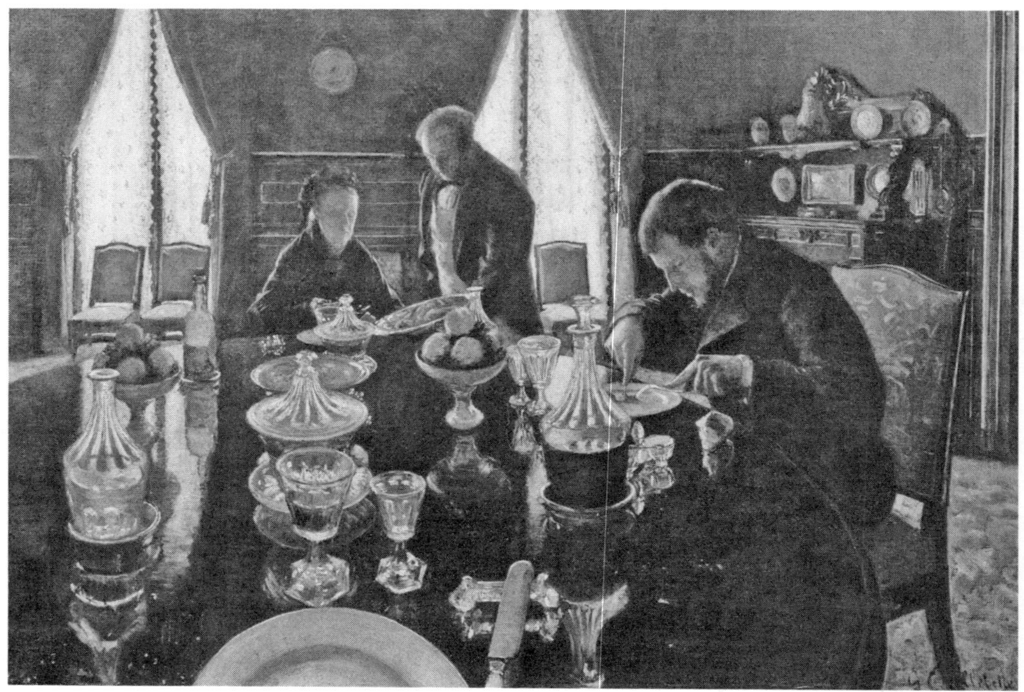

Figure 4.3. Gustave Caillebotte, *Le Déjeneur*
(1876), private collection. Note the upward
tilt of knife and plate, as if the viewer were
peering down at his own place setting,
then glancing up at the rest of the room
and its contents.

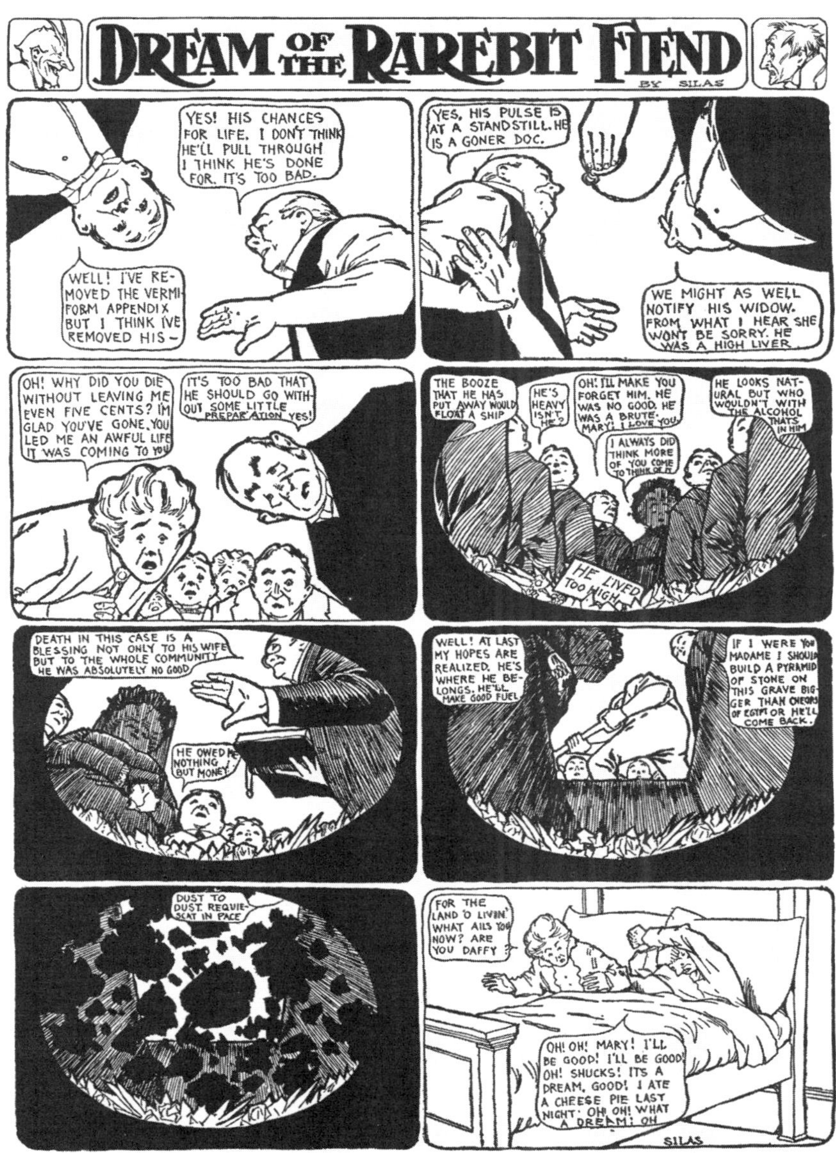

Figure 4.4. Winsor McCay, *Dream of the Rarebit Fiend* (New York: Frederick Stokes, 1905), n.p. The first-person point of view is a provocative way to imagine death — or the nightmare of being buried alive.

view of the world through his left eyehole in the *Antimetaphysical Prefatory Remarks* of his 1886 *Contributions to the Analysis of Sensations* (Figure 4.5). Mach reports having executed the drawing some seventeen years earlier, that is to say, before 1870, and as a matter of fact a far more atmospheric sketch, embellished with such details as the smoke climbing from the end of a cigar and the steam from a tea or coffee cup, can be found among Mach's papers.[12] The more familiar printed picture had attracted recent art historians and historians of science as either evidence of an exaggerated sense of objectivity on Mach's part ("he has tried to render everything *exactly* as he sees it") or of subjectivity ("he has tried to render everything exactly as *he* sees it").[13] It might also be seen as evincing an appeal to the seer's body peculiar to nineteenth-century art, a scientific romanticism or melding of subjective and objective modes of seeing, what Alois Riegl in the previous chapter called our split (*zweispaltig*) approach to nature, seeking it "less and less in an independent objective state and more and more in a subjective state determined by the observer."[14] Riegl, when he did zero in on modern art more specifically, identified this tendency with "secessionist painters" and with "impressionism" in the capacious sense this word had acquired by the fin de siècle. Indeed, though it would be art-historically myopic to mash together French impressionism, Mach's investigations, and the emergence of the first-person perspective at the end of the nineteenth century, the links between these phenomena are real, and they are part of that tendency that I have tried, particularly in the last chapter, to connect with symbolism as a set of theoretical and practical efforts to make subjectivity intelligible.

For Riegl, modern art is a consummation of the history of subjectivity in viewing nature, but a Pyrrhic victory, for we "may not be ripe" for such "reckless subjectivity." Yet this historical progress registers what he cannot help regarding as an objective gain: insofar as objects are shown as they are seen by us, a subjective depiction of them is more truthful, in a physiological and psychological sense, than the "objective" pictures of self-sustaining things Riegl found in ancient pre-Hellenic art. Here Riegl seconds the claim made by

Mach a quarter-century earlier in a popular lecture devoted to the question why humans have two eyes. In that lecture, Mach argued that the Greeks and Assyrians advanced over the Egyptians precisely by making images that are constrained perspectivally, rather than naïvely rendering objects "as they really are":

> There are naïve natures that hold the pretend-murder onstage for real murder, the pretend-action for real action, who wish to fly to the aid of those persecuted in the play. Others cannot forget that the stage contains only painted trees, that Richard III is none but actor M, whom they have often see in public. Both mistakes are equally large.
>
> To see a drama and a picture aright, one has to know that both are appearance and mean something real. With this comes a certain dominance of the inner mental life over the life of the senses, so that the former is no longer killed by immediate sensation. A certain freedom to decide one's standpoint also belongs here, a certain humor, I'd like to say, which the child and young nations decidedly lack.[15]

In this evolutionary perspective, which is making a questionable return in the twenty-first-century humanities, Mach's diagram gains special significance as a particularly high stage of development, exhibiting both the "inner" life and that "of the senses." It shows us objects like books, boots, and vests much as they appear in that repository of Oriental objectivity, the illustrated *Larousse*, but from a standpoint so radically subjective that its very legibility must be doubted outside of the philosophical discourse that informs it. A kind of fictional collapse threatens: in Mach's self-portrait, the world is enclosed by the arch of his eyebrow, much as in Greek myth it rests on the shoulders of Atlas. Mach himself admits the result would have been tricky to render, not to say graphically impossible, for binocular vision. It would have looked like the world seen through binoculars, which is not how the world looks, for in binocular vision the nose, eyebrows, and other features that differ between the two retinal images disappear to some extent, giving us the world more or less without our body intruding into it. The diagram, being objectively subjective in Riegl's hybrid manner, is tendentious. Even in

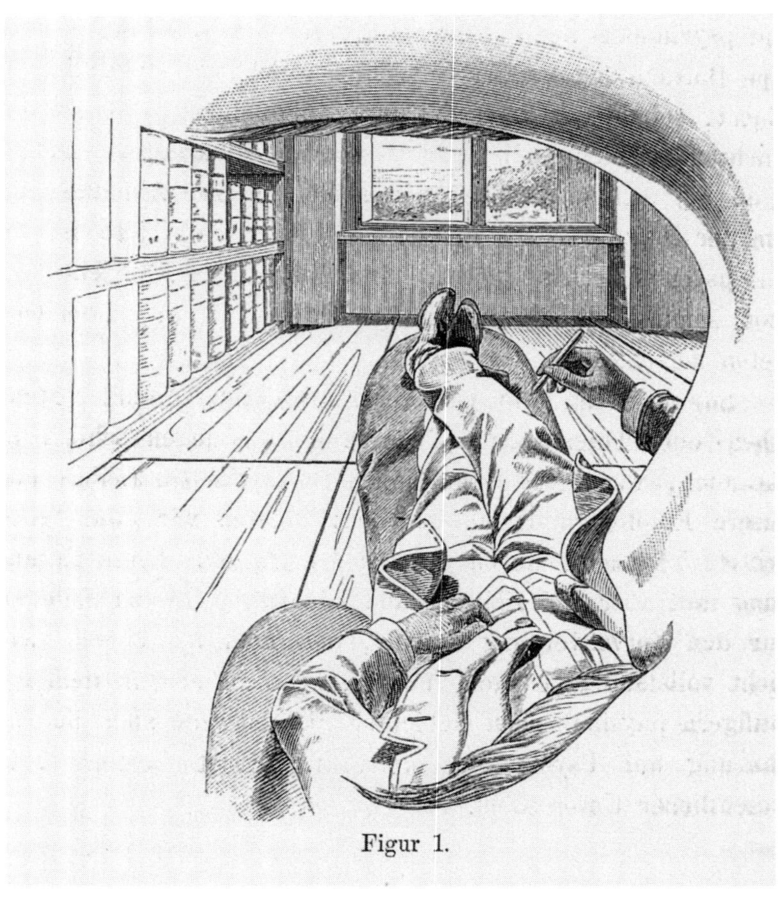

Figure 4.5. Ernst Mach, *Beiträge zur Analyse der Empfindungen* (Jena: Gustav Fischer, 1886), p. 14, pen and wash drawing in the Deutsches Museum, Munich, Archive, CD_62612. *Video ergo sum.* The meticulous preparatory drawing, complete with legible numbered volumes on Mach's shelf, attests to the hard work of letting someone else see what I see.

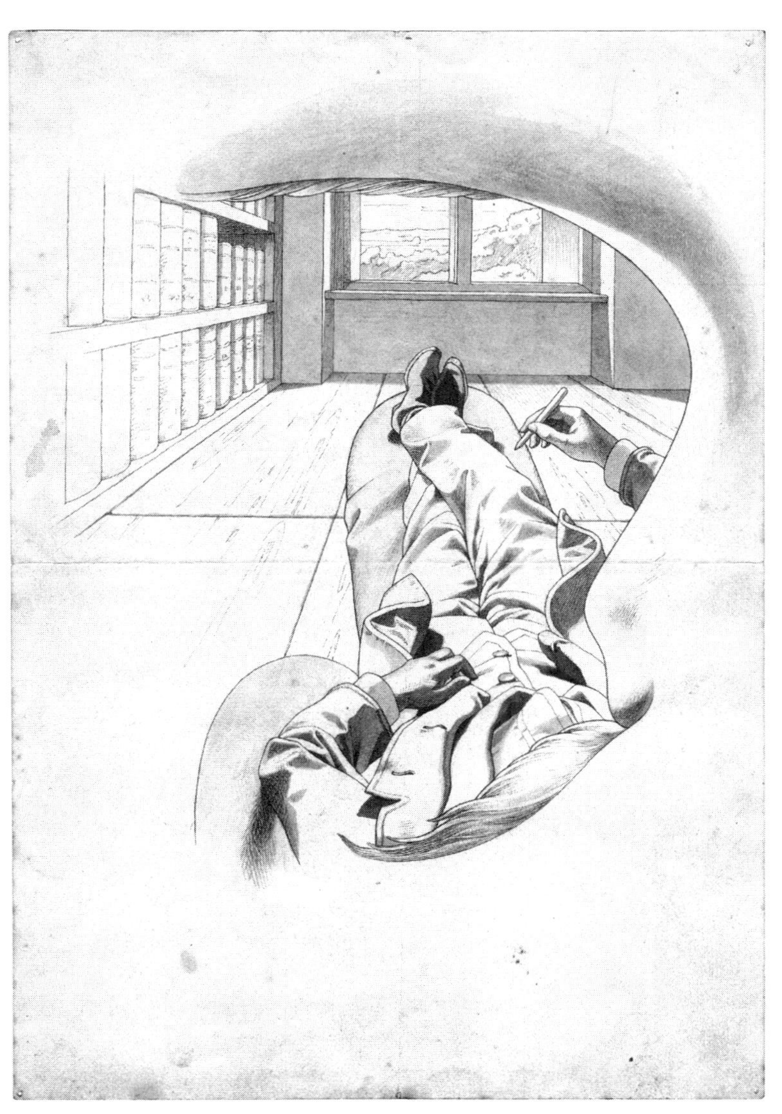

monocular vision, the eye socket would never assume such sharpness of contour, to say nothing of the hatching that makes the nose and inner brow appear to be bathed in shadow — a contribution of the conscious mind, interpreting the uniformly dark tone visible around the eye socket. This tendency is taken to amusing extremes in the diagram's phenomenological "original," the drawing of the late 1860s, where Mach shades in the entire domain from brow through nose to lip, as if he were inspecting the interior of a cave from a vantage point inside it (Figure 4.6). One almost forgets that in both versions the body seen through this aperture — especially the right hand brought up to show off a pen in the printed version — is oddly diminutive in size, or the Mad Hatter effect provoked by the giant moustache.

These, however, are minor imperfections of expression that a more skilled artist attempting this genre (like Henri Matisse in his 1904 *Landscape at St. Tropez*) might avoid. There is by far a greater incongruity in this image: it is Mach's left eye, his moustache, and his body we are seeing. Yet we are not Mach! True, *we* are not any single person, but I am not Mach and you are not Mach, and no one looking at the picture now is Mach. Even Mach, on examining his picture through his left eyehole alone, would have been in the strange position of seeing two left nostrils of Mach (his real one and that depicted), two left moustaches, two left eyebrows. Were we in turn to draw the experience of looking, it might replicate his depiction (Figure 4.7).

And so on, and so on. Yet the oddity of this procedure is by itself nothing more, for we are now dealing with pictures that make no bones about the doubling problem. If we want to know how this impinges on the truth, or aesthetic force, of the image, we must know what the picture is intended to mean or rather, what it does, to which Mach's intention can only be a clue. Is it a pure piece of natural scientific observation to be accepted as true or rejected as false, a piece of meaningful fiction, to be entertained but no more, a thought experiment, or pure nonsense on a par with, say, the view through Mach's large toe? We may orient our questioning after Michael Bax-andall's sane dictum that what needs explaining is not pictures but our accounts of them.[16] This means taking seriously what Mach was doing in his drawing, even if we believe him that he first drew it in the spirit of a philosophical practical joke.

What was Mach's joke, to begin with? The *Antimetaphysical Prefatory Remarks* argue that physics, the pride of the nineteenth century, had unjustly overshadowed the psychological and physiological investigations of self, perception, and mind gloriously carried out by Goethe, Schopenhauer, and Johannes Müller (the last, an important neurologist, is understandably less known in literary and philosophical circles, but we will have reason to return to his work shortly). Arguing that the self is only an artifact of relative psychic continuity, and recalling with relish having just seen himself in an omnibus mirror and thinking "who is that rundown old schoolmaster," Mach offers to spell out the nature of the "I" in less mysterious terms:

> Say I lay on a couch and close my right eye: my left eye shall offer the image in Fig.1 below [i.e., Figure 4.5]. In a frame consisting of the eyebrow's arch, nose and moustache part of my body appears, that part which is visible, and its environment. My body differs from other human bodies, beside the curiosity that lively impressions of motion break out suddenly in its motion, that its touch sets off more conspicuous changes than that of other bodies, through the fact that it is only seen partially and without a head.[17]

In other words, "I" is a word used to describe an unusual kind of perceptual experience, nothing more: a kind of headless horseman.[18] This puppet master theory of the self, according to which we identify our body merely by correlating its movement with other sensations, is proposed not as a serious metaphysical thesis (as it may still have been by Mill), but to get us to see the intent of the picture and its relevance to our own experience. If the picture is how you or I see the world (give or take the facial hair), Mach has made his point, and "self" is only a relation between certain perceived entities. In the original of the late 1860s, Mach jotted alongside the drawing, "Identity more through environment than through psychic identity," and this lesson is, as befits a scientist, presumably supposed to generalize to other people, other selves, as well. In a footnote, Mach goes on to

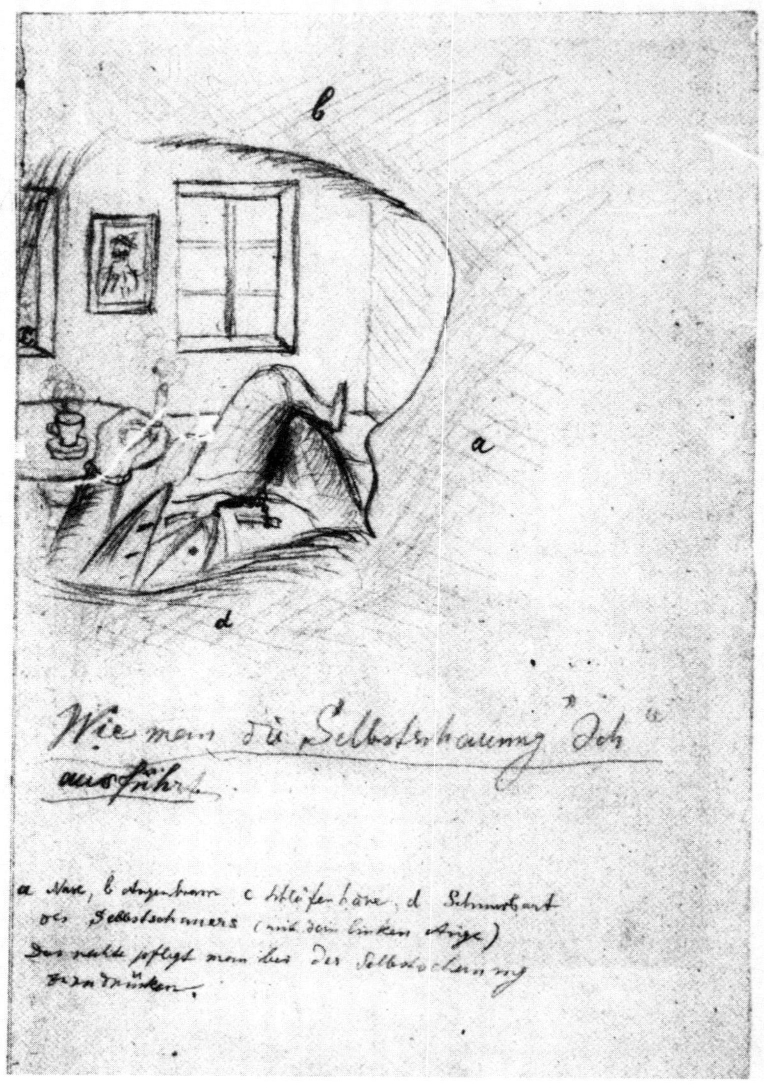

Figure 4.6. Ernst Mach, "How one carries out the self-inspection 'I,'" (c. 1868–69), location unknown. Mach's first draft, with its amusing loose ends—cigar smoke, steaming coffee, and a bearded man in a picture— pasted into a letter to his friend Eduard Kulke, has remained elusive since its reproduction in *Ernst Mach: Werk und Wirkung* (1988).

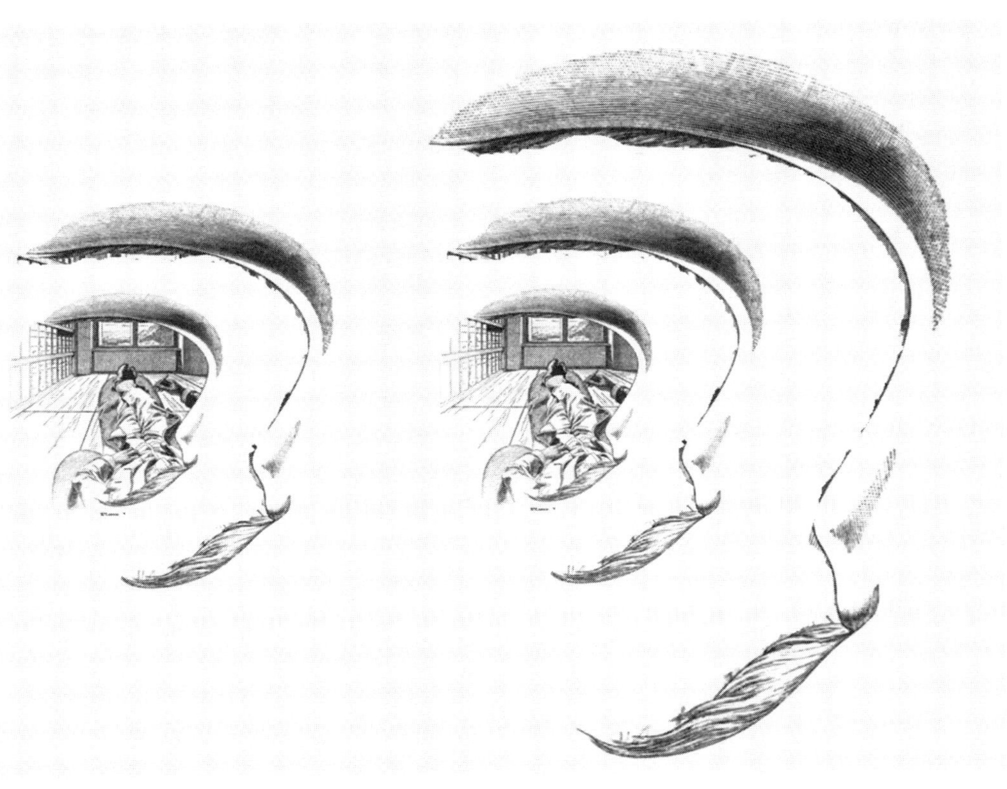

Figure 4.7. Hypothetical drawing of the world through Mach's left eyehole looking at his drawing, and of the world through his left eyehole looking at his drawing of looking through his left eyehole at his drawing.

explain the genesis of the drawing some seventeen years earlier. A friend, long since dead, had given Mach a book to read by the German idealist philosopher Krause: "In this text one finds the following passage: 'Task: Carry out the self-perception "I." Solution: One does it with no further ado.' To jokingly illustrate this philosophical 'Much ado about nothing,' and at the same time to show how really to execute the self-perception 'I,' I designed the drawing above."[19]

Mach mercilessly abridges Krause, whose point, anticipating twentieth-century rediscoverers of the *cogito*, is that the experience of self is fundamental and may be used as evidence rather than requiring supporting evidence itself.[20] But Mach cared little for such metaphysical foundations. He has taken his cue for carrying out the task *practically* from the aforementioned Johannes Müller, who in his 1840 *Handbook of Human Physiology* describes the visual self-experience of the "I" in terms of eyebrow, nose, and cheek — one takes it that Müller was clean-shaven.[21] Consistent with Müller's instructions, the rough draft of Mach's drawing has labels for the nose, eyebrow, temple, and moustache, all of which are identified as belonging to the "self-seer with the left eye" (*Selbstschauer mit dem linken Auge*). Unlike the *cogito* of Descartes or Krause, supposed to be carried out immediately and without outside interference, Mach conceives the perception of his self as the result of carrying out an established experimental protocol — the end of investigation, rather than its cause.

Mach's thoughts concerning the visibility of one's self, or rather one's body, which he thinks comprises all the self that is needed or available, can be thus straightforwardly untangled. In what relation does the picture stand to them? Does it sum them up? It seems unlikely that we can extract all the humor, biography, and scientific references made by Mach from the diagram as printed, which is, in turn, more complex and charming than what the preface has to say. Does it say both more *and* less than what Mach writes? And in any case, just how can one succeed in using a diagram purporting to represent a mental image of the world containing the self in order to dispense with a substantial notion of the self, revealing it to be "much ado about nothing"? Isn't there a circle in using a mental image to show that there are no minds?

Mental Images

The first step to understanding what role pictures can play in such an argument is the nineteenth-century view that pictures are the very stuff of our subjective conceiving — when we think, we see mental pictures (German *Vorstellungen*), sound pictures form upon hearing, and so on. It might seem that a problem with reducing mental activity thus to pictures is that mental pictures are unreliable. Some are fantasies or daydreams or the work of feelings like fear, greed, and vanity. They do not tell us how the world is, but how our mind wants it to be (or *not* to be). The unreliability of mental pictures, and what this meant for their use in scientific work, were very much on the minds of nineteenth-century psychologists and philosophers. Salomon Stricker, in his protobehaviorist 1883 *Studies on the Association of Mental Images* (that is, *Vorstellungen*), before reducing arithmetic, along with all acts of reasoning, to the muscular sensations to which he had already reduced speech and singing, had to admit that our visualizing faculty, and the visual *Vorstellungen* it produced, was hardly up to the task of doing sums:

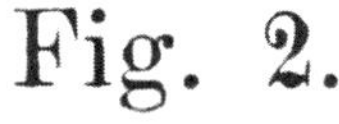

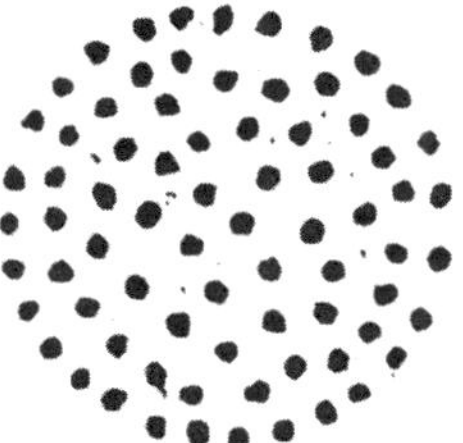

I certainly won't be amiss in assuming that the reader will be unable to tell the number of points contained in Figure 2 from mere visual inspection. However long I look at this pointy [*punktierte*] figure, eyeing it in every direction, the number of points remains unknown to me. If I wish to know the number, I must count the points. But counting involves a labor that can *only* be accomplished with the aid of the musculature.[22]

What we can see right away from Stricker's discussion is that despite the well-founded distrust of our imaging faculty as a tool of knowledge, and perhaps of actual images as sources of information, scientists were fairly optimistic about the prospects of getting smaller, perhaps less conscious sensations or mental images to work with, which, in the manner of atomistic building blocks of our thinking, would do the job of mediating between the conscious mind and the world: they would, as it were, be small and meaningless (because unintelligible?) enough to be material and physical, and yet sufficiently pliable and combinable to result in clear, conscious, meaningful images, feelings, and in the end scientific concepts. We saw in the last chapter how pervasive, and ultimately disappointing, such a faith proved to the quest for displaying subjectivity in painting. And as we saw in the first chapter, some of the most passionate arguments over this use of psychology took place in mathematics. If psychologistic thinkers like Husserl or Stricker found it plausible that there were distinctive mental images accompanying small numbers like 3 or 4, even they had to admit that this is not how we conceive large numbers like 135664 or even 100, not to mention 0, which notoriously does not stand for any particular image or quantity of things.[23] But the point about mental images can be made more general:

> We cannot picture even such a concrete thing as the Earth in the manner in which we *know* it to be; instead we satisfy ourselves with a ball of moderate size, which serves us as a sign for the Earth; but we know that this is very different from it. Though our mental image often does not suffice for what we want from it, we judge nevertheless with great sureness concerning an object like the Earth, even with regard to its size.[24]

Frege's rejection of the mental image discards the ghosts of departed typographical figures (like "C" for 100) that Stricker and Husserl thought necessary to our manipulation of large numbers.

Even an image as vivid as the earth (which we now visualize through color photographs taken from space) is not at all like the thing it stands for—it is a sign, a *symbol* of what we mean by it. This is not just a "symbol" in the minimal Peircean sense of an arbitrary conventional sign, but a symbol as the symbolists meant it: a sensuous object constructed so as to display an intelligible connection between its sense and the object it represents. Thus, although radically distinct in essence from their senses or their objects, images do not impair our ability to think truthfully about them—whether they are images, senses, or real objects. The critique of mental images can thus coexist with a philosophy of symbolism. The latter, however, requires a theory of thought.

Adventures in the Third Realm

What is "thought," in what sense is it related to and in what sense does it go beyond picturing? Frege's answer to this, though reminiscent of a long tradition of Platonic metaphysics, was quite philosophically discordant in its time, the empiricist late nineteenth century; this is familiar to historians of philosophy, but the pictorial aspect of his work, and its proximity to symbolist theory, has been neglected by them and by art historians. In reviewing his work here, I will do my utmost to draw attention to this proximity. It is well to begin where our discussion of his work here and in Chapter 2 left off: in response to the unreliability of mental images as tools of thinking, he located meaning in an objective realm of intangibles called "thoughts."[25] I shall use the term without scare quotes on the understanding that from this point, I mean by thoughts what Frege meant: not the private content of this or that mind, but what such minds grasp in thinking, on the explicit understanding that more than one mind may grasp the same thought.[26] The sentences "2 + 2 = 4," "It rained in Rio on 23 June 2012," and "I am not yet dead" express thoughts. They are *not* themselves thoughts, since the statement of the same thoughts would require different words and perhaps symbols in another language, and their full spelling out requires knowledge of arithmetic in the first case, of geography and chronology in the second, and information about time of utterance and the speaker in the third. Interpretation, often very elaborate and holistic, is thus needed to go from sentences to thoughts: once we have done it, we may "grasp" the thought (a pictorial metaphor, as Frege admits). But still, we can only handle the thought as some sentence or picture or other tangible entity that satisfactorily conveys it to self or to others. It would take angelic beings — with which many a symbolist poet and painter flirted, but which Frege was clear we are not — to commune in pure, unsymbolized thought.[27]

It follows that thoughts and *not* sentences are entities that can be true or false, and they remain eternally so: "it rained today" is not a thought whose truth wavers with each passing day, but a time-indexed statement picking out different thoughts each day it is used. This character of truths, noticed almost two millennia earlier by Augustine, is often misunderstood as a grandiose Platonist ontological thesis, namely that thoughts are eternal, immutable, indestructible entities. The latter qualities are indeed correctly attributed to thoughts, but this is no onerous idealism, since thoughts take up no space, time, or in any other way impinge on the natural world they describe.[28] Frege was not alone among late nineteenth-century thinkers in positing a radically objective realm of "subsisting" entities distinct from physical and psychic reality. His Austrian contemporary Alexius Meinong developed observations on art and play first proposed by his brilliant student Mila Radaković, pointing out that already a child playing at Siegfried, the epic hero who wears the Tarnhelm to be invisible and bathes in the blood of the dragon Fafnir to be invincible (alas, he misses a spot), possesses concepts, like immortality and invisibility, that have no basis in the material world. Nor are they, for similar reasons, mental images.[29] It is typical of late nineteenth-century philosophy that the logical domain of thoughts is argued for on the basis of familiar human activities, rather than in the theological or mystical terms familiar from ancient and medieval Platonism.

For his part, Frege insisted that "a third realm [*that of thoughts*] must be recognized," beyond that of material objects and (psychic)

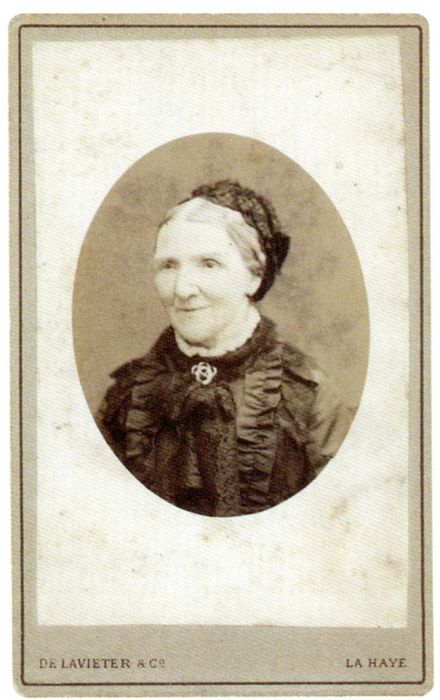

Figure 3.1. Anna Cornelia van Gogh, carte-de-visite, Van Gogh Museum, Amsterdam. Note that Van Gogh reversed the photograph's orientation and simplified his mother's dress, but not her headgear. He made her chin smaller, too.

Figure 3.2. Vincent van Gogh, *The Artist's Mother* (1888), oil on canvas, Norton Simon Museum, Pasadena. Both images were taken in September 2015 with a digital Leica. The image at left is (for me) the most accurate.

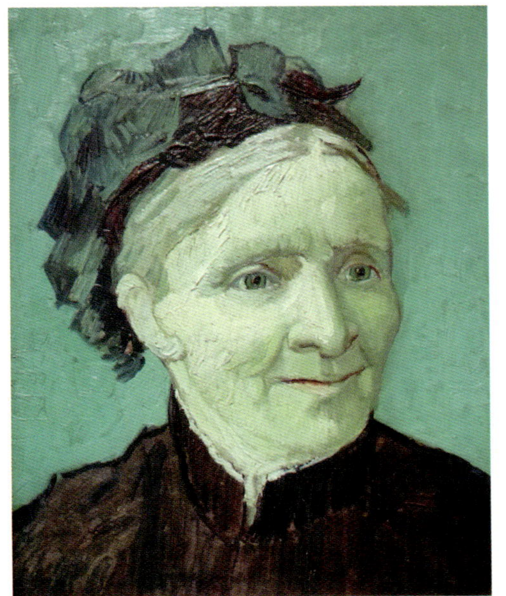

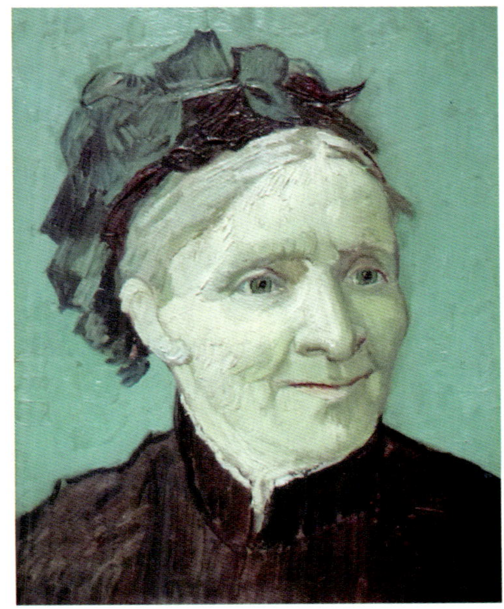

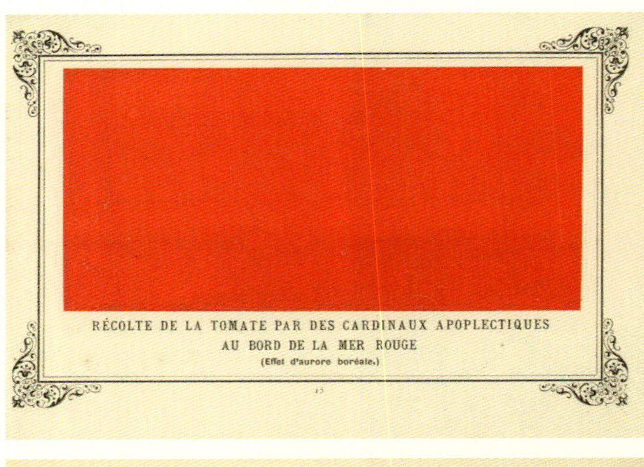

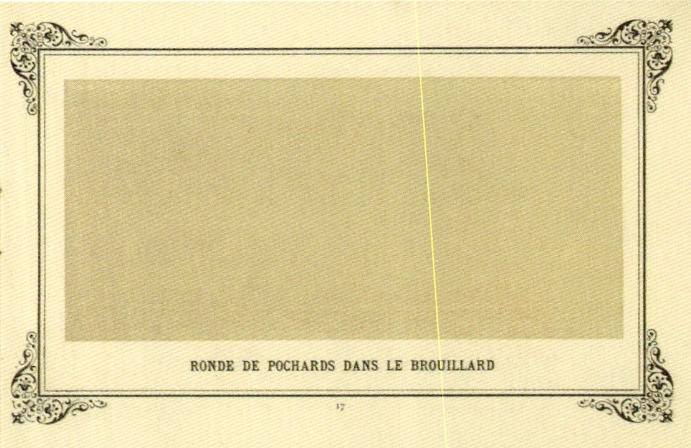

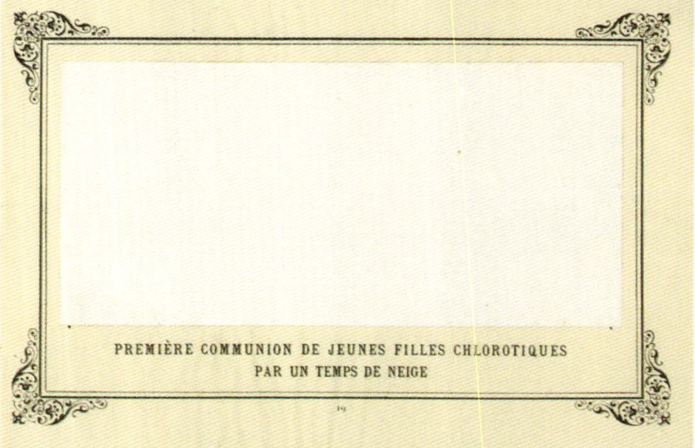

Figure 3.3.a–c. Alphonse Allais, *Album primo-avrilesque* (Paris, 1897), pp. 15, 17, 19. Note the rococo flourish of the frames.

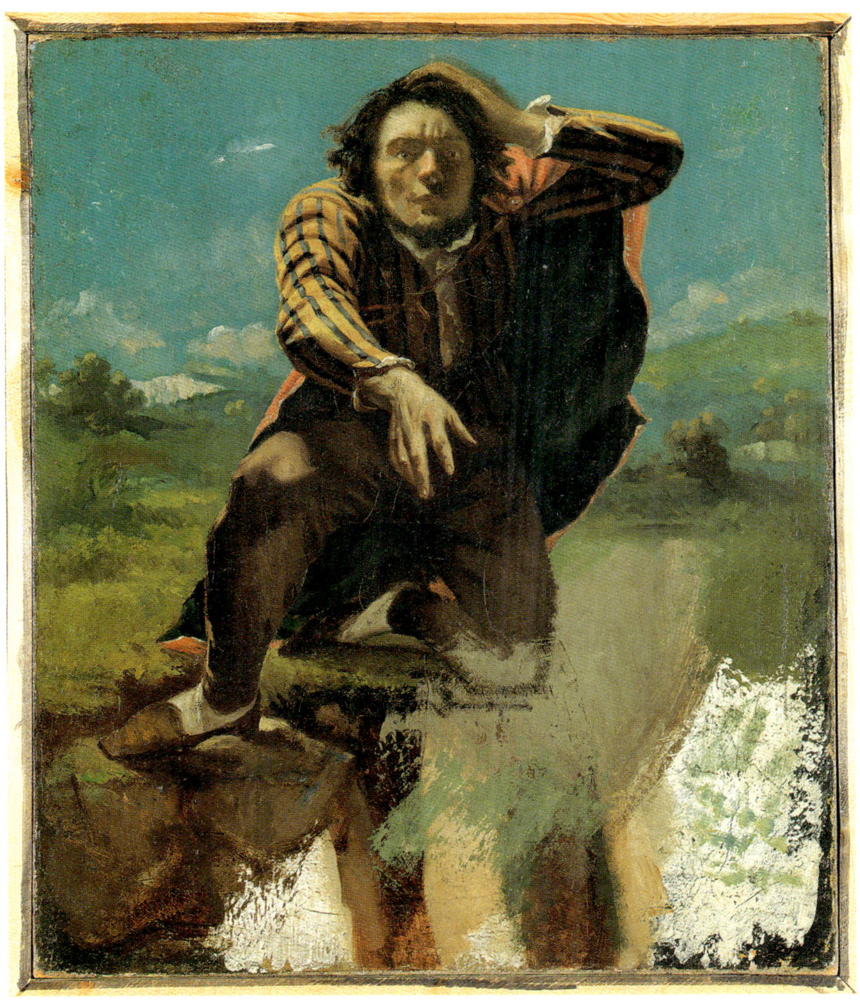

Figure 3.5. Gustave Courbet, *Man Mad with Fear* (1844?), oil on canvas, Nasjonalmuseet, Oslo. Whether Courbet intended us to imagine in the foreground something like an open grave or a precipice akin to those visible in the background, what he shows in the picture is the visual derangement of his frantic alter ego.

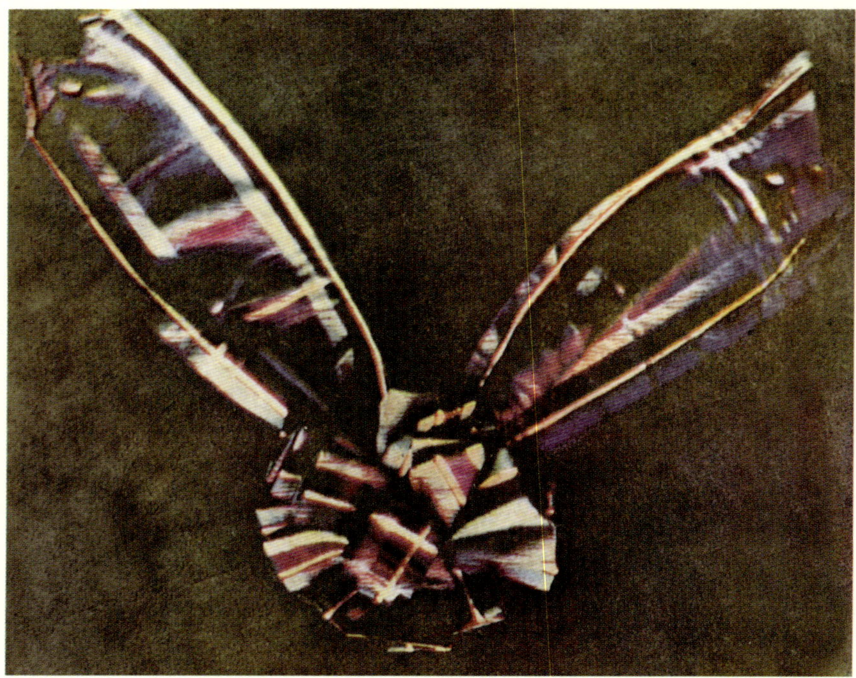

Figure 3.12. James Clerk Maxwell, Tartan ribbon, 1861, Vivex print (1937), National Media Museum, Bradford, UK. The image superimposed from three negatives was only "sort of a photograph" according to Sutton; we do not see it, nor can we see it in a book: what is reproduced here is a twentieth-century polychrome print from the three negatives.

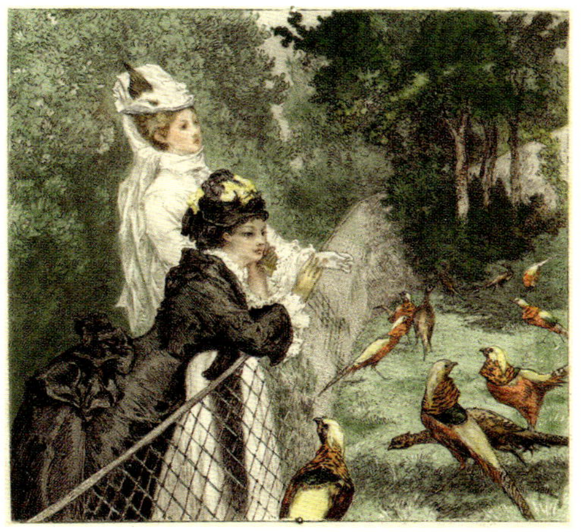

Figure 3.13. Félix Bracquemond, *Au Jardin d'acclimatation* (1879), etching, drypoint, aquatint, final state, British Museum, London. The etching was printed in pink, yellow, green, and black; the birds' orange and the yellow-green are mixtures.

Figure 3.14. Félix Bracquemond, color trial plates of *Au Jardin d'acclimatation*, etching and aquatint, British Museum, London. The two trial plates are printed in green and yellow, respectively, in the second state of the finished etching.

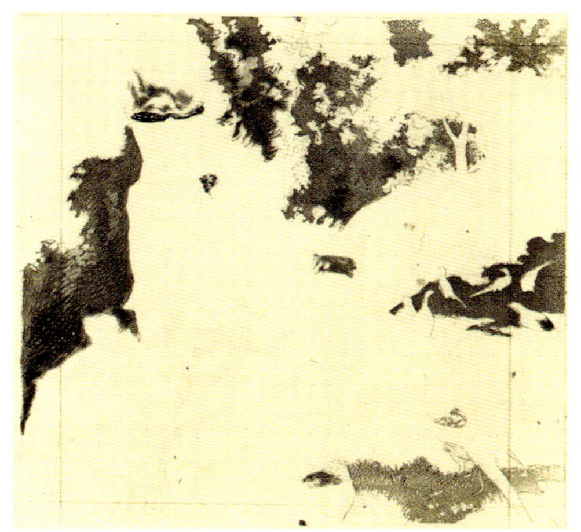

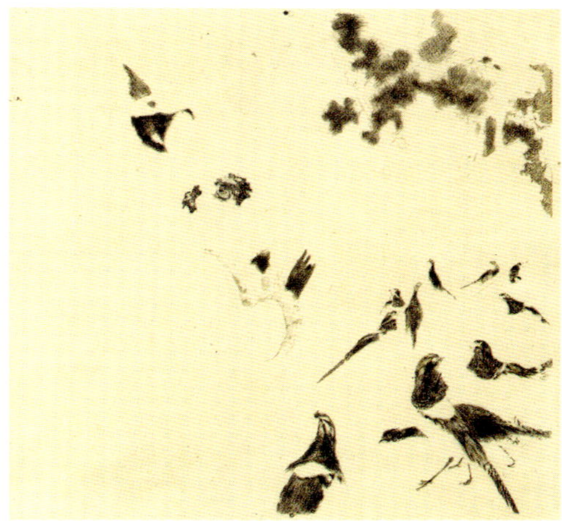

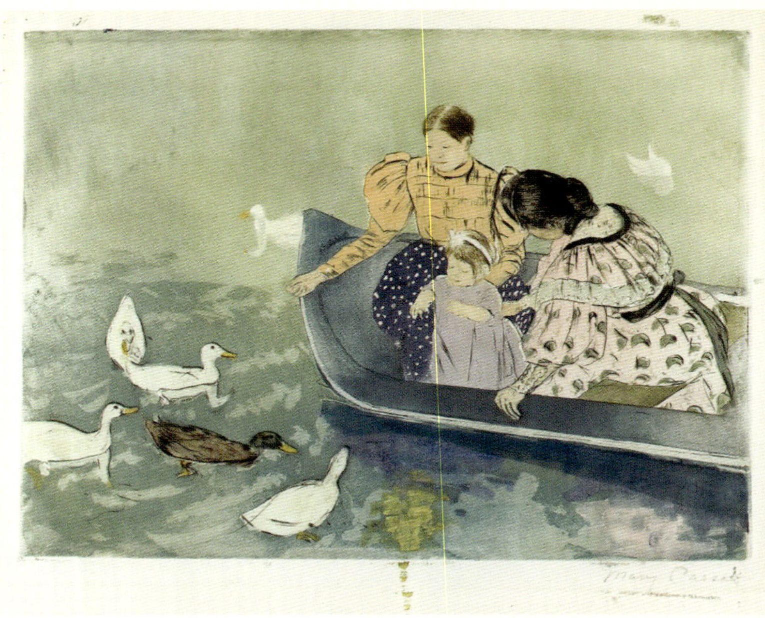

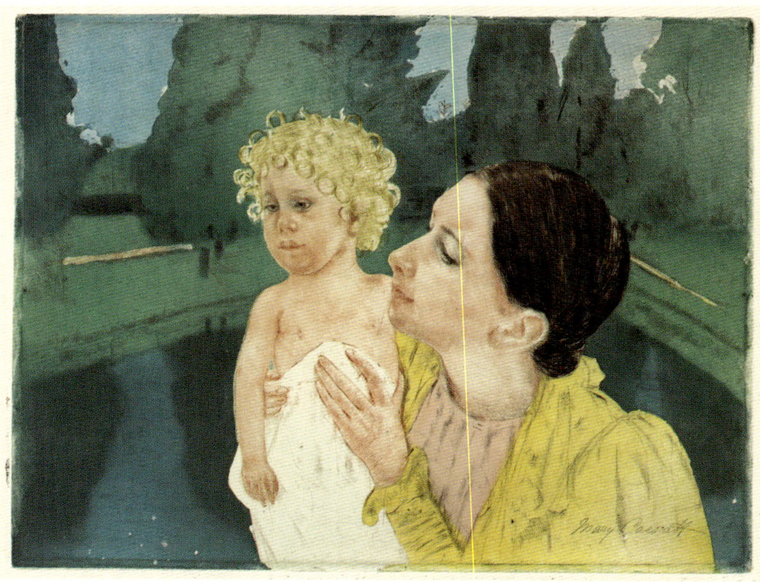

Figure 3.18. Mary Cassatt, *Feeding the Ducks* (c. 1895), drypoint, aquatint, monotype inking, Art Institute of Chicago.

Figure 3.19. Mary Cassatt, *By the Pond* (c. 1898), color drypoint and aquatint (from three plates), Art Institute of Chicago.

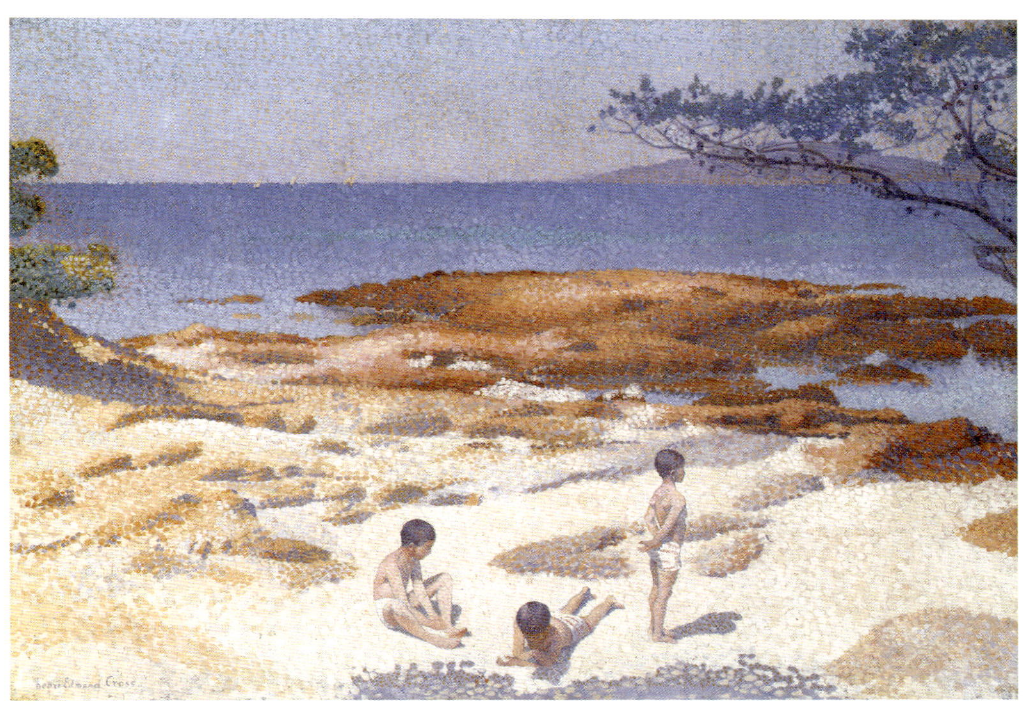

Figure 4.1. Henri-Edmond Cross, *Beach at Cabasson (Baigne-Cul)* (1891), oil on canvas, Art Institute of Chicago. Note how Cross's round brushstrokes, layered on the land and water, are lined up in a neat grid (not unlike Taine's "copper scales") in the sunset sky.

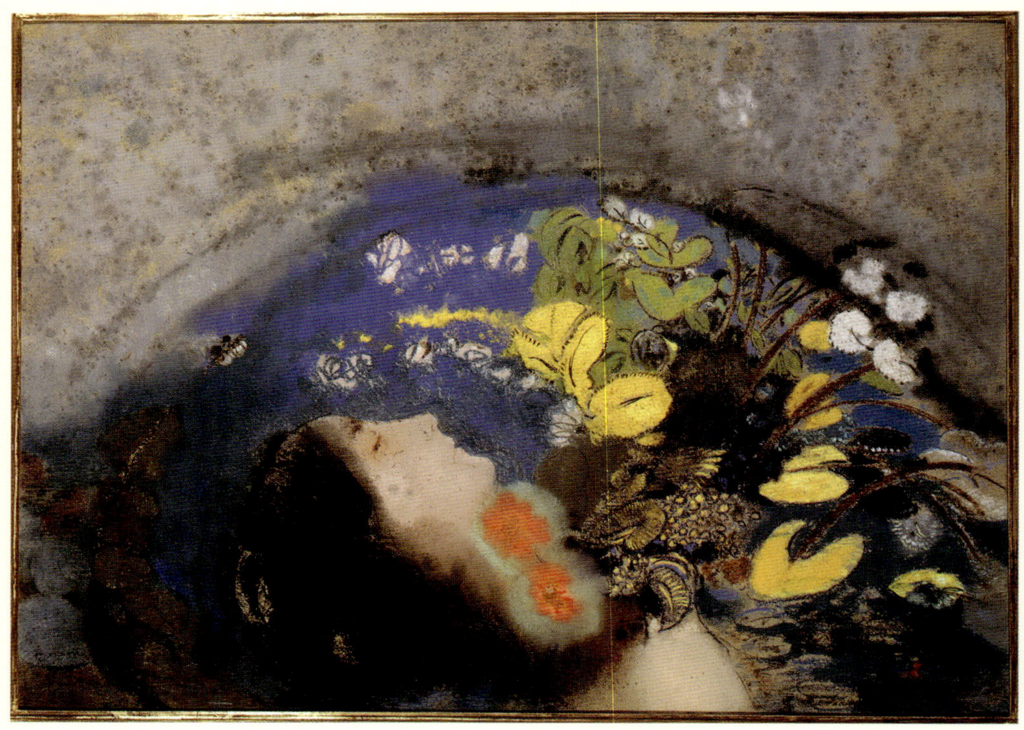

Figure 4.10. Odilon Redon, *Ophélie* (1900–1905), pastel on paper, Woodner Collection, New York. Ophelia and her watery bed are contained as if in the blink of an eye—but see how the white bouquet and the trace of one lonely bloom above it reach into the mottled space enclosing her.

experience; but he was not tempted into sketching a Platonic realm of truth, since he considers thoughts not *real* or *existent* but merely objective.[30] To say they have always been true is simply to acknowledge that, in the case of a matter of fact, the specification of time and place makes something (the fully specified thought) definitely the case, no matter on what occasion it is thought. To talk of thoughts becoming true (or false) is in fact to talk of more or less complete sentences that specify different thoughts at different times.

It is intrinsic to the nature of thoughts that they can be true or false. There is no third option. But, strangely enough, thoughts do *not* seem to possess this property in every context. In particular, the aesthetic, with which this book is concerned, seems to offer a "third option":

> Why is the thought not enough for us? Because and only insofar as the truth value matters. This is not always the case. On hearing an epic, for instance, we are bound, beside the harmony of language, by the sense of the sentences and the feelings and imaginings they evoke. With the question of truth, we would leave behind aesthetic pleasure and turn to scientific investigation. Thus we are indifferent, whether the name "Odysseus" has a reference, as long as we take the poem as a work of art.[31]

It might seem typically Kantian, or Nietzschean, or Wildean, to deny that art has anything to do with truth. Yet Frege's transition to science distinguishes him from these allies in denying truth to art: as we will see, it gives the social context of art its due. The differentiation between aesthetic and other uses of signs (be they scientific, political, or the like) is indeed not internal to the signs themselves, as formalists might think, but depends on how those signs are put to use, what practices they are caught up in. According to Frege, to inquire into the truth of a thought is to embark on a scientific investigation. Various other contexts, which Frege sums up as "poetry and myth" (*Dichtung und Sage*), give rise to no such investigation.[32] Frege says that all thoughts are true or false, and he *seems* to hold the traditional view of art as devoid of truth. Are we then to conclude that fictions, and pictures, are not thoughts? Or is he being inconsistent?

I believe that Frege is not inconsistent, nor that he holds the traditional view; the absence of truth in fiction is a subtler phenomenon than mere thoughtlessness. To show this, I will return finally to Mach's pictorial puzzle, on which Frege himself commented. In figuring out what there is to say about the truth and fiction of this picture, we will come to spell out what it is to be a picture (the relation of the picture to the written word is examined in more detail in the next chapter).

But first, just what is a picture for Frege? Does he ever use the word, apart from his critique of mental images? It turns out that he does make a point of using the German word for actual concrete picture (*Bild*), not just *Vorstellung*. He goes so far as to offer a definition of it: "It would be desirable to have a particular name for signs that should only have a sense [and *not* a truth-value]. Were we to call them pictures, then the words of the actor on the stage would be pictures, indeed, the actor himself would be a picture."[33]

The words quoted point out our goal, and that of any symbolist picture theory: sense without truth is intelligible fiction, independent of truth but not of meaning. Yet how do we reach the goal? Recall that for Frege thoughts never *change* truth-value: a fictional thought cannot "turn" true just because we decide to investigate it scientifically, any more than identical sentences in truth and fiction stand for parallel thoughts in two languages or universes of discourse.[34] Of course we might call a thought lacking truth-value an "apparent thought" (*Scheingedanke*), as Frege does in an unpublished text, but that does not tell us how it differs from a "real" thought.[35]

The reader may wonder whether Frege intended his word *picture* to accord at all with common usage. After all, a fictional sentence and a picture, fictional or documentary, seem like wildly different things. Yet I think Frege never lost sight of pictures in the literal sense. In discussing fiction, Frege considers a variety of visual scenarios, like anatomical illustrations, history paintings, and the actor acting. But his deeper point about thoughts is that they are not *ever* linguistic: words serve (imperfectly and contingently, as in the case of "today") only to pick out *some* thoughts; others are accessible visually, acoustically, and in other sensible guises. Applying the word "picture" to sensemaking objects, whether visual, auditory, haptic, or linguistic, allows us to see what they have in common. Recall Eder's x-ray: what makes it a picture, as opposed to a trace of radiation on photosensitive paper, is the form it has, the sense it makes as a rendering of internal organs. It may but not need be true — or false — of some body.

A picture, whether history painting or actor onstage, is for Frege a coherent thought apart from its truth-value.[36] To make sense of this striking suggestion, we have to consider two of Frege's revolutionary ideas, the doctrine of assertion and the duality of meaning. According to the latter, words, sentences, and other signs don't just mean one thing (linguistic or conceptual sense, their contribution to the thought), they also pick out or refer to (*bedeuten, bezeichnen*) the object they stand for. "Jabberwocky" produces a character (its sense), while Richard III does that, and also names a person; predicates like Lewis Carroll's "brillig" suggest a way "'twas," while real predicates, like "gyre," do this and also describe the way some things may be (or behave — namely spin); finally, whole sentences may convey thoughts (Frege's example, "In deep sleep, Odysseus was brought ashore to Ithaca," regardless of whether or not there was an Odysseus).[37]

Frege's great innovation, which not even Platonists who believed in the independent existence of concepts had anticipated, was to insist that the sentence as such also names something. If subject and predicate refer, the whole sentence does too, and is true or false. You can see this by considering a conditional sentence, like this offer to my son: "If you brush your teeth, you may stay up late." The ten-year-old logician may wonder what makes my sentence true. Surely there's no causal connection between brushing and staying up late? I clarify my point by making its beginning more explicit: "If it is *true* that you brush . . ." Another way of putting this is Frege's: the thought that "you brush your teeth" stands for Truth, or Falsity, which is why we can use it as a part of more complex thoughts that depend on its truth-value. If I had said instead, "If 'twere brillig, you can stay up late," we would grasp an amusing thought, but no consequence would have been forthcoming, unless we agreed about the meaning of Carroll's adjective.[38]

That very possibility should make us pause. If sentences are true and false regardless of what we think about them, how can the "brillig" sentence be truth-valueless until we settle on a concept for it to represent? With this, we are approaching Frege's contextual, deeply pragmatic theory of assertion. Fictional or stage utterances may be identical to true ones, may even be attached to sentences advertising their truth, yet fail to be true in the world: "One could certainly say, 'The thought that 5 is a prime number is true.' ... [But] there, where it lacks its usual force, in the mouth of an actor on stage, the sentence expressed is only a thought, indeed, the same one as '5 is a prime number.'"[39]

In other words, what is said onstage expresses exactly the same thought but means no thing or person or truth in particular. As Frege sardonically puts it, "A physicist wishing to investigate thunderstorms leaves stage thunder unexamined."[40] The causal connections one expects in reality are simply not assumed to attend fiction, no matter how convincing the stage thunder.

In Frege's writings on assertion, from his first mature text, the *Begriffsschrift* ("Concept Script") of 1879 to his last published text, "Compound Thoughts" of 1923, he is at pains to point out that, while there are clues that a thought is being aired, not asserted (the "if . . ." and "then . . ." of a conditional, the question mark), there is no way to guarantee the opposite. The added "I assert" or "it is true that . . ." is equally at home in fictional contexts. Frege is fascinated by the fact that the same truth-value deprivation can be achieved linguistically or institutionally: "To make it clearer that a thought is only expressed, not asserted, I phrase it in subordinate form: 'that the seawater is salty.' Instead I could let it be spoken by an actor onstage, for one knows that the actor in his role only appears to speak with assertoric force."[41]

Frege has an explicit assertion symbol in his sign language, but he knew that, were his biography filmed, such signs would be made on a blackboard by a bushy-bearded actor who did not mean them at all.

As he put it pungently in a 1917 letter: "The actor only makes as if [*tut nur so*, what children do when they pretend] he asserts something, just as he only makes as if to stab someone, and one can accuse him of lying just as little as of attempted murder."[42] This is no deconstruction or anything else subversive; it shows only that the force of assertion does not lie in any symbol or object, but in the context shared by creator and recipient. But the fact that truth can turn into fiction at the drop of a hat, or vice versa, is important if pictures, artistic or scientific in character, possess the power to traverse realms of fantasy to (sometimes) reach truth.

A Picture Language

Having spelled out some Fregean theses on the role of pictures in thought, one may still wonder how the philosophical sense of "picture" accords with actual artifacts and their use, and particularly, with the conceptual primacy accorded to the picture in symbolist aesthetics. To see this, we must examine the use of symbols in Frege's own practice as a mathematician: He notoriously invented a picture-language for proofs, the *Begriffsschrift* or concept-script. In considering it, we will not leave the realm of pictures or interpretation for a more mechanical, automatic, objective, or "linguistic" realm of signs, symbols, and so on. This warning has to be sounded because Frege rejected the claim that mathematics was mechanical, "aggregative thinking" (a charge leveled by his neo-Kantian teacher Kuno Fischer). Against this, and against claims by Frege's friend Wittgenstein and his circle that signs needed only to refer, sense being only a kind of psychological window dressing, Frege insisted that we need this more exploratory, pictorial level of meaning if we are to build novel thoughts out of our supply of signs: "We could well agree that certain signs should express certain thoughts; like signals on the railway (*track clear*) but in this way we would always be constrained to a very narrow domain and we could not build a new sentence that another could understand without a previous agreement being particularly made for this case."[43]

Frege was dissatisfied in a similar way with the symbolic logic of his time, notably Boole's, in which arithmetical symbols were used for new logical purposes, so that "+" meant "or" and "." meant "and." For Frege, this was an unfortunate collision with mathematical practice, bound to run into trouble when actual content, say, equations, was combined with the new symbolism. In contrast, his signs, though reputedly incomprehensible (he himself admitted they were strange), build an elegant kind of network that, beside assertion, negation (|), and generality, consists of a dash (—) indicating content, bound with others through vertical dashes that represent inference.

The basic unit of reasoning, $\vdash\!\!\begin{array}{l}\text{—}A\\[-2pt]\rule{0pt}{6pt}\!\!\lfloor\text{—}B\end{array}$

asserts simply that if B is true, A must be true.[44] The inference symbol (*Bedingung*, literally "dependence") itself is suggestively constructed out of the negation and content signs: if B is false, the vertical duct carries negation into A, making A false if the whole conditional is true; if A is false despite B being true, the whole conditional is false (if B is false, A may be either). In the first, simpler form of the language, Frege had five primitive symbols (generality, content, negation, assertion, dependence — he would later add machinery for handling sets), but the action boils down to constructing and substituting into inferences (*Bedingungen*), an astoundingly simple and elegant way of cutting through the elaborate logical taxonomy built up since Aristotle. The trade-off (at least until he switched to two columns of text in 1893) is a liberal way with the printed page. It is probably this arrangement that led contemporaries like Russell and modern historians of science like Lorraine Daston and Peter Galison to judge Frege's system too "opaque and cumbersome." Daston and Galison see a philosophical scruple, if not prejudice, motivating Frege's adoption of this graphic system: "Like the photograph that checked the impulse to project sharp outlines and pleasing symmetries onto an imperfect specimen, the *Begriffsschrift* held all seductive pictures and equivocations at bay. Both served as sentries against subjectivity, but the one embraced images while the other repudiated them."[45]

Given the scope of Frege's thinking about pictures, the imputation of iconoclasm, though well taken, is perhaps premature. Is an image that "holds at bay seductive images" really a repudiation of images? Consider a slightly more involved sentence of Begriffsschrift:

$$\vdash \begin{array}{l} a+1 > b+1 \\ a > b \end{array}$$

This says that regardless of what a and b are, if a is greater than b, then $(a + 1)$ will be greater than $(b + 1)$. We are not told that a is greater than b: these letters are dummies, they do not refer to any single number. The only truth asserted is about what happens when one is added to both. The visible form of the ducts, the way they order the formulas, shows the dependence of one family of cases on the other.[46] Not that this is self-evident in the diagram. When are pictures ever so? My claim is more modest: that formulas of Begriffsschrift, viewed as pictures, show a thoughtful reader what they explicitly say, exposing as ungrounded any resonating thoughts that extend past it, in a way that, for instance, Mach's suggestive drawings do not. Daston and Galison are right to see a vigilance against equivocation here, but it is not a vigilance against pictoriality or subjectivity as such. Consider the *Begriffsschrift* passage (§15) they use to clinch Frege's iconoclasm (Figure 4.8).

What this tells us is that the deduction in the upper part of formula 1, which depends upon the deduction in the lower half of formula 1, is valid by itself, because that lower deduction (formula 2) is itself independently true. Putting this true formula below its replica in the conditional, Frege is able to eliminate it from the inference, leaving the upper part free under the horizontal stroke (his formula 3). One can look closer and figure out the details of dependence, which if written out would make quite a long sentence (not page-long, but much harder to keep straight than the image). The beauty of the Begriffsschrift is that, once the procedure is mastered, one simply *sees* that parts substitute into one another, true bits eliminate identical bits that are antecedents of a conditional, and so on. In reading, linguistic details might blur as the reasoning gets intricate, but one continues to make right judgments by *looking* attentively at the pattern of dependence. And so the Begriffsschrift by no means repudiates images.[47]

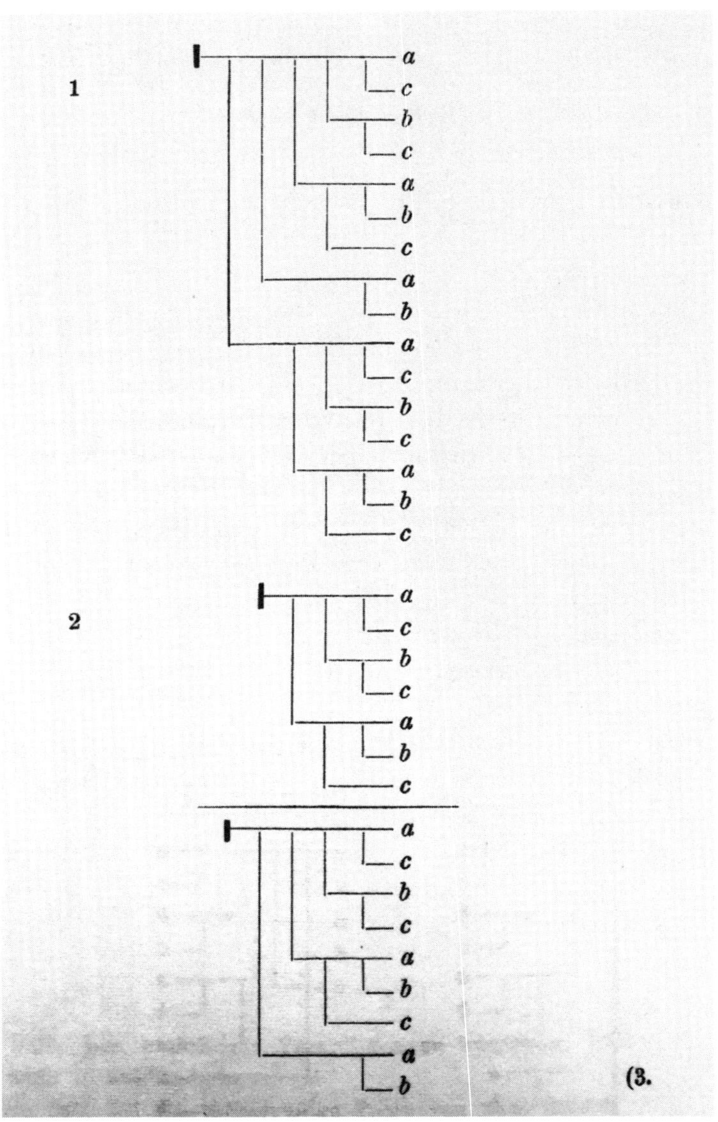

Figure 4.8. Gottlob Frege, *Begriffsschrift*
(Halle: Verlag von Louis Nebert, 1879),
p. 30. The formula at the bottom is #3.

We could go further, for the very next deduction in the book demonstrates its radical pictoriality (Figure 4.9). The main mode of producing new formulas is substitution of letters: into our Figure 4.8's formula 2, Frege proposes to substitute three short formulas for *a*, *b*, and *c* respectively, so that the resulting antecedent (lower part) of the new formula matches formula 3 and may be eliminated, leaving a consequent (upper part) that is his formula 4. Once one gets the hang of substitution, passages expand and mutate in one's mind as they take their place in deductions. The resulting image-work is dynamic, making use of our power to make and manipulate mental images, taking as its paradigm the algebraic practice of substituting equivalent values and in the process vindicating Frege's claim, otherwise less than obvious, that the script was designed "in imitation of arithmetical symbolism." Indeed, the unfolding of formulas resembles the kind of overarching tension Mallarmé built into his page-spanning sentences in *Un coup de dés* (see Figure 2.22). Far from a puritanical rejection of images, concept-script is *all* image, so that if its name referred to how it operated rather than what it operated with, it might be called a *Bildschrift*. Its task is not to eliminate subjectivity so as to automate logic, for Fregean figures are made to be seen, unlike compiled computer code. Nor are they really, as Russell quipped, meant to be such a nuisance as to get the thinker to slow down. Instead, they prevent the skipping of steps by yoking the subjective effects of diagrams to just those thoughts the diagrams validate. It is pictorial thinking more precise than the visual proofs in Euclid, not a mathematics purified of images.

Though Frege made no systematic study of images, their rationality impressed him — as did the importance of precise tools for gains in science.[48] Nor, given Frege's theory of thoughts and pictures, to say nothing of his symbolic practice, can we rest with a distinction between the logical or conceptual realm as a radically anti-pictorial one, and the empirico-psychological realm of *Vorstellungen* as the natural home of pictures. So we must abandon the myth that the reactionary Frege fought a rearguard action against the revolutionary Helmholtz, who invaded "Frege's sacred preserve of arithmetic"

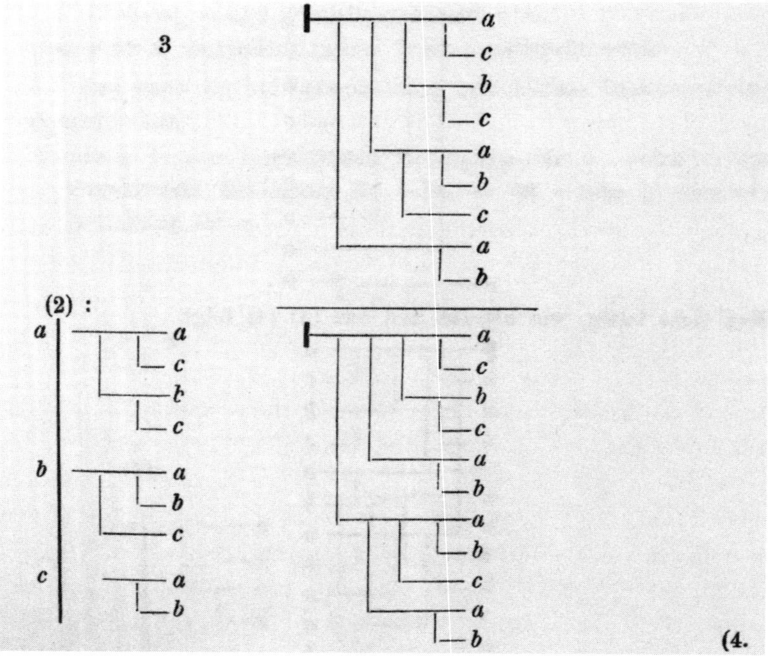

Figure 4.9. Gottlob Frege, *Begriffsschrift*
(Halle: Verlag von Louis Nebert, 1879), p. 31.
This deduction follows the preceding one
(see Figure 4.8) and is reproduced here
as an example of the substitution of variables
to transform a formula into another.

under the banner of experience.[49] Frege began his career as a geometer, comfortable with the role of intuition in the study of space. It was his interest in "imaginary figures, to which we attribute qualities inconsistent with every experience" that led him to question the role of conceivability, experience, and mental images in reasoning.[50] His dissertation, which this quotation introduces, busied itself with developing methods for picturing imaginary numbers in geometrically intuitive ways. In the course of his career, he came to the conclusion that what is shared and reliable in such intuitions cannot be the purely subjective impressions they have on us, for "a phantasm contradicts another as little as a whirlpool in a stream contradicts another."[51] Again, these are not the words of an iconoclast, though an image-critical tendency breathes in them, as it does in Aurier and Mallarmé. Nor was Frege alone among scientists in thinking a two-dimensional notation useful for abstract thought. Just three years after the publication of *Begriffsschrift*, the American logician Charles Sanders Peirce wrote to a friend about the utility of "spreading formulas over two dimensions." To experiment with this, Peirce sketched two figures,

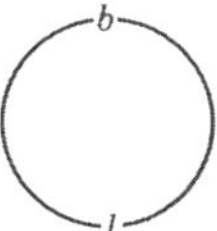

"to express the proposition that something is at once benefactor and lover of something" and

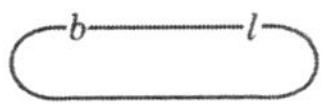

to show "that something is at once benefactor and lover of something, that is, something is benefactor of a lover of itself."[52] Peirce also gave the familiar Boolean notation for both cases: $\Sigma_x \Sigma_y\, b_{xy}\, l_{xy} > 0$ and $\Sigma_x \Sigma_y\, b_{xy}\, l_{yx} > 0$, which are indeed concise, but evidently lack something — not something merely subjective, but as it were logically intuitive or subjectively logical — possessed by his diagrams. Here too one sees that a visual logic, whatever its virtues, involved expenditure of the printed page, a fact decried by the Boolean logician Ernst Schröder in an review of the *Begriffsschrift* that charged Frege with "paying homage to the Japanese manner of writing vertically!"[53]

If Frege's symbolic logic was hard on printers, it had the advantage, no minor one to our purpose, of bringing out the formal structure of picturing: it is no great stretch for someone thinking along these lines to acknowledge fiction as consisting of analogous structures of sense, with reference left out of account.[54] But we also have to know when fiction can diverge from the forms of reality without losing meaning: what would it mean, even in a fictional context, for Odysseus, non-metaphorically, to leap off the page? To be able to attribute sense to the universe of discourse, with its fantastic beings and actions, we have to be able to bind the linguistic or pictorial structures that we use in daily discourse to subjective experience in such ways that thoughts resonate in as determinate a manner as the fiction demands. That is, in saying that "Odysseus leaps off the page" I have not said much, besides rehashing a tired phrase: a brilliant postmodern novelist might use the phrase to represent a fictional world of considerable lucidity. Frege engages in such extensions in our ordinary understanding of sense in his polemics against psychologistic and formalist theories of numbers. One has to think out one's fictions. "Numbers are a row of similar things," runs one hasty theory. Very well, Frege replies, let's try it out:

> A train is a row of similar things, capable of moving themselves with wheels on tracks. One might think the locomotive rather special. But that makes no real difference. And so a number of this sort comes rushing out of Berlin. Obviously the science of such numbers will be quite different than that in which numbers are supposed to be marks on paper.[55]

If each wagon is "one," can we make trains by "repeating similar elements"? "I don't think railroad administrators are aware of this means of producing trains."[56] Or, granted that the train is a number: can we multiply it by the "number of books" on a shelf? What will the product be made of? Book-wagons? Wagon-books? Mental images?

The drollness of Frege's prose should not distract us from its consistent way of using images to disprove claims about them, whether it is that numbers are a kind of habitual picturing or arbitrary marks on paper. The point is not that we cannot identify numbers arbitrarily with some object or other (that is, an object functioning as some symbol or other), but that if we do, we have to abide by the consequences:

> One takes an object, maybe the moon, and declares: the moon multiplied with itself is –1. Thus we have in the moon a square root of –1. This explanation seems permissible, because out of the previous meaning of multiplication no such product can be obtained, and thus in an extension the definition may be fixed arbitrarily. But we also need the product of a real number with the square root of –1.[57]

The trouble with definitions is having to honor them: if the moon is to serve as *i*, it must answer to the accepted properties of imaginary numbers, like giving -1 when multiplied "with itself." Pictures, far from being cast out by a logical puritan, may show us when we have theorized our way from useful to inconsistent fiction. But to see that, we take their structure seriously.

Let us apply this hint about the logical articulation of subjective acts of picturing to Mach. We know there are thoughts of Mach's to which the "left eyehole" image contributes: most important, that the self is a particular way of perceiving the world and nothing else. There are other Machian theses, such as that volition is shown by vivid mental images issuing in action, to which his drawing, being a still image of a body in repose, cannot contribute much. What of the "special perception" thesis: the idea that the self is, and is only, the correlation of a number of sensations, among them the image of a headless body that is always part of the visual field? Frege considers just such an argument in his late (1919) essay on the objectivity of thoughts.[58]

> But I seem to hear a strange objection. I have several times assumed that the same thing I see can also be observed by another. But what if everything was

> a dream?... Something that can no more exist independently of me than my feeling of tiredness, a mental image can be no human being, cannot see the same meadow as me, cannot see the strawberry I am holding.[59]

As Daston and Galison note, the worry is not that the world as product of the mind is not as it seems, but rather that, as product of the mind, it grants no access to other persons.[60] Could one then really know anything about oneself and one's own experiences? A comic interlude follows:

> All is mental image? All needs a bearer, without which it cannot subsist? I have seen myself as bearer of my mental images; but am I not myself a mental image? It seems to me as if I lay on a lounge-chair, as if I saw a pair of polished boot tips, the front of pants, a vest, buttons, parts of a coat, especially sleeves, some beard hairs, the vague outlines of a nose. And this union of sense impressions, this total image is me? It seems to me also that I see a chair. It is a mental image.... How do I come to pick one of these images out and call it the carrier of the others? Why should it be the one, which I am used to calling I? Could it not just as well be the one I am tempted to call chair?[61]

Frege's point is not that the chair might really be me, but that on the Machian reduction of the world to a particular way of seeing, all perceived things become (subjectively) equal, so that one might as well take a chair as a human body to be the carrier of mental images. If on the other hand one discards the carrier, one also discards the mental images, for "without a ruler, [there can be] no subjects." Images would fly about freely. Yet that corresponds as poorly with our experiences as it does with Mach's picture. Is it better, returning to the moon and the judgment that "I see it" to assign a mental image to the moon and another to the "I" who perceives it? That mental image would have, as components, mental images of the moon and itself. The resulting "endless nestedness in myself" is untenable, for then there would be no single "I" but infinitely many, as in our reinterpretation of Mach's diagram (see Figure 4.7). And so, we must conclude, against Mach, that "I have a mental image of myself but I am not that image."[62]

Frege goes further: the doctor can treat a patient's pain, possessing only an idea (really, a thought) representing the pain. The doctor may err, as we do all the time. In erring, "we fall against our will into poetry." But if we do not venture outside our private sensations and risk the fall into error, we have no outer world at all. We can get a more vivid sense of where Mach went wrong, and how worthwhile was his effort, in looking at one of the artworks that arguably gets Mach's gambit *right*: I mean Odilon Redon's breathtaking rebuke to the nineteenth century's literary painting, and its manifold attempts to depict Ophelia's death in *Hamlet* with something approaching Shakespeare's combination of brutality and delicacy (Figure 4.10; see color plates). What artist-readers as thoughtful as Delacroix and Millais envisioned flatly as a woman adrift in a forest landscape is reinterpreted by Redon as a muted *yeux-clos* scenario: we see the supine profile of the delirious Ophelia, flowers clutched about her, in a blue pastel swirl of river and reflected vegetation. Beyond it, gray and mottled like the void, or the interior image upon a closed eyelid, arches a world indifferent to the dreamer, one which she no longer heeds, to her doom. This arch around her is almost a first-person window like Mach's left eye.[63] But though it gives us various subtle hints about her detached state of mind, it does not purport to *make us* into Ophelia, whom we see. A philosophically acute picture does not do the impossible, but it makes us vividly aware of the work of reason and of imagination.

Pictures as Functions and Beyond

In what remains of the chapter, I want to draw the moral of the logical analysis of Mach's image and Frege's symbolist picture theory and practice, asking what implications it has for art history more generally. Before we do, it is worth throwing a backward glance at the history of artificial languages, in order to be able to appreciate the distinctively symbolist quality of Frege's own. Frege knew and admired Leibniz, who was the source of the Boolean symbolism he disliked; and he must have known the method of representing sets by overlapping circles or squares pioneered by Leonhard Euler and refined by

John Venn and Lewis Carroll, though these too, like Boole's, were more useful for demonstrating rules than for any kind of sustained reasoning. He probably didn't know, and would have despised, such ad-hoc visualizations of complex metaphysical theses as Richard Jack attempted with letters and shapes in his "mathematical theology" of 1747; the case of John Wilkins, the celebrated seventeenth-century English Copernican and inventor of an artificial language, a project that was reprinted and further worked out in the nineteenth century, is more interesting (Figure 4.11).[64] Frege, whose parents ran a school and were very interested in language (his father published a book on grammar), may have known of the performance of the Anglican bishop.[65] In any case, Wilkins's procedure was the opposite of Frege's; he invented a large set of basic symbols, which could in turn be inflected to play various grammatical roles, but also to pick out objects from exhaustive lists classifying metaphysical notions, plants, animals and minerals, and various human products. Wilkins made it clear in his prefatory remarks that the classification was not exhaustive, and that arbitrary terms would have to be added to the language, as the categories grew, outrunning the space available for inflection on the little symbols. Being a pastor, he entertained wishful notions of overcoming the "confusion of Babel," but also, like Frege, he hoped the philosophical language would help expose "several of those pretended, mysterious, profound notions, expressed in great swelling words... [which when] examined, will appear to be, either nonsense, or very flat and jejune."[66] The similarity to Frege's observation that politics contains phrases of dubious reference like "the public will," in his essay "Sense and Reference," is striking. But Wilkins never got as far as exposing any fallacies.[67] Looking at his rendering of the Lord's Prayer, one cannot help but expect the language, indexed to Wilkins's tables of concepts, to mirror the assumptions and prejudices of its learned inventor. For it is a language of natural nouns and predicates, not of their logical articulation.

Of course, there are languages and languages. Wilkins's script surpasses in mad ambition any that Frege attempted. It is a utopian performance, like the invented visual signs of a Humbert de

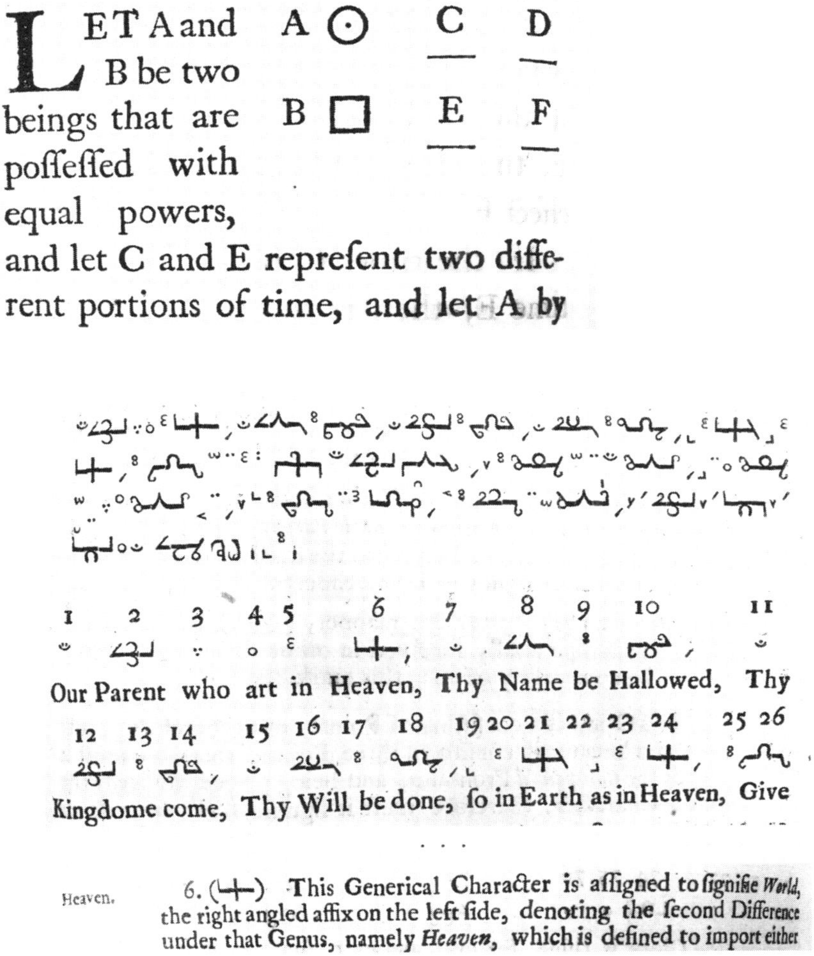

LET A and B be two beings that are poffeffed with equal powers, and let C and E reprefent two diffe-rent portions of time, and let A by

A ⊙   C̲   D̲

B ☐   E̲   F̲

1 2 3 4 5 6 7 8 9 10 11

Our Parent who art in Heaven, Thy Name be Hallowed, Thy

12 13 14   15 16 17 18 19 20 21 22 23 24   25 26

Kingdome come, Thy Will be done, fo in Earth as in Heaven, Give

Heaven.   6. (⊥) This Generical Character is affigned to fignifie *World*, the right angled affix on the left fide, denoting the fecond Difference under that Genus, namely *Heaven*, which is defined to import either

Figure 4.11. Diagrams from Richard Jack, *Mathematical Principles of Theology* (London: G. Hawkins, 1747), p. 122, and John Wilkins, *An Essay Towards a Real Character and a Philosophical Language* (London: Royal Society, 1668), pp. 395–96. Jack used letters and shapes to stand for objects and underlined letters for their states. Wilkins's written language (he invented a phonology, too) involved reading the tiny hooks as marking modifications of general categories: thus "the second" item in the genus "world" is heaven.

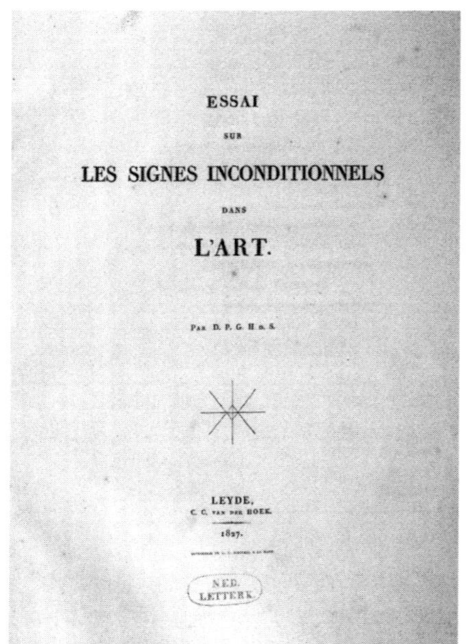

Figure 4.12. Title page of D. P. G. Humbert de Superville, *Essay on Unconditional Signs in Art* (Leyden: van der Hoek, 1827).

Superville or Paul Klee, who hoped that humans would eventually find intuitive access to their synthetic modes of perception (Figure 4.12).[68] Frege, on the other hand, aimed to bring out perspicuously the logical properties of the contents of thoughts, the latter preserved in their familiar guises (not the linear network of Begriffsschrift, but the scientific statements connected by the ducts). Like the contours, solid colors, and other framing devices favored by Aurier, his techniques result not in a transfigured, unrecognizable world, but in a conceptually clarified one.

This deeply contextual approach to logic and meaning is not unique in the late nineteenth century. The period abounds in logical pictures that are both diagrammatic and mimetic, which stand as it were between Begriffsschrift and Aurierian symbolism. In the case of Mach's partly visible body, the context of a Cartesian "cogito" taken from Krause injects the scene with the air of a comical *reductio ad absurdum*. Equally remarkable is the image introduced by William

James to track the time-dimension of a thinker's self-awareness in formulating a particular thought (Figure 4.13). This performance of the Cartesian ego is certainly very different from Mach's. Making James's soft sculpture, or just gazing upon its diagrammatic form, perhaps while thinking, "I am the same I, that I was yesterday," we coordinate an image with the experience of a reader-viewer. Though not intended to prove anything, as Mach's picture had been, this gridded blob is a lucid formulation of the introspective claims about the unity of consciousness advanced by James in his lectures. It is as if the stream of consciousness had been arrested and allowed to crystalize in the form of a concept made visible.[69]

The link between pictures and concepts, palpable when pictures represent abstractions like the Seven Virtues or Deadly Sins, gains in legibility in Frege's logical theory. As we have seen, Frege identified thoughts with entities capable of truth and falsity under the right circumstances. Concepts occupy a humbler place in the hierarchy. They are not, as for many Platonists (including Meinong), autonomous objects, but possible parts of thoughts. An object falls (or does not fall) under a concept, and the resulting thought is true or false. The concept is compared with a mathematical function: as $(2)^2 = 4$, so does Capital_of (Germany) = Berlin. Moving on to truth functions, which are central to logic, as the mathematical function $x^2 = 4$ is true for $x = 2$, so is the proposition "Berlin is the capital of Germany" true.[70] The introduction of truth functions links mathematics to reasoning in all domains of human inquiry. In a metaphor taken from chemistry, Frege called concepts, and functions generally, unsaturated, for only in tandem with an object to which they may apply (or an input value, producing an output value) do they produce the self-subsistent objects that interest us, be they thoughts or truth-values.[71]

This limpid view of concepts as functions is complicated in real life and conversation. In "The pig is a mammal," we take the pig as an object falling under the concept "mammal," but in a different context the sentence might say of the "[concept of] the mammal" that it "is exemplified by [the concept of] the pig." Thus, without confusing concept and object, we can take opposed parts of a thought to function as

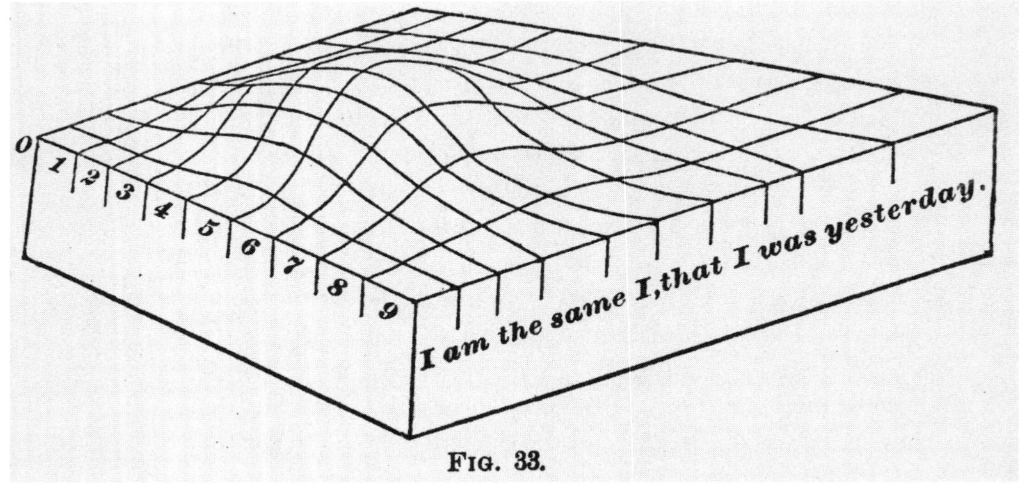

FIG. 33.

Figure 4.13. William James, *The Principles of Psychology* (New York: Henry Holt, 1890), vol. 1, p. 283. James even offers manufacturing advice: "If we make a solid wooden frame with the sentence written on its front, and the time-scale on one of its sides, if we spread flatly a sheet of India rubber over its top, on which rectangular co-ordinates are painted, and slide a smooth ball under the rubber in the direction from 0 to 'yesterday,' the bulging of the membrane along this diagonal at successive moments will symbolize the changing of the thought's content in a way plain enough, after what has been said, to call for no more explanation."

concept and object in typographically identical sentences.[72] And divergences can afflict sense, as when one thinks of Aristotle as the "philosopher from Stagyra" and another as "the teacher of Alexander the Great."[73] Frege did not think names were shorthand for descriptions, but he sometimes regarded them in this way to sharpen the possibility of misunderstanding. He considered two friends who use the same name to speak of the same man, Dr. Lauben, but know different things about him; given the identity of sign, and of reference, divergence of sense leads them to affirm different thoughts. With regard to Lauben, Frege fears that the two friends "speak different languages."[74] This is a stark way of putting it, and it has led to various amendments, some in line with Frege, some revising or discarding his categories. I find it a reminder that an adequate theory has to face the tribunal of practice. Thoughts are what we aim for in talking and looking at pictures; that we can fall short is a fact of life, not a refutation.[75]

Context, which throws up these difficulties for a philosophical theory of language, might seem familiar enough to art historians to require no apology. Yet the context principle by which Frege sought to overcome ambiguity, according to which "only in the context of a sentence does a word mean something," is notorious for an ambiguity about its application. This is so because Frege introduced it before he had made the distinction between sense and reference.[76] Which one is context-dependent, or are both? From the preceding arguments, we might expect sense to be: there is no fact of the matter whether "buns" refers to hair or bread, aside from the intelligibility of the resulting candidate sentences. The realm of reference might seem a bit more resistant to context: names label us, for better or worse, sentence or no sentence. But here too the practice of naming, of holding names constant, abbreviating, distinguishing between family and given names and between people with identical names, can be construed as the context needed for reference to be fixed.[77] Indeed, even subjective effects and resonant side thoughts parallel to the official sense are shaped by context. This emerges clearly in Frege's discussion of tone (*Färbung*), the subjective coloring of signs. The word "horse," noted Frege, produces very different mental pictures in two such sentences as "How gladly he rides his noble horse" (epic?) and "I just saw a horse fall on the wet asphalt" (realist fiction?).[78] Context, then, runs throughout the phenomenon of meaning, from naming to mood. It is in pictures too.

What I am driving at is that pictures, though they have myriad uses, do not have built-in functions; they *are* functions. As physical objects, supplemented by us with an intricate network of conventions and assumptions, they take as variables our subjective contribution — the feelings and thoughts we bring to perceiving them — and give as output thoughts, and sprouting about these, a veritable forest of resonating side-thoughts, feelings, and conjectures.[79] That is why artworks are not identical to the lumps of matter called paintings and statues, but also not to our ideas, much less our texts: artworks are our interactions with all these things and our interest in their senses. This interest can be limited neither to brute physical things nor to their interpreted aspect as functions, which require objects in turn to make sense of them.

Is there a further step to truth-value? There must be, for to proceed from sense to inquiry into truth is a further activity that, like the transition from the senses to sense, is not contained in a thought or object considered, however much it may provoke in us the desire to do it. Yet some art invites assertion. Thus John Ruskin, in his account of the verisimilitude of Renaissance art:

> As soon as art obtained the power of realization, it obtained also that of *assertion*. As fast as the painter advanced in skill he gained also in credibility, and that which he perfectly represented was perfectly believed, or could be disbelieved only by an actual effort by the beholder to escape from the fascinating deception. What had been faintly declared, might be painlessly denied; but it was difficult to discredit things forcibly alleged; and representations, which had been innocent in discrepancy, became guilty in consistency.[80]

What makes pictures illusionistic, idolatrous, or humble, or scientific, is not their appearance, but how appearance is put to use. Frege's observation on these assumptions complements Ruskin's:

> As an artwork, the history picture does not in the least make the claim of bringing the real course of events before our eyes. A picture meant to

> represent a historically significant moment with photographic fidelity would
> be no work of art in the higher sense of the word, but comparable rather to an
> anatomical illustration.[81]

The phenomenon of implicit assertion in pictures does not contradict Frege's definition of a picture as an unasserted thought; it is the cultural forces brought to bear on every picture that determine us to remain with the sense, or to use it for some purpose private, scientific, political.

In sketching an analogy between pictures used as art (in our modern sense, which is to say, as fiction) and Fregean functions, I have followed Frege in talking of fiction and pictures as senses apart from reference. How one actually gets from sense to reference, and from sense-data to sense, did not interest Frege; yet it is just such psychological, historical, and cultural questions that interest art historians, and that have tempted me to extend Frege's use of functions to transform sensation to sense, and sense to reference. How does this extended "functional" view look in practice? And what can it add to the theory of pictures as logical structures?

To appreciate pictures as both logical and full-blooded sensuous beings, it helps to regard a shocking image like (what may be) James Gillray's last drawing (Figure 4.14): a blast of furious ink defacing an elaborate self-portrait of the artist as beggar, hung with a plaque around his neck approximating an eighteenth-century sentimental poem by the clergyman Thomas Moss ("Pity the sorrows of a poor old man! / Whose trembling limbs have borne him to your door"), thrusting forth his hat for alms and clutching a cane, his mouth open in pain or entreaty. It has elicited powerful prose from historians struggling to come to grips with it. Gillray's biographer Draper Hill, himself a cartoonist, wrote that "the caricaturist's pen seems to have pulsated electrically across the surface, leaving in its wake a meaningless trail of short broken lines."[82] "He has three visible teeth" and the "eyes . . . are absent," according to a more recent author.[83] This interpreter may be right that the drawing is a "portrayal, in a moment that strives for visual clarity, of a mind reappearing through

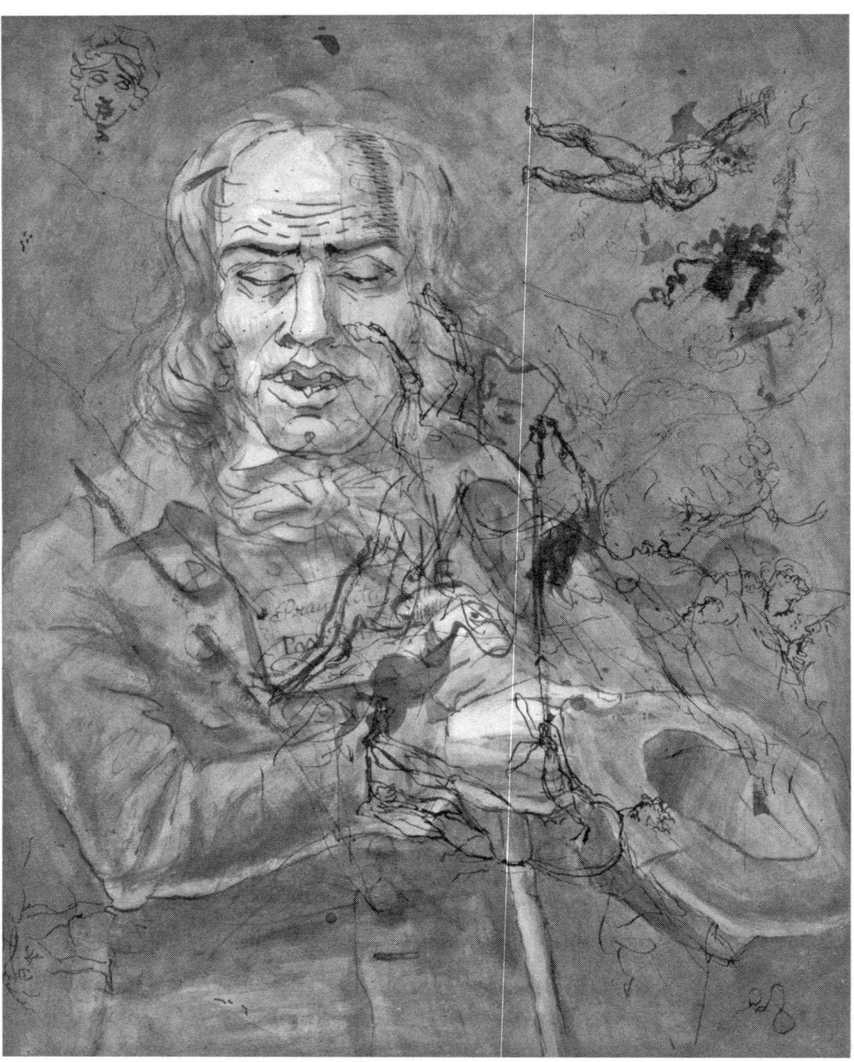

Figure 4.14. James Gillray, *Pray Pity the Poor Blind Man* (1811), black and brown ink, Fitzwilliam Museum, Cambridge University. Note the precision of the ink scribbles, from the Oedipus-like head with ruined eyes at upper right, to the infant touching an adult's mouth just above the beggar's hat, and the suggestion of pince-nez at bottom right.

uncertainty, invasion, and stark unrelenting oblivion," but we can do better as far as looking is concerned.[84] The blind beggar has at least five visible teeth, one under the upper left lip, two at the upper right corner of his mouth, and two lower incisors where the lower lip curves in the center; that is, setting aside uncertain passages at the corners of his mouth. And the eyes, though hooded, peek through the lids, or rather roll listlessly to the left (his right), pursuing (or is it evading?) the figments that afflict him. As for the figments themselves, teeming in the space around the figure, they are not just ink stains solidifying here and there in a legible head. There is a hard-driven horse, shooting vertically upward, his hoofs kicking in the vicinity of the blind man's nose, and a haloed child reaching out to touch an adult's face farther right. These extremes of violence and tenderness coexist in a compressed space before the portrait. We cannot come to grips with the physical image of the blind man (or of the ailing artist, if they are not one and the same), nor with the volatile flashes of his inner states, without first attending in such detail to the work.[85] Indeed, it is as if the sense of the whole demands that we stop and respond to the texture of the very thing: a sensory function returning the sense of the drawing, which in its turn seems to take our experience of what we see and return, if we have looked carefully, a truth or at any rate a thought about what it is like to be Gillray, or *any* person, in an intense state of suffering.

Despite its unassuming position in the history of art, this drawing takes it place in the grand drama of the painting of subjectivity in which Courbet's *Self-Portrait Mad with Fear*, Manet's *Shadow*, or Redon's *Ophelia* figure. Symbolism avant la lettre? Perhaps. But in any case a sobering reminder of the *difficulty* of the symbolist enterprise, of the fact that understanding how an image works, and how the mind works, is no triumphalist venture, but grinding and unfinished, perhaps in the end unfinishable work. An important lesson is also contained in this contrast: the thought genuinely contained in a picture, objective sense, is *not* objectively available to just anyone, but reaches us, as do the senses that Frege's friends have about Dr. Lauben, only through the fallible channels of sense that are the senses.[86]

What is missing from Frege's account of sense, and is indeed uncongenial to his whole way of thinking, is an account of how subjectivity — imagination, memory, perception — touches sense, which is in turn the road to reference. In Frege's view there must be such an account, or perhaps many, for how human beings grasp thoughts. They are just not the logician's business, but they are very much the artist's. Nor do they threaten the objectivity of sense. However the two friends came to their different senses of Dr. Lauben, they can compare them, discover that they attach different senses to his name, and thus learn each other's language, which they never could do with each other's mental images, even if they possessed sci-fi machinery making telepathy possible.[87] This suggests that beside sense functions (thoughts) taking as arguments objects (referents) and returning truths, there are material functions taking as arguments mental acts of reading or looking at pictures, and returning thoughts. The resonating thoughts characteristic of picturing find here their raison d'être. Consider the first-person picture from Gillray and Courbet to Caillebotte, Mach, and McCay: it is reasonable to suppose that the very understanding of the thought(s) these images comprise requires us taking them as views through the eyes of a particular seer (or in the case of Gillray and Courbet, as glimpses of a troubled visual faculty, overlaid on a portrait of the bearer of that faculty). In order to take an image as *a representation that I see things in a particular way*, that is to say in order to reach that very sense with its first-personal content, an act of seeing must occur — it must be passed along as object to the function manifested in the physical artwork, if the right first-person thought is to be grasped. Whether the resulting view is in some substantial sense true — whether "to be blind and a prodigious visual artist is thus," or whether "seeing the world thus is all there is to being an 'I,'" and so on — is in turn a sense function given objects of reference — the self, the blind man — which are by no means parts of these images, but are connected to them by the thoughts the images express.[88] The passage from a question to an assertion, from saga to science, is the product of such resonance resulting from experiences with image and language. We already know that there is nothing else, nothing in the thought or the picture or any of its parts, that constrains a movement to truth-value.

In using Frege's concept of a function to articulate the link between mental image and sense on one hand, and between sense and truth on the other, we obtained notions of fiction, picturing, and inquiry that are naturalistic without depending on any particular theory of biology, psychology, or neuroscience. Frege himself thought an account of how humans came to reason would not rely on some divine dispensation of verbal language — something that twentieth-century phenomenologists and language philosophers still seem to assume implicitly. In an unjustly neglected early essay on "The Scientific Justification of a Concept-Script" (1882), he reflected on animals' developing ability to gain control over their mental life (at first, by the simple expedient of running away so that they might have other feelings and sensations), a process of manipulation and articulation that in the end issued in the linguistic, pictorial, and other kinds of symbols among which reasoning beings live and think. This vision of the forest of signs offers a glimpse of how, in the mute work of imagination and fantasy, beasts come to gain autonomy in their inner lives and their dealings with one another through image-making: "So we penetrate ever deeper into the inner world and move there as we wish, using the sensuous to break its bonds. Signs are for thinking of the same import as learning to use the wind to sail against the wind was for navigation. And so let no one despise signs!"[89]

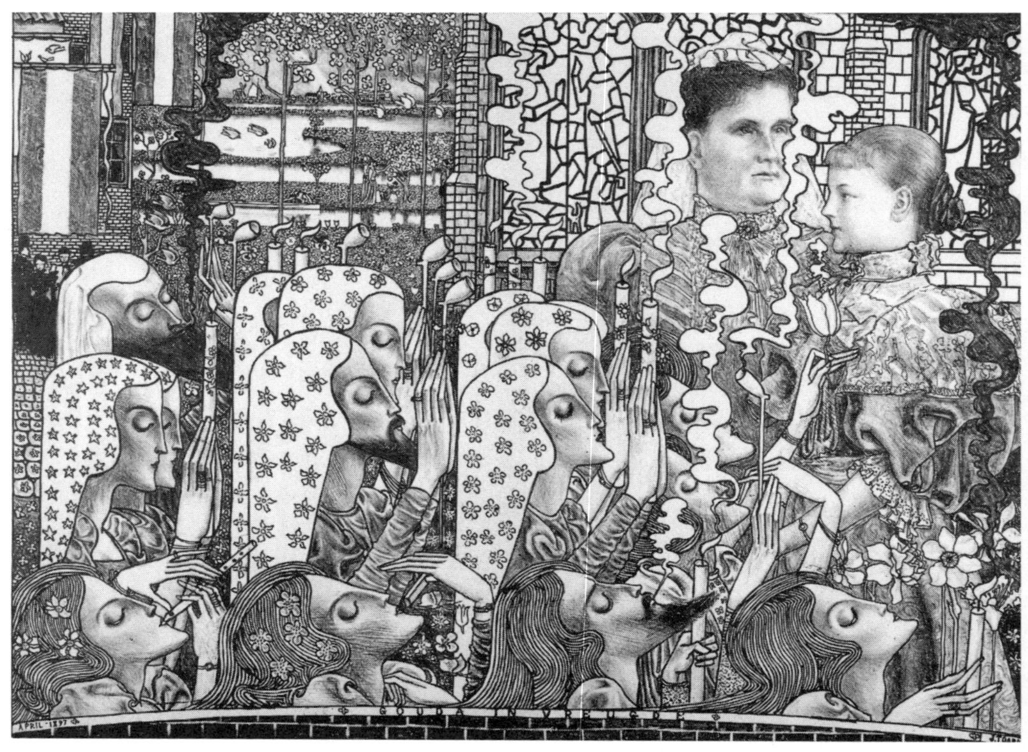

Figure 5.1. Jan Toorop, *Happy Gouda* (1897), lithographic supplement to newspaper *De Kroniek* (May 2, 1897). Several years later, Toorop made naturalistic drawings of the workers (male and female) of the Kaarsen-fabriek Gouda.

CHAPTER FIVE

Where Are We Going?

Consequences of Symbolism

This final chapter is devoted to the consequences of symbolism in both the historical and logical sense: to art and theory that took symbolist insights as its starting point, but also to phenomena that, before or after 1900 (in some cases, before symbolism as an official movement), embody practical results of symbolist modes of thinking. This will bring us to an "upper" limit of the concerns of symbolism — language, the mystical, the incommunicable — as well as to its "lower" limit in fin-de-siècle empiricism and the arts, from pointillism to film, that seem to embody it.

To begin with the esoteric, the reader may notice how little of the iconography of pale consumptive virgins, suicidal consumptive students, pale consumptive Marys, and Polynesian girls who at least don't look consumptive — in short, the mainstream of symbolist imagery — is found in these pages. It is not that this imagery, to say nothing of the considerable symbolist ventures in theater and music, is less central to symbolism than the tradition I have been sketching. I do not deny the mysticism, from theosophy and Rosicrucianism to the French Catholicism of Maurice Denis and the interest in antique paganism and non-Western religion of many fin-de-siècle artists; nor can I deny the political and social importance of the aesthetic *Kulturkampf* heralded by Aurier, whether it leaned right (orthodoxy, Frenchness) or left (Pan, Polynesia, syncretism).[1] What I do deny is that it is these concerns that constitute the more religiously and mythically inclined symbolists *as* symbolists. Against this, I urge that the transcendence they explored as symbolists, be it poetic or traditionally religious or visionary or merely obfuscatory, cannot be separated from the theory of meaning.[2]

Take, for instance, Jan Toorop's *Happy Gouda* (Figure 5.1), printed in a supplement to the Dutch literary weekly *De Kroniek* in May 1897, the occasion being a visit to a candle factory by Regent Emma and Princess Wilhelmine, who would be crowned Queen of Holland a year later.[3] There is almost a campy, tongue-in-check quality to the obeisance paid. Identical young men and women with closed eyes and sharp chins carry lilies and emit black and white smoke, from candles, pipe, or mouth, while the royal pair sits stiffly admist the billows, their faces modeled in the manner of the sentimental portrait photographs of them then in circulation. Happy Gouda indeed. The extent to which piety, pride, and (self-)parody mix in this lithograph is no greater than the extent to which decorative line, photograph, and caricatural simplification intertwine to form "those two dogmas symbol and synthesis, that is to say expression of ideas and aesthetic and logical simplification of forms," as Aurier summed up the variety of symbolist practice.[4] This pictorial discipline led naturally to decoration, as Aurier was farsighted enough to observe: "they evidently lack only walls [to paint on]."[5]

Symbolists did not get many walls, but the turn to ceramics, furniture, book design, and other occasions for decoration is conspicuous (Figure 5.2). As with Toorop's print, many awkward questions may be raised about Gauguin's most flamboyantly rustic collaboration with Bernard: who contributed what, how the panels and the cultures they depict (Brittany left and above, Martinique at right, with biblical elements at center and below) relate to one another, and above all the work's relation to Gauguin's *Vision after the Sermon*, which the center panel boils down to two Bretonne heads, atop which a pair of nudes wrestle with a tree. These nudes are as much Jacob and the Angel as Adam and Eve. Questions of theology, of the relation of the various arts, and of image to text can hardly be consigned to

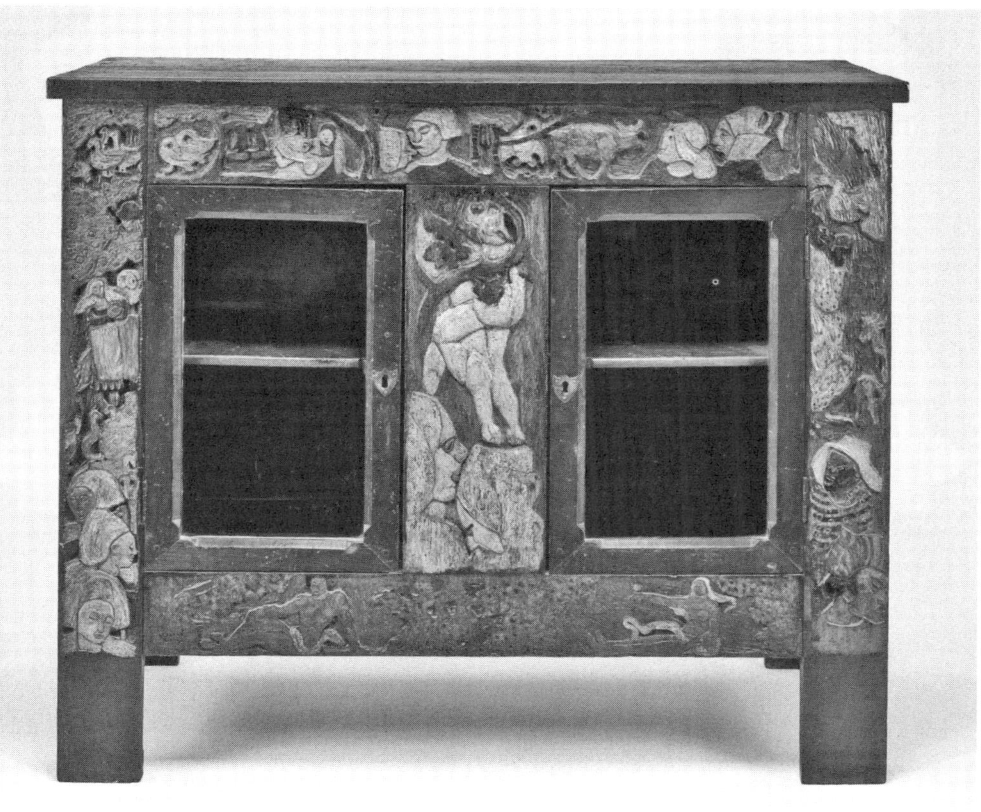

Figure 5.2. Paul Gauguin and Émile Bernard, *Earthly Paradise* (1888), glass, metal, chestnut and pine cabinet, carved and polychromed, Art Institute of Chicago. Bernard carved the left panel and perhaps the decorative upper and lower panels; Gauguin almost certainly carved the center and right panels.

the work's conceptual background. It is as if Gauguin were asserting rather their inextricability — quite a tall order for a cupboard.[6]

A work that wears its semantic conditions of possibility most insistently on its sleeve is Georges Rodenbach's *Bruges-la-Morte* (1892), the premier symbolist novel, and also the first novel illustrated with photographs (Figure 5.3). In the opening "Avertissement," Rodenbach insists that the story (a melodrama of a man falling in love with the doppelgänger of his dead wife, who spurns him and finally drives him to strangle her with a lock of the wife's hair, after which he catatonically repeats "Bruges-la-Morte...") stars the city itself, the way its quays, deserted streets, nunneries, and so on exert a psychic influence, until the reader feels "the shadows of high towers across the text."[7] A metaphor of the photograph as shadow projects the city, through its images, into the text. But how does this work? Must we rest content with an inarticulate notion of psychic "contagion," as Rodenbach himself suggests?

The works of these full-blown symbolists (post-Manet, post-Mallarmé, and post-Frege for that matter, though he was hardly read at the time) raises issues of commensurability between pictorial and discursive thought on several levels: between photography and drawing (Toorop), between photography and writing (Rodenbach), between painting and sculpture, and between visual images arranged spatially into an ensemble (Gauguin and Bernard). All of them lead back to an old art historical conundrum, that of the link between word and image. If we approach it through symbolist art criticism, as we did in the introduction, we meet with ringing assertions like "It is pure logic," and learn that it is a matter of "the idea imposed integrally in all its tyranny." Émile Verhaeren, the inventor of these slogans, derives them from the very nature of language:

> The sentence considered as a living thing in itself, independent, existing through its words, animated by their subtle, savant, sensitive position, and upright, and supine, and walking, and carried along, and shocking, and dull, and nervous, and flaccid, and rolling, and stagnating: organism, creation, body and soul pulled out of the self and certainly, perfectly created, more immortal than their creator.[8]

This sounds like an animistic version of Frege's doctrine of thoughts, articulated and self-sufficient, and, in Verhaeren's active fancy, alive to boot.[9] But the anthropomorphism implicit in treating sentences as living things, for all its belletristic charm, disguises a real difficulty: Is language in its descriptive capacity up to making legible a symbolist concentration of form and content, or is the symbol opaque to language? To begin to answer this question in a way consistent with the symbolist devotion to ineffability *and* logical clarification, I explore an extension of Frege's logical theory of pictures, brilliantly but too concisely expounded by Ludwig Wittgenstein in his *Tractatus Logico-Philosophicus*. It was first published in 1922, but was started before the First World War, and in many respects a late symbolist/ Fregean philosophy of signs. According to its doctrine, language and pictures alike are logically articulate, and thus in principle comparable. The twist Wittgenstein gives to this doctrine — some think it mystical, others antimetaphysical — is that certain things cannot be said informatively, for a relation common to saying and showing *conveys* meaning, and this cannot be isolated for literal transmission. I shall illustrate this theory, appropriately I hope, through the three remarkable pictures actually printed in the *Tractatus*.[10] In the second part of this chapter I apply the insights gained thereby to the worldlier heirs of Manet's painting and Frege's philosophy — the radically empiricist, but logically rigorous vision of the world as a set of images found in the work of Bertrand Russell and Georges Seurat. Thus we will round out the conceptual portrait of symbolism by addressing its reputed opposite, neoimpressionism.

Symbols and the Words that Fail Them

Are there thoughts we must think in responding to a picture, or to a picture plus context? A rhetoric of necessity may bother humanists, but it just sharpens the prosaic question, "What makes one interpretation better than another?" In practice we agree that some interpretations are better than their competitors. They may be more powerful, subtle, concrete. If this were not so, art historians would produce not scholarship but only works of (interpretive) art. But that, say the

Figure 5.3. Ch.-G. Petit, similigravures after photographs by the house Lévy et Neurdein, in Georges Rodenbach, *Bruges-la-Morte* (1892), pp. 1, 169, 221. I reproduce the first and last photograph, as well as the sole image to contain a pedestrian. Is he the alienated hero of the novel?

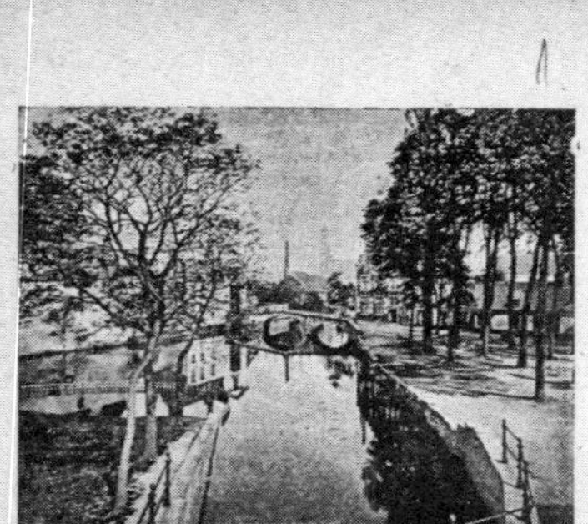

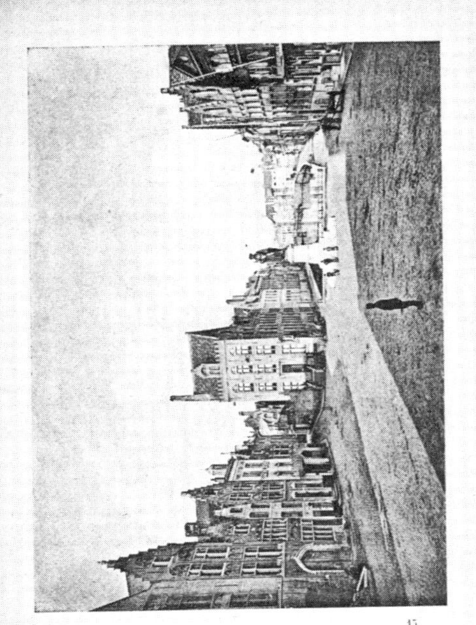

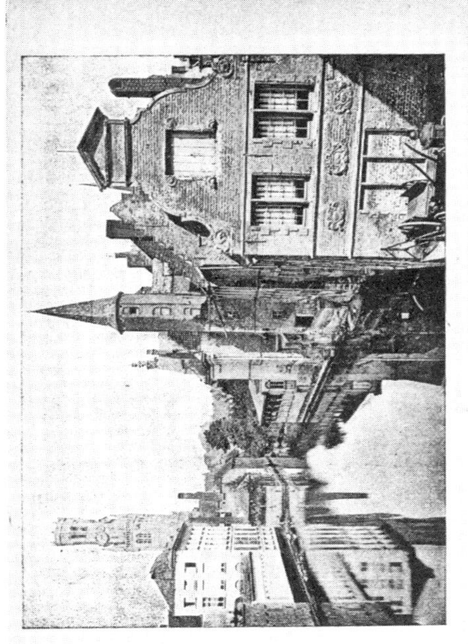

iconoclasts, is precisely what we do. In an essay that has not received the attention it deserves, Jaś Elsner has argued that the dissimilarity of word and image makes art writing "at best a parallel work of art" and "inevitably a betrayal of the original."[11] Why?

> The act of translation from one medium to the other undertaken by art historians is central. We conduct it with such ease. And yet the conceptual apparatus into which the object has been rendered, and its transformation from a thing that signifies by volume, shape, visual resonance, texture into one that speaks within the structures of grammar, language, verbal semiotics (call it what you will) and can be appropriated to numerous kinds of argument or rhetoric, are quite simply vast.[12]

Elsner's points range from the plausible to the tenuous: "however good the approximation in words . . . it can never fully be or fully replace the object." Well, who expected it to? As Blake saw, two distinct things cannot be one, "Nor canst thou ever change Kate into Nan."[13] What troubles Elsner is not so much perfect identity (proffering a text "equivalent" to an image) as epistemic overload: translation of images into words is "quite simply vast" and not helped one bit by technology. Whatever tools we bring to the table — not just photographs, which Elsner regards as personal acts of visual ekphrasis, but also archival documents, specialist knowledge, and so on — they are hooked to the picture by temperamental acts of description. The resulting assertions are not such as can be either accepted or denied solely on the basis of the picture.[14] For Elsner, what results is not unacceptable, but deeply tendentious. The picture determines nothing; it only affects us subjectively, and we mount an elaborate defense of what we intuitively feel.[15]

Elsner's argument is important. I will not try to prove him wrong, since one cannot have one's worries proved away, but only to show why it dissatisfies me. If all writing about art is ekphrastic, so is Elsner's text. Conscious that his own text does not read thus, he appends a very personal account of Michelangelo's *Rondanini Pietà*. If that transmutes the text as a whole into art, we are left with no reason to accept its skeptical argument. If Elsner is right, he is wrong.

Where is the rub? Like some romantics and symbolists, Elsner is impressed by the ineffability of images. But are pictures special in this? No less a formalist than Heinrich Wölfflin expressed doubts: "Indeed one feels a pictorial work in general to be a far more definite message than the written word, to which ambiguity adheres in greater degree. Schiller somewhere says he stands on a precipice whenever he thinks of the indefiniteness of verbal expression."[16] And feelings, perceptions, memories can be just as rich or ambiguous. The thicket of the imagination and mental imagery are, as we have seen, parallel objects of skepsis. The difficulty, then, does not lie in the essential poverty of one medium respective to another, but in the question of how two complex phenomena, writing and picturing, correspond. If essential differences between pictures and words can be found, it must be at the level of matching them. Wittgenstein's distinction between saying and showing is an account of such matching; I will try to show that it doesn't have the consequences for pictorial uniqueness that one might expect.

A Cardinality Theory of Pictures

The mystic marriage of word and image is first announced in Wittgenstein's writings as a direct figurativeness of all thinking. "We grasp the facts through pictures," reads the passage in a rough draft of Wittgenstein's *Tractatus* where pictures are first mentioned.[17] The same sentence in the finished text rearranges the words: "Wir machen uns Bilder der Tatsachen": we make for ourselves pictures of the facts, we picture the facts.[18] Is this a "metaphysical correspondence" between the structure of symbols and that of the world, or suggestive nonsense we are forbidden by Wittgenstein to believe, but which we need to help us to get over our illusions? Though Wittgenstein notoriously concludes the book by declaring its sentences nonsensical—a ladder to be thrown away after one has climbed its rungs—neither of these dominant views should appeal much after study of the picture theory of Frege, whose work Wittgenstein admired and thought he was extending.[19] The "homemade" quality of pictures, their rough and ready use as tools to navigate the world, is what one finds again and again in his writings, rather than some mysterious correspondence between symbol and world.[20]

Do Wittgenstein's dicta have any literal applications? In art history, can we say that we make pictures (art historical texts) of the facts (artworks)? I think we are not there yet, but on the right track, for this is just the point at which Elsner would object: we have no guarantee that our pictures match the texts! Wittgenstein anticipated this complaint. The picture we make of a fact is not a mental image copied from reality by the senses. It is a picture in the normal external way, one that stands for something else. So the concept of matching needed is a logical, not a crudely mimetic one. In a notebook entry that precedes the *Prototractatus* by over a year, Wittgenstein discusses the source of his insight: in a Parisian newspaper, he had read of a court case wherein an automobile accident was visualized three-dimensionally (with "dolls").[21] What struck Wittgenstein was the reversal of the usual scheme of pictorial interpretation: instead of turning pictures into words, a three-dimensional picture replaced language, with a gain in legibility. If this is possible, so is the reverse: "In the sentence a world is put together experimentally [*probeweise*: literally, one tries it out]."[22] Without such contingent links between model and world, not even Elsner's "betrayal" could take place, since manifest nonsense deceives no one.

My object is not to pit Elsner against the most lionized philosopher of the previous century: what is striking is how Wittgenstein, from familiar Fregean and symbolist starting points, echoes the contemporary art historian concerning pictorial meaning. Wittgenstein's note on the Parisian dolls contains a small doodle enclosed in quotes:

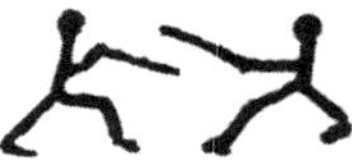

He comments: "If in this picture the right man stands for person A, and the left for person B, the whole might say 'A fences with B.' The sentence in picture writing can be true and false. It has a sense independent of its truth or falsity. Everything essential must be demonstrable in it."[23]

Given the simplicity of the picture, even Elsner might concede that some description of this sort is right or wrong. Not that it would exhaust the picture: it may well be that what the author meant was rather "B lunges at A." How many such statements must be made is a matter of tact, part of the "art" of describing pictures. What Wittgenstein, like Frege, emphasizes is that whatever our degree of success in this connoisseurial venture, there is something in the picture on account of which this venture succeeds or fails: "One could say, we don't indeed have the certainty of being able to bring all states of affairs in pictures on paper, but definitely the certainty of being able to picture all logical features of the state of affairs in a two-dimensional script."[24]

What all effective translations of a picture will have in common is logical form: Fregean sense. What did Wittgenstein add to this? The thirty or so aphorisms in the *Tractatus* following "we picture the facts to ourselves" (*T* 2.1–2.225), though typically oracular, struggle to locate pictures in space and time, pointing out how correspondences can be found between disparate domains. Later in the text, he gives a concrete example concerning music:

> The gramophone record, the musical thought, the notes, the sound waves, all stand in that internal pictorial relation to one another that holds between language and world.
>
> They are built the same logically. (As in the fairy tale the two youths, their two horses, and their lilies. They are all in a sense one.)[25]

These are heterogeneous things, few of which would count as pictures even to a visual culture theorist (though, as we saw, Frege is just as broad). If we are willing to regard the record, the sound it produces, and the notes played by the musicians as pictures, we thereby admit that pictures needn't look like the things they depict. The pictures we make for ourselves (*die wir uns machen*), a pun Wittgenstein explained to his English translators as "we imagine," are thoughts.[26] The section on pictures (*T* 3) accordingly ends: "The logical picture of the facts is the thought."[27]

Here we might suspect, as we did in Frege's case, that Wittgenstein leaves far behind the ordinary sense of the word "picture." But he defends himself hotly against this imputation: "On first sight the sentence — perhaps as printed on paper — seems no picture of the reality it deals with. But notation on first sight also seems no picture of the music, and our phonetic (alphabetic) writing seems no picture of our spoken language. And yet these sign languages prove, even in the ordinary sense, to be pictures of what they represent."[28] Of course, Wittgenstein never said a record is a picture in the sense a snapshot is: instead, as for Frege, they are logical pictures, carriers of the same thought: "Every picture is also a logical one. (By contrast not every picture is spatial)."[29]

What makes a picture logical is precisely its connection with other things: "That the elements of the picture stand (behave) in a certain way to one another represents that the things stand (behave) thus to one another" (*T* 2.14). A conventional picture of a tree might show a trunk with a crown of leaves above it. Inverted, with the trunk above the leaves, it might show a tree uprooted, or from a bat's point of view. These sentences are, for Wittgenstein as for Elsner, only new pictures about the tree picture. The relations "above" and "below," on the other hand, can be seen in the pictures that exemplify them. "The sentence shows the logical form of reality. It points to it" (*T* 4.121). This is not a peculiarly linguistic view of pictures (a "picture theory of the sentence"), but rather a wide net thrown around pictures; even language shows what can be shown in the same wordless way pictures do — that is, through logical relations between words. This is the simple point of one of the most comically abstruse of Wittgenstein's pronouncements: "Not 'the complex sign "aRb" says that a stands in the relation R to b' but rather: *that* 'a' stands in a certain relation to 'b' says *that* aRb."[30]

The doctrine concerning "saying" and "showing" was for Wittgenstein "the main point," even "the cardinal problem of philosophy."[31] That point, boiled down to one sentence, is: "What *can* be shown, *cannot* be said."[32] This is memorable. Unfortunately, ineffable showing is bound to sound suspicious.[33] Frank Ramsey, Wittgenstein's friend and critic, joked: "But what we can't say we can't say, and we can't whistle it either."[34] In other words: if something is showable, how can it be unsayable? If we can show, can we not also say, "I show you *this*"? Did Wittgenstein believe in a categorical distinction between picture and language? His *Notebooks* seem to indicate that he did: "Can one deny a picture? No. And there lies the difference between picture and sentence. The picture can serve as a sentence. But then something is added, which makes it say something. In short: I can only deny that the picture fits, but the picture I cannot deny."[35] Alas, from "wrong way" signs to canceled plates in printmaking, the world is awash in negated pictures, and what is more, in pictures negated not verbally but with pictorial means.[36]

There is not a hint of this bad argument in the *Tractatus*. True, we are told, supposedly in criticism of Frege, that a sentence has a sense, and can thus be true or false, prior to assertion or negation.[37] But by this token, a picture and a sentence are analogous: what one asserts, when one says, "The picture fits" or "The sentence is true," is the ready-made sense. In *Tractatus* theses on language, objects, and music, there is no hint of any essential difference between picture and language: just the contingent "Kate and Nan" differences of medium, history, and use. On the other hand, the whole book is devoted to an essential difference between saying and showing. If this were a strict disjunction, would Wittgenstein accept the resulting absurdity that "what can be said cannot be shown"? Hardly. *T* 4.022 announces that "the sentence *shows* its sense" and adds cheerfully: "The sentence *shows*, how things are, *if* it is true. And it says, *that* they are so." Paying attention to the stresses, it seems that it is only on the added condition "*if* it is true" that a sentence shows what it says. But does it not show "how things are if it were true," that is, does it show the same thing if true or false (only, if false, things aren't that way), or does it show this *only* if true? The comma and "*if*" unfortunately suggest the latter, implausible view — unfortunate too since Wittgenstein was committed to the thesis that sense is independent of truth value; had he used the subjunctive, we could be sure he held the plain Fregean view that a sentence says its sense and shows its reference (truth-value).[38] Of course, there are cases, like Frege's "the concept 'horse' isn't a concept" and Wittgenstein's own candidates for nonsense ("'1' is a number") in which only showing and not saying is possible: logically simple cases may be noticed, but they cannot tell us that they are such (for doing so isn't simple).[39] But even in such cases, showing has no special content beyond saying, for if it did, *that* could be said in a parphrase. Hence Ramsey's point about not being able to "whistle" the unsayable. What then is the distinction?

○ ○ ○

Consider three circles. On Stanisław Leśniewski's count, the drawing represents seven things, because every combination of the circles is counted as a thing (first; second; third; first and second; second and third; first and third; first and second and third).[40] Such divergence (3 or 7) might satisfy a relativist that even arithmetic is conventional, devoid of language-transcending truth. But that is a linguistic illusion. 3_{NORMAL} and $7_{LEŚNIEWSKI}$ are not autonomous facts, but the same number pictured two ways. Whatever counting system we use, the picture conveys *one* state of affairs. If we take one circle away, the descriptions "3 things" and "7 things" both become false. Wittgenstein would have agreed: "In our notations something is indeed arbitrary, but *this* is not arbitrary: that, *if* we have determined something arbitrarily, something else must be the case" (*T* 3.342). We can shift gears as much as we like, writing 1, 2, 3 or *a*, *b*, *c* or 1, 10, 11 (binary code). But in so doing, we don't switch from an ontology of numerals to one of letters or bits, only to different signs. The logical specificity of a fact (which Wittgenstein calls by the Kantian term *Mannigfaltigkeit*, "manifoldness") may be the same in various representations. But this common denominator, according to Wittgenstein, cannot be independently represented: "My basic thought is that the 'logical constants' do not represent. That the *logic* of the facts cannot be represented" (*T* 4.0312). There is here indeed a problematic slide from logic not representing to its not being representable. But on the picture theory, the two go together: a relation like "and" or "3" does not represent anything, since it is just "showing-cement" for substantial things that have the logical property shown. 1–2–3, *abc*, III might all show "threeness," but "threeness" cannot be isolated.[41] Indeed, it could not be "shown" if showing were an independent activity: I may well say "I show you that!" but I do not succeed in showing it to you unless you see it. It seems fitting to call this a "cardinality theory of pictures" because it presupposes not any special metaphysical relation between world and symbol, but the same *number* of significant logical aspects in two representations (two Fregean senses) of the same referent: "*that* 'a' stands in a certain relation to 'b' says *that* aRb." What is hereby equated is not sentence and reality, but two kinds of images.

Pictures of the Unsayable

To what use did Wittgenstein put his picture theory? In the last third of the *Tractatus*, he argues three times using images of his own devising. The passages are digressions, but they deal with classics of metaphysics, and thus, for Wittgenstein, strongholds of error: complexity, the self, and space. They show the extent to which a symbolist practice of picturing which, as Wyzewa and Aurier insisted, consists in logical elucidation, could be applied to classical philosophical problems. The first picture in the *Tractatus*, *T* 5.5423 (Figure 5.4) reworks the famous Necker cube.[42]

We can see this cube as the original "duck-rabbit" (*Hase-Ente*, literally "hare-duck" in Wittgenstein's German), a picture that may be seen in two ways, only not both at once.[43] On seeing that picture, some see *only* a duck, others *only* a rabbit, which it depicts slightly more plausibly. Of course, a careful viewer is given evidence of both: What makes us adopt one view or the other, switching between the two?[44] In his late work, Wittgenstein worried away at this question for pages, wondering what would happen if we met the monster in a cluster of reeds by the waterside, and so forth. If his main interest was the role of context in overcoming ambiguity in perception (and meaning), the duck-rabbit has become a symbol of the cultural ambiguity of all seeing: on being told to see a rabbit we duly see one, ditto the duck, and visual facts are supposed to play along meekly without determining anything.

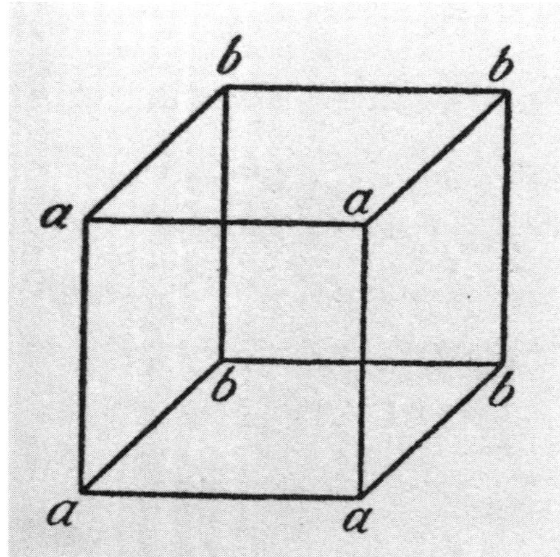

Figure 5.4. Ludwig Wittgenstein, *Tractatus Logico-Philosophicus* (London: Kegan Paul, Trench, Trubner & Co, 1922), pp. 144, 145 (*T* 5.5423) and anonymous illustration in *Die Fliegende Blätter* 2465 (October 23, 1892), p. 147. The august duck-rabbit was originally filler on a page dedicated to a quite unfunny hunting story about a bear in an eagle's nest.

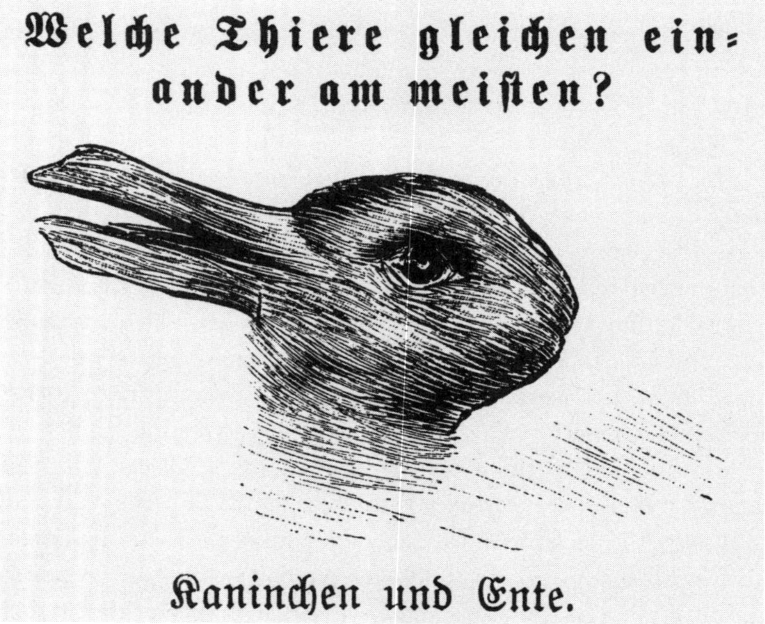

The reader is by now sufficiently familiar with this book to be able to guess its reply. Playing at the cultural relativists' game, we might say that, far from visual context guiding what we see, we approach duck-rabbits with a *parti pris* (Wittgenstein! aspects! social construction of knowledge!) and for this reason see what we are expected to see. If the image of the duck-rabbit is really indifferent between its two interpretations, nothing can prevent a well-informed viewer from grasping the dilemma by the horns and seeing the two together: a rabbit and duck joined at the eye. It is a biological improbability, but not logical impossibility nor any other kind.

The Necker cube, being less of a visual prodigy, lends itself less easily to inflated claims. It is at the same time more effective in presenting genuinely incompatible aspects: the two cubes it can be taken to stand for, one seen from above and to the right (with square *a* being the front), the other from below and to the left (with square *b* being the front) differ spatially, if the viewer is assumed to be in the same position, or they project the viewer at two different points in space, if the cube is assumed to occupy the same position. We have two genuinely incompatible facts. Wittgenstein calls the whole resulting picture a complex. In accordance to the cardinality theory of pictures sketched above, a complex is not a thing possessing the quality "complexity" apart from its parts; their coexistence *is* what we call complexity. The total really is nothing but a sum of its parts: or better said, the sum is the parts. In the Necker cube this is dramatized by the fact that the parts are not things that can exist at once: Wittgenstein could have illustrated the aphorism with the law of contradiction, "*a* or not(*a*)." But this he did not do, since he regards tautologies as senseless: "I know, e.g., nothing about the weather if I know that it rains or does not rain."[45] In the form of the Necker cube, however, he is willing to acknowledge a logical fact: "To perceive a complex means to perceive that its components behave in such and such a way to one another. This may well explain as well why the figure can be seen as a cube in two ways; and all similar phenomena. For we really do see two different facts."[46] If "a cube seen two ways" shows "two different facts," we can conclude that this is a complex of two possibilities.

The prejudice that pictures are too rigid to represent negative facts, conditional facts, counterfactuals — what might be or could not be (if something else is true) — is thus elegantly refuted.[47]

Just before the Necker cube, Wittgenstein had noted in one of his irritated asides that "the soul — the subject, etc. — as handled in today's superficial psychology — is a nonthing. For a composite soul would be no soul at all" (*T* 5.5421).[48] The cube does not show this directly, but it does play its part in Wittgenstein's psychological doctrine, according to which we ought not to think of the subject as a homunculus directing our body but as the stage setting of our whole view of the world, indeed, its boundary (*T* 5.632). This "solipsism" of Wittgenstein's, which he insists is nothing but the purest realism (since the data received by this world-spanning self is accepted as the world, rather than as mental images), sounds on first hearing like Mach's: there is no separate self, but the self can be inferred from the first-person aspect of the world. Though drawn to Arthur Schopenhauer, Wittgenstein rejects this stance, for it sins against the theory of meaning. The relevant passage foreshadows the dialogic style of the *Investigations*, Wittgenstein wrestling with himself:

> *Where in* the world is a metaphysical subject to be noticed?
>
> You say this is just like the eye and the field of vision. But you really do *not* see the eye.
>
> And nothing *about the field of vision* allows the deduction that it is seen by an eye.[49]

The next paragraph, 5.6331, which according to Wittgenstein's numbering system is supposed to serve as a gloss on the former (5.633), offers a diagram of the field of vision (Figure 5.5) with commentary: "For the field of vision does not perhaps have [*hat nämlich nicht etwa*] a form of this kind."

It is hard to impress on a non-German speaker how evasive this sentence introducing the picture is. Just what is Wittgenstein denying here? Surely not the general shape of the field of vision, therefore not the silhouette he has drawn as a whole. He did not believe he had revolutionized physiological optics from the armchair — or, to

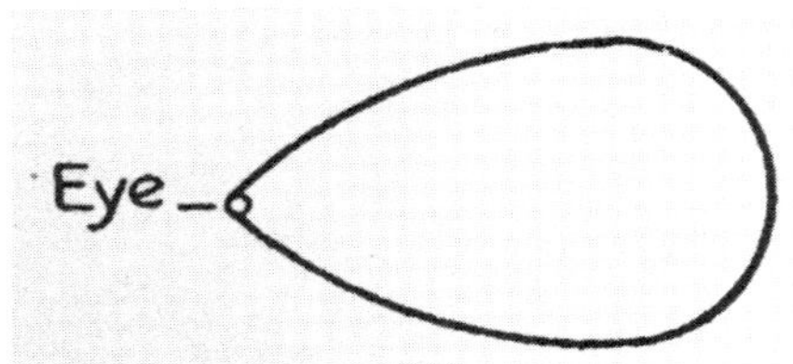

Figure 5.5. Ludwig Wittgenstein, *Tractatus Logico-Philosophicus* (London: Kegan Paul, Trench, Trubner & Co, 1922), pp. 150, 151 (*T* 5.6331).

be more precise, the officer's school.[50] Rather, the weight of his negation bears on the labeled eye, which, it should be observed, lies *in* the field of vision of the drawing. If a doctor were to reply that from his own field of vision he can identify blind spots, distortions, and other features that allow him to deduce the structure of a human eye, Wittgenstein would not disagree. For all his reservations about causality ("belief in the causal nexus is superstition," *T* 5.1361), he does not doubt that the eye does some kind of work in seeing. Nor does he doubt that we can see the eye: in a mirror image, for instance. What he is getting at is that such pictures remain external pictures of the world "around us" (another nonsensical phrase whose spirit we should be able to grasp), they are not the products of a self-seeing eye. But the latter was the analogy that the idealist interlocutor needed in order to deduce the self from one's experience on the analogy of the eye and the field of vision: not as a possible cause, but as the logically necessary complement of what is seen.[51]

Wittgenstein's argument, even if sound, is deeply misleading. The field of vision as a picture could certainly be produced otherwise than by an eye, but as *seen*, that is, as consciously seen, as I am doing at the moment of writing and you are at the moment of reading, it seems indubitably the product of an eye (or a pair of eyes). Such objections are well taken, but what Wittgenstein is trying to get at is nevertheless important, and it is falsified only in his words, and my own that seek to elucidate his picture. Here is a case where the showing done by the picture is really a matter of pragmatic success in making a convincing criticism: for the picture does not show something false, as the hesitant words introducing it suggest, but something impossible. Now, according to Wittgenstein, doing *that* would be impossible: "from the picture itself one cannot recognise if it is true or false" (*T* 2.224). But he is speaking of meaningful, informative pictures; a tautological picture would be necessarily true, a contradictory one necessarily false. Indeed, before writing the *Tractatus*, its author had used the image of the eye as a metaphor of negation: "The comparison of language and reality is like that of a retinal image and visual image: to the blind spot nothing in the visual image seems to correspond, and thereby the boundaries of the blind spot determine the visual image — just as true negations of atomic propositions determine reality."[52]

The metaphor, though intricate, can be unraveled: reality is like the visual image (what we experience) while language or picturing is like the retinal image (the logical fact). There is a real blank space in the latter (corresponding to a negative statement), but the former is as seamless as reality, which doesn't contain nonexistent things. Returning to the eye diagram, we can no more deduce the eye from the field of vision (a positive fact) than we can see the blind spot (a negative fact, which determines our experience only by its absence). A picture of *one's own* blind spot is an impossibility. We grasp this by recognizing such a picture (which looks like a picture of someone else's blind spot) as necessarily false.[53]

We have in the Necker cube and the self-seeing eye respectively pictures of impossibility and a priori impossibility. These are shown to us to be such, and we can *discuss* them as such — about this Wittgenstein, who denied that we can represent any substantial metaphysical facts, was in the wrong. But his distinction between showing and saying remains valuable in tracing just what our thinking consists in: whereas in learning facts we assemble a picture that may or may not correspond to some object(s), in the cases of showing we notice something (often in a flash), then make some effort to formulate it for ourselves or others. This is always the case with logical truths and often the case with pictures, which strongly suggests that the logical content of pictures is richer than generally supposed. The third and last picture in the *Tractatus* drives this home: it is the most exotic of them all, a picture of something unimaginable.

Once again, the explicit doctrine is not original, but made Wittgenstein's own through his imaginative intensity, which extends to the picture as well. In *T* 6.36111, unusually, Wittgenstein names a classical philosopher, Kant, and criticizes his doctrine of space. Kant thought spatial relations real but only subjectively so, applicable to phenomena as perceived by mind, but not to things in themselves. Most of Kant's arguments depend on Euclidean geometry being the only one possible, but one is still of philosophical interest: the left and right hand, identical in every detail, which cannot however be brought to overlap perfectly, because of their "left" and "right" orientation. How does this show space to be subjectively real? Leibniz had defended the (now orthodox) view that space consists of relations between things: Kant could not come up with a better positive account of space, but he came up with a puzzle, at least on a crude view of relativity according to which congruent objects in such a space, consisting purely of relations, should be susceptible to superposition. Since the hands seem to violate this requirement, Kant declared an antinomy, and consequently, the absoluteness but subjectivity of space.[54]

To this Wittgenstein replies, following Russell, that the effect is generic, requiring nothing so picturesque as hands. It can be found in one-dimensional arrows, such as the ones he draws (Figure 5.6). Russell had noted with deceptive placidity that given two line segments AB and BA, the left-right difference doesn't look as mysterious as Kant thought, even if they can't be superposed.[55] Wittgenstein duly provides a picture of the inconvenient segments, but he also proceeds to do the impossible, or at least to think it. He says superposition would in fact be possible if one could move the oppositely oriented objects through a space of one dimension more than the space they occupy. In the case of his arrows, it is enough to fold this sheet of paper (rotation through a second dimension). Even this had been said by Russell.[56] Yet how could this be done in Kant's case of the hands? In response, Wittgenstein turns futuristic, considering not hands but gloves: "One could put the right glove on the left hand, if one could turn it in four-dimensional space."

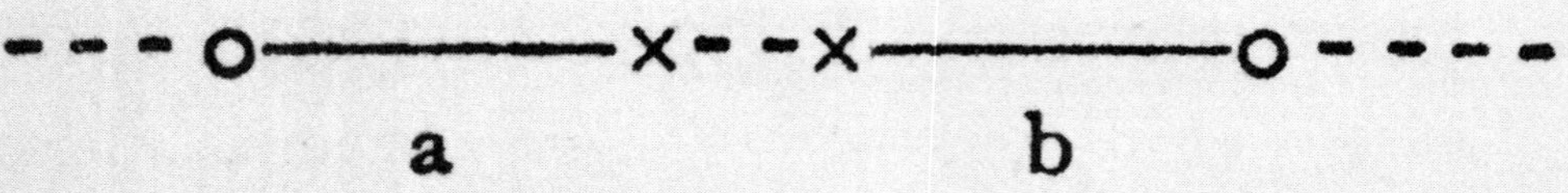

Figure 5.6. Ludwig Wittgenstein, *Tractatus Logico-Philosophicus* (London: Kegan Paul, Trench, Trubner & Co, 1922), pp. 178, 179 (*T* 6.36111).

Commentators, notably Max Black, have chided Wittgenstein for bringing in a "fantastic" fourth dimension. It is not clear that a physical impossibility is being proposed. Mathematically, there is no more wrong with it than with seven-dimensional space; it is a matter of the number of coordinates that objects in such spaces are taken to possess (Russell, as we will see, considers perceptual space six-dimensional). But, Kant might protest, four-dimensional space is not conceivable. It may be that our three-dimensional brains can only entertain analogical notions of a fourth dimension. But since such a space shares a logical property (the rotability of figures possessing fewer dimensions) with lower-dimensional spaces, the principle applies, even if we cannot picture it. The unimaginable is possible (at least logically). An image can help us at least grasp this bare pivot of the argument; in 1952, a brilliant young philosopher, Honor Brotman, showed with diagrams how, as one might gain a sense of the third dimension by looking at a two-dimensional persepective drawing, one might gain a sense of the fourth by looking on an analogous three-dimensional figure (Figures 5.7 and 5.8).[57] The key is to realize that the "small squares" are in fact the same size as the large (in perspective), and thus, that apparent trapezoids are themselves squares — in the third or fourth dimension, respectively.

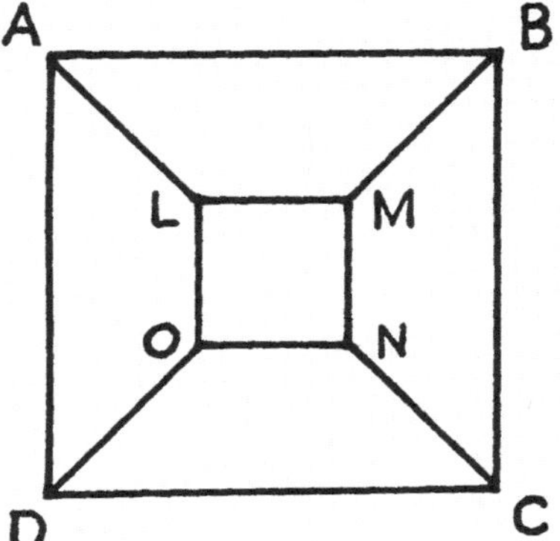

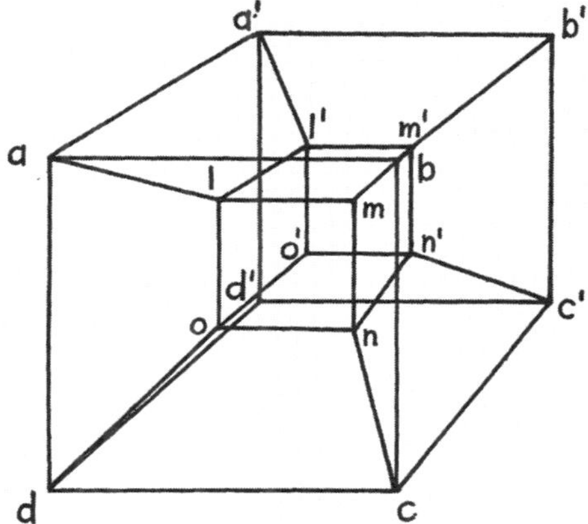

Figure 5.7. Figure 4 in Honor Brotman, "Could Space Be Four Dimensional?," *Mind* 61.243 (July 1952), p. 321.

Figure 5.8. Figure 5 in Honor Brotman, "Could Space Be Four Dimensional?," *Mind* 61.243 (July 1952), p. 323. Claude Bragdon had devised very similar diagrams, probably unknown to Brotman, in his popularization of Hinton's four-dimensional geometry, *A Primer of Higher Space* (Rochester, New York: The Manas Press, 1913).

Setting aside Ramsey's whistle, we are thus on firm ground with Arthur Conan Doyle's sardonic comment on Sherlock Holmes: "It is not easy to express the unexpressible."[58] Wittgenstein's doctrine of showing and saying, if it is neither obscure metaphysics nor a deflationary view of philosophy as nonsense, can be understood as an extension of the symbolist study of pictures. Its exploration of incompatible, impossible, and inconceivable objects in lucid images is one highwater mark of the mathematically pure art dreamt by Aurier. No wonder the *Tractatus* emerged as an aesthetic and philosophical ideal of postwar concrete poetry, and conceptual art, which one practitioner, Joseph Kosuth, defined as "inquiry into the foundations of the concept 'art.'"[59]

Naïve Impressionism

Having reached the midpoint of the final chapter, it is a good place to survey the path traveled. Gottlob Frege's positing of a realm of sense in which the work of the scientist as well as that of the artist comes to its meaning was tested in a variety of images, most dramatically in efforts to picture the self directly through the first-person viewpoint. The peculiar prominence of pictures in the realm of thought was then explicated through a rereading of Ludwig Wittgenstein's analogy between the structure of pictures and of the world. Wittgenstein's own mystical ethical and aesthetic commitments put pictures on an austerely transcendent plane, which our earlier discussion of down-to-earth works like Rodenbach's photographic novel was designed to extend: to say that Wittgenstein's pictures of logical states of affairs often constitute very precise arguments is helpful in reading *Bruges-la-Morte*, whose abandoned streetscapes and photographs of artworks convey its hero's alienation and fetishism more precisely and evocatively than what Rodenbach actually wrote. This applicability of pictures — when logically articulated — to our experience of the world more generally was explored most thoroughly by Wittgenstein's friend and teacher, Bertrand Russell, who admitted to having learned much from his pupil.[60]

The symbolist alternative to psychologism, whose uneven development among artists and some theoretical scientists we have been observing over the course of this book, was only ever truly popular among logicians. And the thinker responsible for this respectability, Frege's most famous correspondent and Wittgenstein's friend and logic teacher, was Russell. The irony is that throughout his lengthy and eventful intellectual life, Russell was often an unwilling or a lapsed Platonist, the doctrine of his mathematical youth. The author of *The Principles of Mathematics* (1903, composed largely in 1900) could exuberantly affirm that "Numbers, the Homeric gods, relations, chimeras and four-dimensional spaces all have being, for if they were not entities of a kind, we could make no propositions about them."[61] The preface of that book went even farther: "The discussion of indefinables which forms the chief part of philosophical logic is the endeavour to see clearly, and to make others see clearly, the entities concerned, in order that the mind may have that kind of acquaintance with them which it has with redness or the taste of a pineapple."[62]

This quest for a sensuous experience of abstract, logical entities recalls the modernist poet's decree that words be "as fully flavoured as a nut or apple."[63] Russell admitted that "it is often easier to know that there are such entities than to perceive them," but acquaintance with numbers and other logical entities was a distinctive feature of his early philosophy, so much so that he worried already in the preface of the *Principles* that he had not been able to attach such gritty familiarity to his notion of classes, which were furthermore vulnerable to contradictions.[64] These paradoxes, coupled with the lack of intuitive availability, led Russell, through his theory of descriptions, gradually away from the hospitable ontology of his first great work. In the 1937 preface to the second edition he observed: "At the time when I wrote the *Principles*, I shared with Frege a belief in the Platonic reality of numbers, which, in my imagination, peopled the timeless realm of Being. It was a comforting faith, which I later abandoned with regret."[65]

Certainly there were later important philosopher-logicians, foremost among them Kurt Gödel, who retained young Russell's heroic view of a kind of "mental telescope" trained by the mathematician on the numbers. But numerical intuition is foreign to Frege; Russell himself never *relies* on intuition, but his willingness to put logical objects on a plane with the concrete furniture of the world, curiously, did not leave him when he came to believe that they could be paraphrased away. In offering a reductive account of numbers, and later of matter and mind, as logical fictions, Russell was to make profound use of Fregean and Wittgensteinian notions of the logical picture. His work parallels that of modern art and mass imagery as it diverged from symbolist hopes for transcendence into the kind of vernacular materialism typical of film.

To understand how Russell used pictures to get a logical grasp of the world, one freed as far as possible of the abstract entities that mattered so much to the symbolists, it is worth dwelling a moment on his milieu. Though educated in the reigning idealism of late Victorian philosophy, Russell preferred the sophisticated naïveté of his Cambridge friend G. E. Moore, whom in *Principles* Russell professed to follow in all things metaphysical.[66] Russell and Moore were also intimates of the Bloomsbury group of artists and poets, including such luminaries as Virginia Woolf, Lytton Strachey, and John Maynard Keynes.[67] Russell was too intellectually independent to owe much to them (with the exception of Moore and Keynes who shaped Russell's thinking about scientific induction), but it is clear that, though he had no interest in aesthetics, he absorbed a modernist attitude toward the visible.[68] The traffic was two-way. Woolf's own view of painting as making "hard, tangible, material shapes of bodiless thoughts" owes as much to Russell's "taste of a grapefruit" as it does to the nuts and apples of imagism.[69] And this modernist vindication of appearance was of especial interest to the philosopher. Russell called his first bestseller, the 1912 *Problems of Philosophy,* a "penny shocker" for the way it thumbed its nose at commonsense opinion, most dramatically perhaps in Russell's dissolution of a table into a cloud of sense data:

> To the eye it is oblong, brown and shiny, to the touch it is smooth and cool and hard; when I tap it, it gives out a wooden sound. Any one else who sees and feels and hears the table will agree with this description, so that it might

> seem as if no difficulty would arise; but as soon as we try to be more precise our troubles begin. Although I believe that the table is "really" of the same colour all over, the parts that reflect the light look much brighter than the other parts, and some parts look white because of reflected light. I know that, if I move, the parts that reflect the light will be different, so that the apparent distribution of colours on the table will change.[70]

Russell draws an immediate conclusion — no two people will see the same distribution of light, since they do not occupy the same space — and raises similar difficulties for the other senses, sharpening the divergence between appearances in order to give them their due in mental life: "For most practical purposes these differences are unimportant, but to the painter they are all-important: the painter has to unlearn the habit of thinking that things seem to have the colour which common sense says they 'really' have, and to learn the habit of seeing things as they appear."[71]

This declaration of pictorial modernism is *impressionist* in its rhetoric. The idea of the visible world as a motley of hues makes no concession to the symbolist emphasis on outline and unity of tone, used to endow sensible objects with conceptual coherence.[72] Russell is of course not talking about what artists *make* of what they see but about the *raw materials* they go on to arrange, cohesively or kaleidoscopically. His attitude, it should be stressed, was widespread at the turn of the century. It can be found in formalist art history, notably Franz Wickhoff's 1895 comparison of Pompeiian frescos to modern French and Japanese art (a text translated into English in 1900).[73] The very framing of the project as "learning the habit of seeing things as they appear" suggests a triumph of the ephemeral over the permanent. For Russell, it means seeing them impressionistically, that is to say, as subjective through and through.

This may explain Russell's weakness for the psychologistic philosophy he had renounced early on under the influence of Frege and Moore. In particular, the relation between perception and the world was for him, for Frege's hapless "physiological psychologist," a fraught one:

> The observer, when he seems to himself to be observing a stone, is really, if physics is to be believed, observing the effects of the stone upon himself. Thus science seems to be at war with itself: when it most means to be objective, it finds itself plunged into subjectivity against its will. Naïve realism leads to physics, and physics, if true, shows that naïve realism is false. Therefore naïve realism, if true, is false; therefore it is false.[74]

This conundrum was propounded by Frege as motivation for questioning empiricism. For Russell, it functions quite differently: puzzles about skepticism thrown up by science can be met only by expanding the logical picture to bridge the divide between the objective and the subjective, into the psychology that Frege neglected. In a review of Alexius Meinong, Russell noted that theory of knowledge "may be approached either through psychology or through logic, both of which are simpler than it is."[75] The approach from psychology, or rather from both sides, was part of the symbolist program in art and science: an investigation of how subjective realities fit together, both in the relations between individual minds and between minds and the world. Russell pursued this project from the implicitly impressionist standpoint of an atomized perceptual world, on which order may be imposed, but which is not ordered to begin with.

In calling Russell's aesthetic attitude, which determined to a large extent his project of understanding the shared physical world in terms of private experience, naïve impressionism, I do not attack it, in the manner typical of discussions of naïve realism.[76] Rather, what is crucial about the adjective "naïve" is that it identifies that aspect of experience that the theory takes as basic, not requiring (nor allowing for) further justification. A naïve realist assumes that objects cause perception, a naïve phenomenalist like Mach, that perceptions are the basic stuff of the world; the view, or rather philosophical mood, of Russell is subtler, since though it basically agrees with Mach about the primacy of our experiences, it aims, with William James, to grant equal reality to all aspects, both experiential and logical, of the world.[77] This will turn out to be of particular importance to art, even to that art which ostensibly puts sensation first.

Wild Particulars

From his early, Platonistic phase to the skeptical scientific philosophy of later years, Russell democratically believed in the reality of dreams, daydreams, hallucinations, and various other mental states that philosophers had distrusted.[78] He is delighted to poke holes in the assumed connection between such anomalous perceptions and lack of rationality:

> What makes the patient, in such cases, become what others call insane is the fact that, within his own experience, there is nothing to show that the hallucinatory sense-data do not have the usual kind of connection with "sensibilia" in other perspectives. Of course he may learn this through testimony, but he probably finds it simpler to suppose that the testimony is untrue and that he is being willfully deceived. There is, so far as I can see, no theoretical criterion by which the patient can decide, in such a case, between the two equally satisfactory hypotheses of his madness and of his friends' mendacity.[79]

From this leveling outlook, Russell concludes that "wild" particulars are simply those that do not have "the usual relations by which the classification is effected; perhaps dreams and hallucinations are composed of particulars which are 'wild' in this sense."[80] Far from being unreal, mental events of this sort merely give rise to false inferences. But they belong to reality, and, in Russell's evolving doctrine of sense-data, they are themselves physical, for otherwise they could not participate in the world and provide a window upon it.[81]

This affirmation of the reality of mental events, however "wild" their constituents, may be characteristic in Russell of a lingering Poeian fascination with the vividness of psychic life, from everyday sensation to dream, extrasensory perception, and mental illness.[82] What should not be missed in this psychic pluralism is its rationalist motivation: Russell set aside the metaphysical distinction between appearance and reality, according to which the former is illusory, in order to be able to give a logically full account of as much of the world as possible.[83] The breaking down of objects, and of minds, into a motley of sense-data or "experiences" arranged thingwise or mindwise was by no means original, being outlined by Mach in 1886 and elaborated by William James in his essay "Does 'Consciousness' Exist?" (1904) and other texts published posthumously by Ralph Barton Perry as *Essays in Radical Empiricism* (1912).[84] The fact that the same sense-datum — my seeing of a tree, for instance — is part of both my mental life and the tree itself was regarded by James and Mach as explicable only by a "neutral monism" able to overcome traditional metaphysical distinctions between self and other, mind and matter, physical and psychological, subject and object.[85] This theory had an obvious appeal for the fin-de-siècle empiricist, simplifying psychology by ditching the self and physics by discarding the thing. Russell, by contrast, did not seek simplicity for its own sake, but in the service of explanatory power: to "construct the world" out of classes of ordered particulars that comprise the aspects of an object, both along the space and the time axes (producing what he calls biographies).[86] The corresponding sets of the three-dimensional "private" perspectives of perceivers in a three-dimensional public space, resulting in a six-dimensional world, were meant to reconcile psychology and physics, solving ancient puzzles about how various conflicting appearances can belong to one object.[87] Though he sometimes encased "things" in scare quotes, Russell was less interested in eliminating them than in transforming everyday concepts in a way consonant with modern science. And art. The most telling passage of "The Ultimate Constituents of Matter," a lecture given in Manchester in February 1915, draws a link between ontology and film:

> My meaning in regard to the impermanence of physical entities may perhaps be made clearer by the use of Bergson's favorite illustration of the cinematograph. When I first read Bergson's statement that the mathematician conceives the world after the analogy of a cinematograph, I had never seen a cinematograph, and my first visit to one was determined by the desire to verify Bergson's statement, which I found to be completely true, at least so far as I am concerned. When, in a picture palace, we see a man rolling down hill, or running away from the police, or falling into a river, or doing any of those other things to which men in such places are addicted, we know that

there is not really only one man moving, but a succession of films, each with a different momentary man. . . . Now what I wish to suggest is that in this respect the cinema is a better metaphysician than common sense, physics, or philosophy. The real man too, I believe, however the police may swear to his identity, is really a series of momentary men, each different one from the other, and bound together, not by a numerical identity, but by continuity and certain intrinsic causal laws. And what applies to men applies equally to tables and chairs, the sun, moon and stars.[88]

Russell must have been an avid cinemagoer to rack up such a catalogue of "attractions," as Tom Gunning has called the antinarrative thrills of early cinema. I know no film containing them all, and indeed the great poets of film athleticism, Douglas Fairbanks and Buster Keaton, had yet to enter the arena. Perhaps Russell was recalling such classics as *The Great Train Robbery* (1903); or perhaps he had recently seen Charlie Chaplin's screen debut in *Making a Living*, with its climactic chase down a busy Los Angeles street, camera tracking before him (Figure 5.9).[89] Russell's point about cinema is not that its flicker make objects look impermanent. Film does not so much show us a state of flux as symbolize it lucidly. Early film excels in making motion itself the cause of character development, overshadowing the hero's — or antihero's — motives.[90]

This is no postmodern paean to self-invention: to be a "different momentary being" at every flicker of the screen is only to be in motion. Russell expressed the point as early as *Principles of Mathematics*, a decade or so before his encounter with cinema, with reference to Zeno's paradoxes of motion:

In this capricious world, nothing is more capricious than posthumous fame. One of the most notable victims of posterity's lack of judgment is the Eleatic Zeno. Having invented four arguments, all immeasurably subtle and profound, the grossness of subsequent philosophers pronounced him to be a mere ingenious juggler, and his arguments to be one and all sophisms. After two thousand years of continual refutation, these sophisms were reinstated, and made the foundation of a mathematical renaissance, by a German professor,

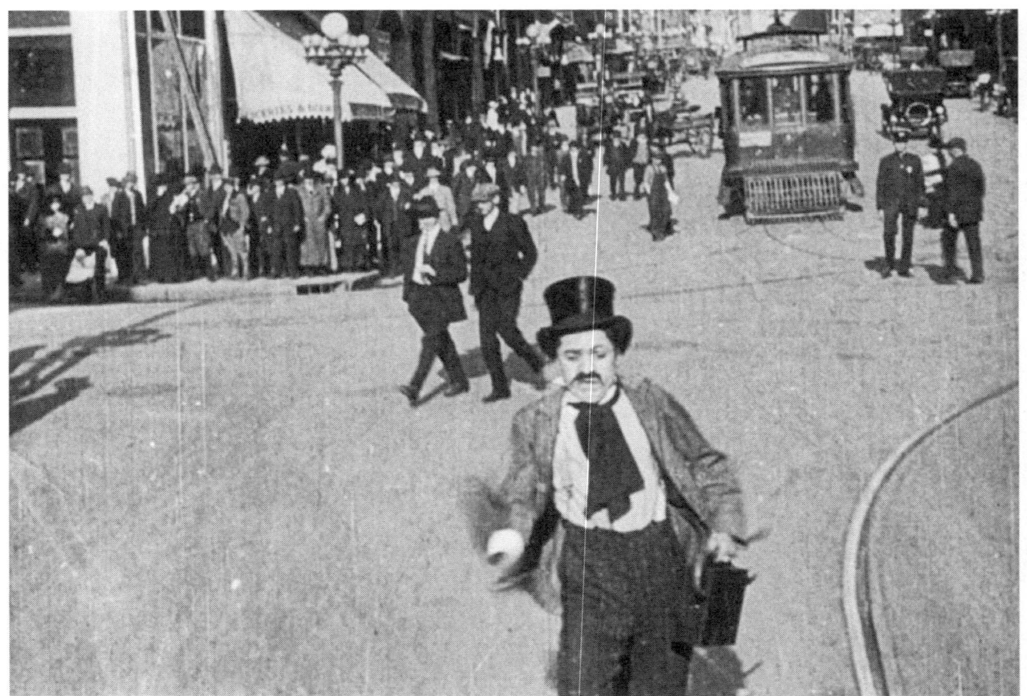

Figure 5.9. Charlie Chaplin in *Making a Living*, dir. Frank Lehrman (Mutual Film Corporation, 1914). Chaplin disliked this first role playing a thief; by 1915's *Kid Auto Races at Venice*, he was already the sympathetic Tramp. The frontally filmed chase scene, resembling in composition nothing so much as Manet's *Races at Longchamp*, prefigures the elaborate race down the streets of Los Angeles in Buster Keaton's *Seven Chances* (1925).

who probably never dreamed of any connection between himself and Zeno. Weierstrass, by strictly banishing all infinitesimals, has at last shown that we live in an unchanging world, and that the arrow, at every moment of its flight, is truly at rest.[91]

There is no change, yet there is time: the standpoint is pellucid, if not very Bergsonian, the world viewed as sets of sense-data arranged along the space and time axes.[92] The ultimate goal, and the maxim driving research, a kind of Ockham's razor for the era of mathematical logic, was *"Wherever possible, logical constructions are to be substituted for inferred entities."*[93] Logical constructions, or fictions, as Russell also called them, do not commit us to shadowy entities we may never know; instead, they allow a translation from the language of science "by a sort of dictionary, into propositions about the kinds of things which are given in sensation."[94]

We have come a long way in interpreting Russell's impressionism from the admission of wild particulars to the reduction of the world to sets of sense-data. In the process, the study of experientially or scientifically resistant facts has given way to an ambitious regimentation of these and other realms of experience. Its purpose, as in Mach, is ostensibly to battle traditional dualism, but in practice it issued in behaviorism and the dismissing of subjectivity, agency, and culture — the whole Fregean "third realm" and indeed much of his second.[95] But have not modern visual technologies like film and digital media undergone and provoked kindred transformations, from expansions of our experience of the world to its homogenization?

The question of the reduction of consciousness to sense-data, and more extensively of the world and its contents to sets of pictures, is no idle matter of philosophical speculation. But there are simple and vivid instances that can bring the problem to life. Does the reduction of what I do to my perceptions, or that of a physical thing to the sensibilia it elicits, amount to a verbal definition or a genuine change in perspective?[96] Is a legal fiction, or a political or scientific one, necessarily an illusion? Think of the claim, as current in literary theory as in Mach, that the self is a fiction. Or the fiction of human "races" — which is unfortunately enjoying a resurgence. It matters very much whether showing that one thing is composed of another or multiple others ("reduction") amounts to showing that it doesn't exist ("fiction"). In particular, to show that one may talk about one sort of thing (e.g., sense-data) *instead* of another (a thing or a person) may not mean that one has dispensed with the latter.[97] The question also strikes at the heart of the symbolist project. If we don't need an objective realm of concepts to make sense of the physical world and ourselves, the art and science that thought so may be a dead end.

Russell shows how this question may be brought down to earth by considering pictures. His efforts to rehabilitate wild particulars led him to confront their supposed unreality: "The 'unreality' of images may, on our present hypothesis, be defined as consisting merely in their failure to fulfil the correlations which are fulfilled by sense-data."[98] Is the self, are objects, in this way "wild," detachable from the sets of experiences they are supposed to order? If this hunch about the fictionality of the entities reduced to sets of images is correct, then images themselves could carry out the required enlightenment, as Mach's eye-picture tries to do. Does a verbal account like Taine's sunset, or Cross's painting, explode the perceptual clichés of a Cartesian ego in favor of the teeming complexity of a plurality of perceptions, patiently cobbled together? Impressionism might be thought to be engaged in such a project: painting, in order to see not objects but pencils of light. This tendency was further refined by the neoimpressionists. In a programmatic painting like Georges Seurat's *La Grande Jatte*, we can observe the ordering of sensations into an edifice of objects in space, and perhaps, even, of a subject's vantage on them.

La Grande Jatte, or *Un Dimanche à la Grande Jatte, 1884 (A Sunday on the Grand Jatte — 1884*; the *afternoon* is a later accretion), as the painting was called in 1886, is a large painting, to which I cannot do full justice here (Figure 5.10).[99] I look to it for the pictorial monism we have explored in Mach and Russell: a monism alive in the development of the impressive whole from sketches whose large flat marks radicalize the patchwork of impressionism, at the risk of losing its human presence, or else build up monumental human figures from subordinate patches (Figure 5.11). How confident, how varied in its unity is the effect of the finished painting, despite a disastrous and nearly immediate discoloration of the zinc yellow and emerald ("Veronese") green, which made the overall tone duller and more earthy than it once was or remains in the sketches.[100] I bring in the *Grande Jatte*, an exact contemporary of Mach's book, primarily to test Russell's linkage of fiction to pictoriality, but the deeper methodological rationale for doing this, as we have seen in connection with Mach's drawing, is that pictures may serve as the critique and revision of a philosophical view as much as the other way round. In the case of Seurat's painting, what is still revealing is how the sublimely homogenous paint application clashes with the toy-soldier look of the persons and things in the image. *Even the trees have become cartoonish*: an examination of the central trunk in the studies I reproduce does not reveal the elegant leftward-sweeping fork, with another forking trio breaking from the right tine, that we see in the finished picture.

This distinctive character of the denizens of Seurat's world struck his contemporaries so much that they did not tire of interpreting them in terms of some aesthetic ulterior motive, whether realist, satirical, or eulogistic. The habit is so pronounced that his friend Paul Signac, who after Seurat's death became the veritable pope of neoimpressionism, could disclaim it impatiently in a 1935 encyclopedia article:

> When Seurat exhibited his manifesto painting *A Sunday on the Grande Jatte* in 1886, the two schools that were then dominant, the naturalist and the symbolist, judged it according to their own tendencies. J. K. Huysmans, Paul Alexis, and Robert Caze saw in it a Sunday spree of drapers' assistants, apprentice *charcutiers*, and women in search of adventure, while Paul Adam admired the pharaonic procession of its stiff figures, and the Hellenist Moréas saw panathenaic processions in it.[101]

Signac's own preference, following Félix Fénéon, was to read the picture formalistically as "a luminous, cheerful composition, with a balance between verticals and horizontals, and dominantly warm, luminous colors with the most luminous white at the center." This talk of colors and verticals, far from solving the problem of content,

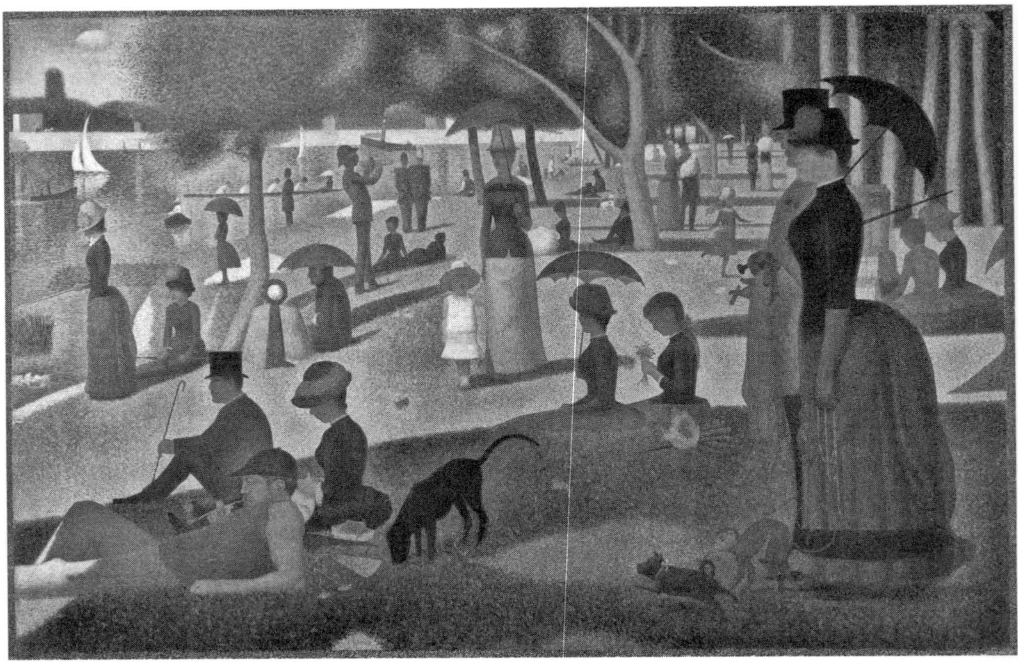

Figure 5.10. Georges Seurat, *A Sunday on La Grande Jatte—1884*, 207×308 cm, Art Institute of Chicago. This massive, cohesive painting not only changed drastically in color in its first two years, but is a result of three painting campaigns: the second provided the consistent web of dots, the third the painted "frame."

Figure 5.11. Georges Seurat, studies for *La Grande Jatte—1884*, oil on wood, National Gallery of Art (top), Washington, DC, and Art Institute of Chicago (bottom). Note the evolution of the forking tree over the studies and into the final canvas (the distinctive curved tree at left and the two very straight trees in between, which change little, suggest that this is one and the same tree, ceaselessly viewed and reordered by Seurat).

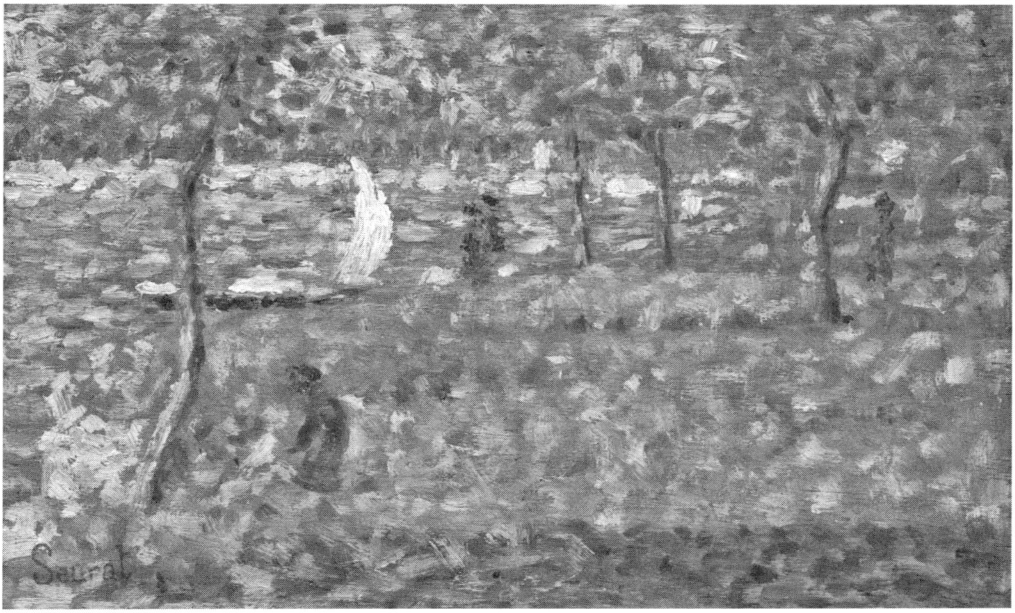

Seurat

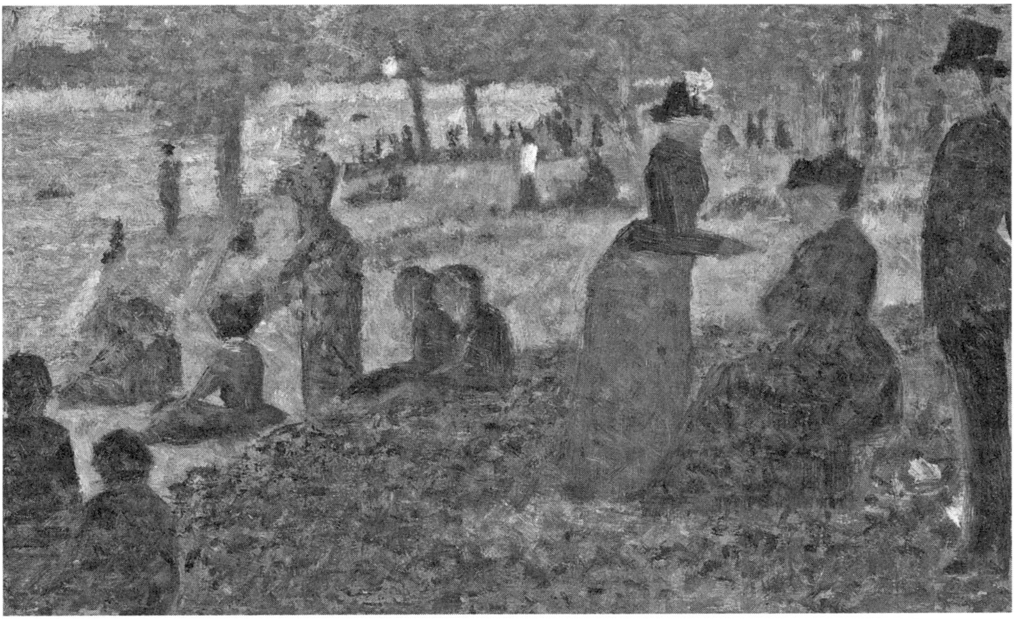

only postpones it. It is the very transition from part to whole and contingent shape accreting from the rounded strokes to represented object that is problematic for "scientific" painting. For if Seurat did not build up his picture out of the small, imperfectly round brush-strokes of the kind he covered the canvas with in the second painting campaign of 1885–86, the figures, objects, and landscape constitute a second level of composition whose relation to the neoimpressionist surface texture may only mask a thinking alien to the bottom-up positivism of dots.[102] One might have to abandon the fiction that bodies are built up out of individual light sensations. And indeed, pink outlines trace the distant figures at upper right, with darker silhouettes around the figures and trees framing them.[103]

Is the visibility of outlines only a matter of the painting's age, the *pentimenti* appearing when a paint layer becomes transparent? The detail of the man with the pipe suggests otherwise: his right hand *does* show a pentimento, where the middle finger crosses the blue outline of his abdomen. But why is his belly outlined in blue? And why all the pink and black outlines on the man sitting behind him, or the left edge of the dress of the woman behind them? Only a graphic emphasis, meant to bring out the silhouettes sharply from the grassy background, can account for these outlines: they are symbolist, if only in the deflated logical sense of Frege and Aurier that we have explored in this book. Seurat defines objects sharply by their concepts: his people are social types he sardonically orders by their buttons and hats and pets. They are constructed by firm outlines, ordering without falsifying the "thousand vibrant combats" of refracted light.[104]

If Seurat as a Platonist about shapes sounds surprising, we can corroborate the conclusion by pursuing intellectual history. Michelle Foa has recently argued that the critics' consensus on some aspects of neoimpressionist practice, such as the time dilation that makes pictures seem like "patient" depictions of the "silhouette of an entire day," in contrast to impressionist instantaneity, betrays a familiarity with the physiological aesthetics of Helmholtz.[105] Correlations abound, from simultaneous color contrasts to the preference for large canvases, allowing one to stand back and counteract flatness caused by accommodation of the eye.[106] But such effects, in particular the juxtaposition of figure and ground that gives the figures in the *Grande Jatte* their distinctive haloes, depend on the firm distinction between an upright object and the more or less horizontal ground visible behind the figure. Helmholtzian rhetoric aside, nearly all that Seurat sought to achieve depended on a robust notion of objects, rather than homogeneous sensations.

Close looking at Seurat's *Grande Jatte*, then, affords us a real critique of Russell and his allies Mach and James. The nature of the "logical fictions" that play the role of solid objects is left deceptively casual by these philosophers: if an object is a linked set of images, there is nothing to prevent us inventing arbitrary objects, such as that comprised by my right toes and a football, or the cane and left hand of Seurat's hatted sitter, excepting his pinky.[107] The continuity of sets of images doesn't help here. A monist theory won't draw interesting bounds between arbitrary pseudo-objects and those that are worth remarking on, like psychologically interesting wild particulars.[108] What of the idea of subjects constituted by sets of possible viewpoints? Seurat's painting is once again helpful in testing overzealous revisionism in metaphysics. It has been suggested that the rightward gaze of the little girl in the white pinafore dress and hat fixes the viewer's position at far right, in front of the large couple facing left.[109] Is that all there is to the subject implied by this painting? Let's see (Figure 5.12).

I have reproduced the painting from extreme vantage points to the right and left. Three considerations matter in weighing which is more suitable: that objects retain the shape they would have on close inspection, that the resulting trajectory for the eye is a plausible sightline through the space of *La Grande Jatte*, and thirdly, that we meet the girl's gaze. Both views meet the third condition, as does a centered view: the girl is so nicely cylindrical that she presents a convincing frontal view, her eyes following us, wherever we stand.[110] On the second count, both views afford convincing depth cues, indeed, the far left gives greater depth, perhaps because of its alignment with the light-green diagonal that runs from lower left to upper right. On

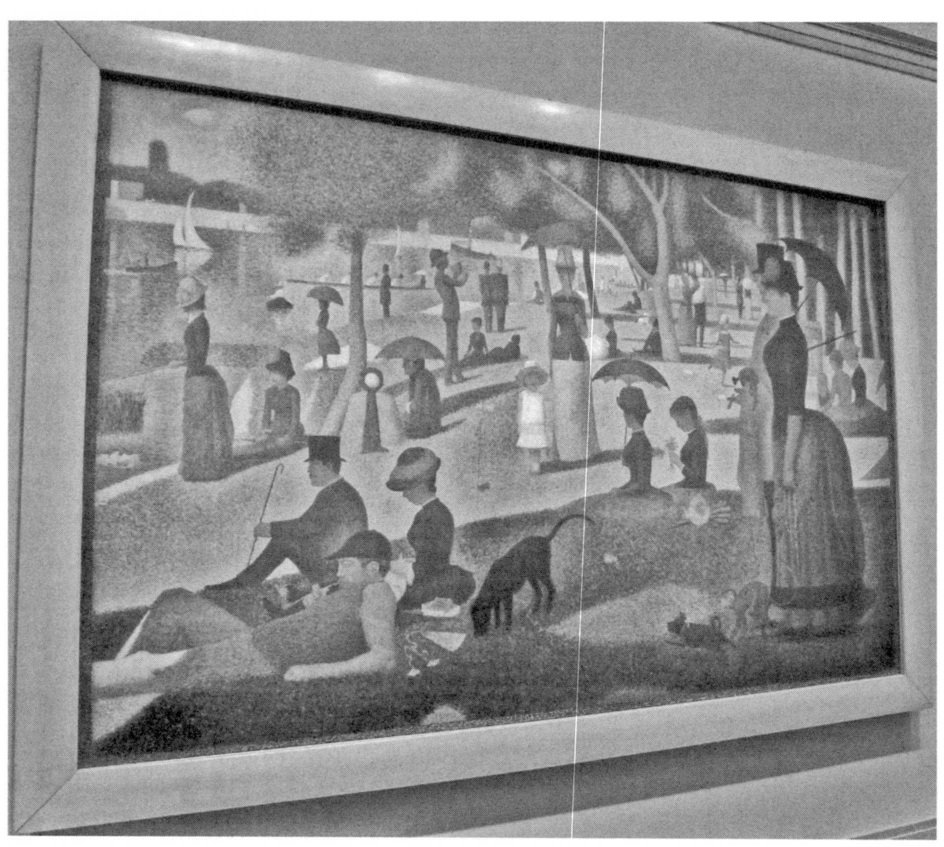

Figure 5.12. It seems to work! If I stand to the left, the girl's gaze seems to fix me in a new location just before the reclining pipe smoker. Or does it? If I stand on the right side of *La Grande Jatte*, the girl's gaze meets me there.

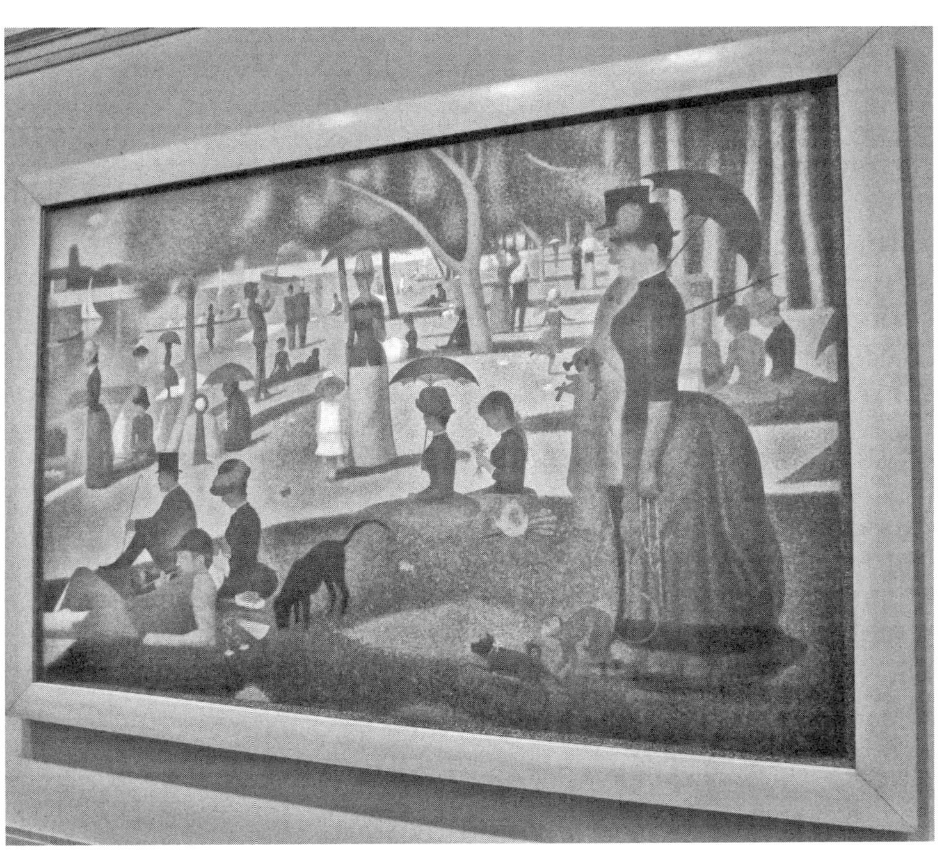

the first criterion, there is more distortion in the right-side view: The dress of the woman fishing looks much thinner than the similarly styled one of the woman with the monkey. This difference is minimized in the view from far left. On the other hand, the greater size of the couple at right is consistent with a viewer situated there.[111]

I conclude that Seurat could not have been too concerned where we stood after all: we were meant to perambulate, or at the least take in the (one) view from a number of perspectives, a state befitting the description of the painting as friezelike ("panathenaic"). Seurat may have painted a subject's trajectory, an official biography as Russell would put it, but by leaving room for us to see the painting from other vantage points without substantial loss, he failed to paint a "subject's view on the world." How does such a failure constitute a critique of Russell? In his 1921 *Analysis of Mind*, Russell toys with the idea of dispensing with consciousness altogether by regarding the images that constitute a subject's views as material — to be specific, photographs. "Thus what may be called subjectivity in the point of view is not a distinctive peculiarity of mind: it is present just as much in the photographic plate."[112] To be clear, Russell is not invoking the photograph to argue that there is no such thing as consciousness: he merely wants to argue that the perspective, and the resulting picture, visible in some particular place from some particular orientation has an objective reality. Knowing Frege's and Wittgenstein's picture theory, we may agree that an objective structure is shared by any appropriately situated photos. But a fatal remnant of the doubling fallacy lurks here. For one thing, a photograph, if it is not merely an imprint of objects in contact with it (a photogram) is an image produced *through* an optical mechanism. Consider another virtuoso work by Josef Eder, the "retinal image" of a firefly, magnifying a hundred times the image projected a millimeter behind a faceted lens, prepared with glycerin on a mica laminate by Viennese physiologist Sigmund Exner in 1889 (Figure 5.13).[113]

There is of course some conceptual confusion among the scientists who made this remarkable photograph: Exner and Eder refer to it as the insect's retinal image, but the notes on the double mount

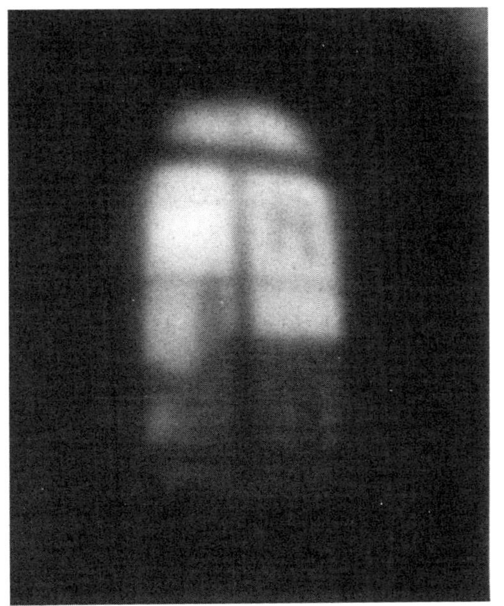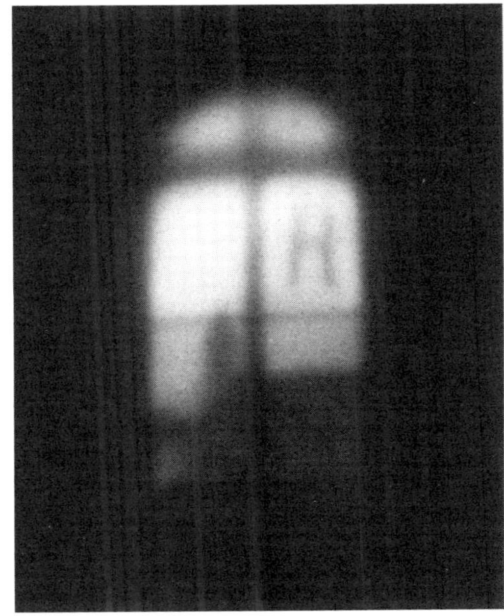

Figure 5.13. Josef Maria Eder and Franz von Reisinger, *Two Enlargements of the Retinal Image of a Firefly* (1890), albumen prints on cardboard, Albertina, Vienna. Through the window (on one of whose panes a large letter R is painted) one can see the tower and nave of the late-baroque Schottenfelder Kirche in Vienna's Seventh District.

call it a "photographic enlargement" of the "little air-picture (*Luft-bildchen*) formed by the firefly's eye."[114] As suggestive as it may be for "what it's like to be a firefly," the production notes are right: this is a particular kind of human artifact, exhibiting lawlike connections with the seeing apparatus of the firefly, and by no means "capturing" its subject position. Russell's own set of hypothetical photographs, positioned so as to capture every possible subjective standpoint, must in turn be *viewed* to match a subject's presence; even their location remains ambiguous, since a photograph takes up space, and is no dimensionless point, like a Russellian subject or the focus of a lens. Russell has only pointed out a similarity between subjectivity and artifacts in recording visual reality. The reduction does not eliminate consciousness, just as Seurat's reduction of things, including people, to points of light requires visual and conceptual armatures for those points, silhouettes which the sets of sensations fill like communicating vessels.

In his old age, Rusell admitted that efforts to eliminate entities in theory, to which he had devoted so much of his life and intellectual energy, are only really interesting when, despite them, something proves indispensable.[115] We are left with our dualism, or rather trinity, of minds, things, and pictures, but Seurat and Russell's patient efforts to reduce the first two terms to the third help us see how they interact logically, how minds read pictures to understand minds (psychology), material things (physics), and pictures themselves (aesthetics). This lesson of divisionism, the indispensability of what resists reduction, informs a picture that might serve as the swan song of symbolism. Odilon Redon's *Cyclops*, painted around the time Russell published *Our Knowledge of the External World*, is far from the uneasy truce between impressionism and abstraction it might at first appear (Figure 5.14). Rather, we have a picture of cognition itself overcoming the lonely reaches (and riches) of perception. Polyphemus, who has never looked this sensitive or this much like an ambulatory vehicle for the sense of sight (and hearing, with his prominent ears), is generally taken to dote on the sturdy classical body of the resting Galatea. Her red contour and weight contrast with

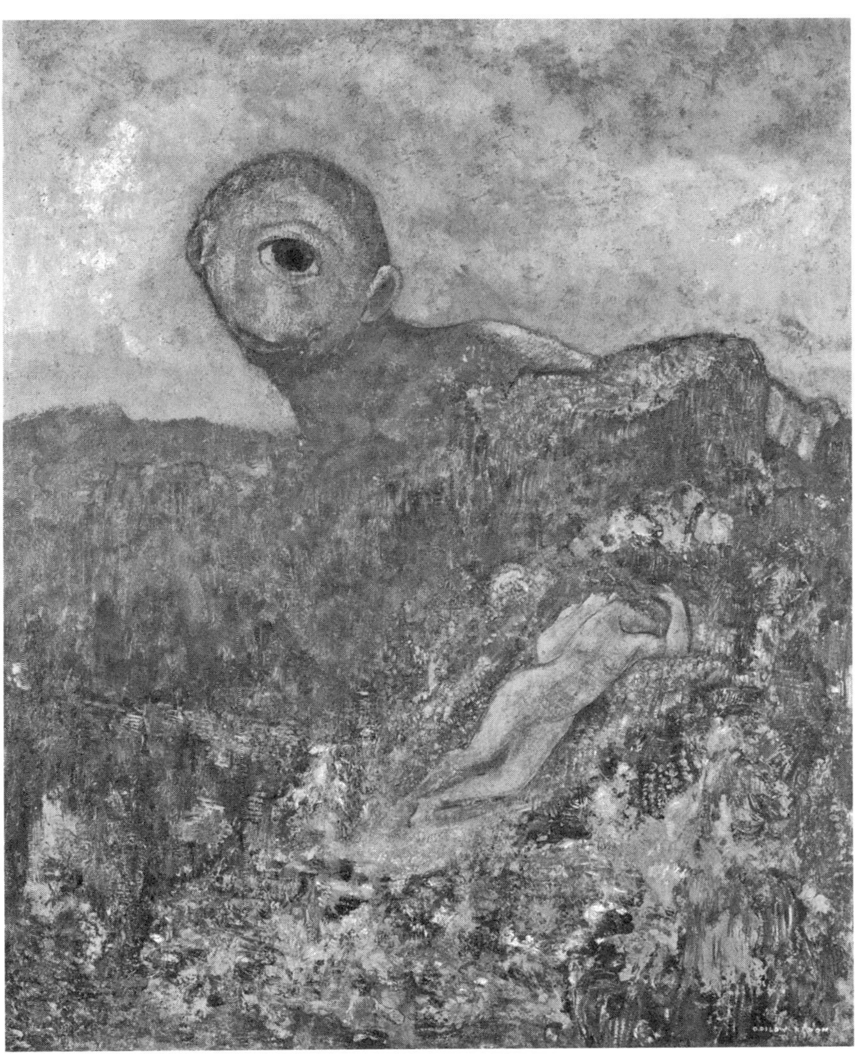

Figure 5.14. Odilon Redon, *Le cyclope* (1914), oil on cardboard mounted on panel, Kröller-Müller Museum, Otterlo, the Netherlands. The cyclops is all sensory equipment: eye and nose and ears and fingers. The saturated aurora-like dabs of paint around Galatea aren't really neoimpressionist or impressionist, any more than Jasper Johns's or Robert Rauschenberg's paint handling is abstract expressionist. They are symbols of such techniques—and of symbolism from Van Gogh and Gauguin to Puvis and Moreau.

the efflorescent, Seuratian vegetation that surrounds her like the sumptuous setting of a precious stone, blurred by indifference. It is as if Polyphemus only has eyes, or rather "eye," for her. But the composition contradicts this insipid valentine. The Cyclops, head raised over the rock he grasps with his left hand, cannot see his beloved; his giant iris, instead of pointing down, turns upward, meeting ours. Or perhaps he is gazing at infinity. But we are in the way, and his gaze meets ours. It is *we* who see Galatea; it is in our eyes, if anywhere, that the Cyclops sees his beloved. Cooperation between thinkers, alive or depicted, is a genuinely shared task of art and science, whether we call it symbolist or not.

Beyond Symbolism

This study of symbolism and its significance to art, science, and modernity is drawing to a close. The narrative could have been extended forward and backward in time, and perpendicularly to encompass more of the world. Friedrich Nietzsche, writing on "Truth and Lies in an Extramoral Sense," painted the same drama in very broad strokes: "In some obscure corner of the countless glimmering solar systems of the universe there was once a star, on which clever animals invented knowing. It was the proudest and most deluded minute of 'world history': but only a minute. Nature had only taken a few breaths when the star hardened, and the clever animals had to die."[1] Nietzsche's point is that there is more to life, and to the cosmos, than human knowledge and its purview. As for that knowledge and its ideal, truth, Nietzsche evinced a withering skepticism toward the reigning psychologistic orthodoxy of his day: "What is a word? The depiction of a nerve stimulus in sounds.... A nerve stimulus, translated first of all into a picture! First metaphor. The picture again reformulated in a sound! Second metaphor. And each time a complete leapfrogging of the sphere one is in to land in the middle of an entirely different and new one."[2]

In light of this arbitrariness, which as we saw struck Russell and Frege as well, Nietzsche finds little sense in an assertion like "The stone is hard," if it tracks only our subjective perception.[3] Reflecting that we have little sense of what it is like to perceive the world from another's standpoint, and that the laws of nature too only track regularities of relations of such perceptions, he famously concluded that truth is a "mobile army of metaphors, metonymies, and anthropomorphisms."[4] The philosophy that grew from this pronouncement, and that has dominated intellectual life in university humanities departments in the latter half of the twentieth century, was blissfully indifferent to science and logic. But that was hardly Nietzsche's intent. In light of a dominant positivistic worldview, he attacked empiricism and empiricist logic of his day with the weapons of logic itself. Of psychological abstraction he wrote:

> Let us think especially of the formation of concepts: each word becomes a concept in that it is no longer to serve as recollection for the unique, fully individualized ur-experience which has caused it, but is to stand at the same time for countless more-or-less similar, that is strictly speaking never identical, that is to say, for a mass of divergent, cases. Each concept originates in the equation of the nonequal. As surely as no leaf was ever exactly the same as another is the concept "leaf" made by arbitrary dropping of individual differences, through a forgetting of what distinguishes them, and it gives rise to the impression that in nature there is something besides leaves, the "Leaf," some sort of Urform, after which all leaves are woven, drawn, measured, colored, curled, painted, but by clumsy hands, so that no exemplar is the correct and reliable, faithful copy of the Urform. We call a person honest; why did he act honestly today? we ask. Our answer tends to be: because of his Honesty. His Honesty! That means once again: the Leaf is the cause of leaves. We know nothing at all of an essential quality called honesty, but certainly of numerous individualized and thus unequal actions that we equate by omission of the unequal and now label as honest actions; finally we formulate out of them a qualitas occulta with the name: Honesty.[5]

All very clever and right, but it tells only against the theory that a concept is a "common name" abstracting from the variation found in real objects. It is Nietzsche's lack of curiosity about the logic of language — and of pictures, as we saw figure in his story of perception — that could make him conclude on the basis of this ailing theory that there is no truth, nor concepts. Indeed, he was classicist enough to know from Plato's *Parmenides* that we need some means of accessing concepts to be able to talk of anything, for instance of leaves, by which Nietzsche of course means objects falling under that concept, not some mental image in his own head.[6]

There were cracks in the empiricist consensus, palpable in its more perceptive skeptics, such as Nietzsche. Though he came to celebrate the "superman" conscious of his desires and willing to act on them, he hated the tyranny of the subject. "In this consciousness he is locked, and nature threw away the key," complains Nietzsche in a text "on the pathos of truth," a dry run for the essay on truth and lies.[7] Yet, complain as he might, he could not deny individual subjective divergence: "if we each had a different sensory apparatus, so that we could perceive only as a bird does, another time as a worm, another time as a plant, or if one of us saw a particular stimulus as red, another as blue, while a third even heard it as a tone, then no one would speak of such regularity of nature, which would be then grasped as a highly subjective notion."[8] The pervasiveness of subjectivity, and the difficulty it brings to understanding the thoughts and actions of others, including their artifacts, and in attempting to act together or build any kind of institution, is the one truth of psychologism. It is my conviction in this book that we can do justice to this insight only by insisting on what it occludes: logical structure. A historian content to track the mobile army of metaphors will miss the forest for the trees.

What does this mean for the art historian? Perhaps nothing that can be put in the form of a methodological maxim, such as those with which Frege began his *Foundations of Arithmetic*: to always strictly distinguish psychology from logic, and object from concept, and to never ask for a word's meaning in isolation from its context in a proposition.[9] These are good maxims, though with the exception of the first, they are not directly applicable to most pictures. Even in the case of the third, it should hardly be taken to mean the familiar thesis in social history that (social) context alone determines meaning. But the context principle might be taken to mean something like this: an artwork means anything only in a context. Often it is its meaning at a particular time for a particular group of people that interests us, particularly in a social history of art. But for the greatest, most interesting art, that is not usually its only interest. What it could mean in its continued existence, both the subtle and unsubtle variation in meaning and — what is more striking yet, and nearly always ignored or downplayed by reception studies — what stays the same or is conserved with the passage of time, are functions of the logical analysis of pictures, which I hope to have convinced you at least through the example of this book need not consist in treating pictures as linguistic utterances. Symbolist art, if never the easiest to interpret, is in any case fairly consistent in insisting that an object's aesthetic value is intimately related to its intellectual content. Of course, to get to the point of considering a picture for its possible role in thoughts, a lot of physical work — looking, walking, even touching or smelling (or hearing, obviously, in the case of music) and a lot of thinking about these experiences must take place. Not that it is all work. Much of this is done with pleasure, and some experts perform it as easily as they breathe. But it is another conviction of this book that no appeal to sensibility or "intuition" will get the art historian over a question of meaning. Beauty and sense are not one.

To engage with pictures intensely is not merely to perceive, but to think and argue about them. Doing so involves, implicitly, the work's context, which is also its history. The Platonist has to be a historian — for we never get beyond appearances without understanding the time-sensitive ways these contribute to meaning. I hope to have shown by example that a historian can also be a Platonist. That an artist may be as well, I hope will emerge from the last image in this book: an early etching by Bracquemond that Jean-Paul Bouillon considers among his most sought after. Bracquemond himself was sensitive to subjective divergence, recounting the story of an old amateur of the prints of Wille and Berwick who could not stand Rembrandt. Once, when Bracquemond exclaimed that only originals could be studied with profit, that "on the basis of engravings, it would be difficult to decide whether Rembrandt or Canaletti [sic] had more genius," the old man lowered his head as if to charge, said goodbye, and never came back.[10]

The *Basket of Vegetables* (Figure C.1) looks at first like a print for

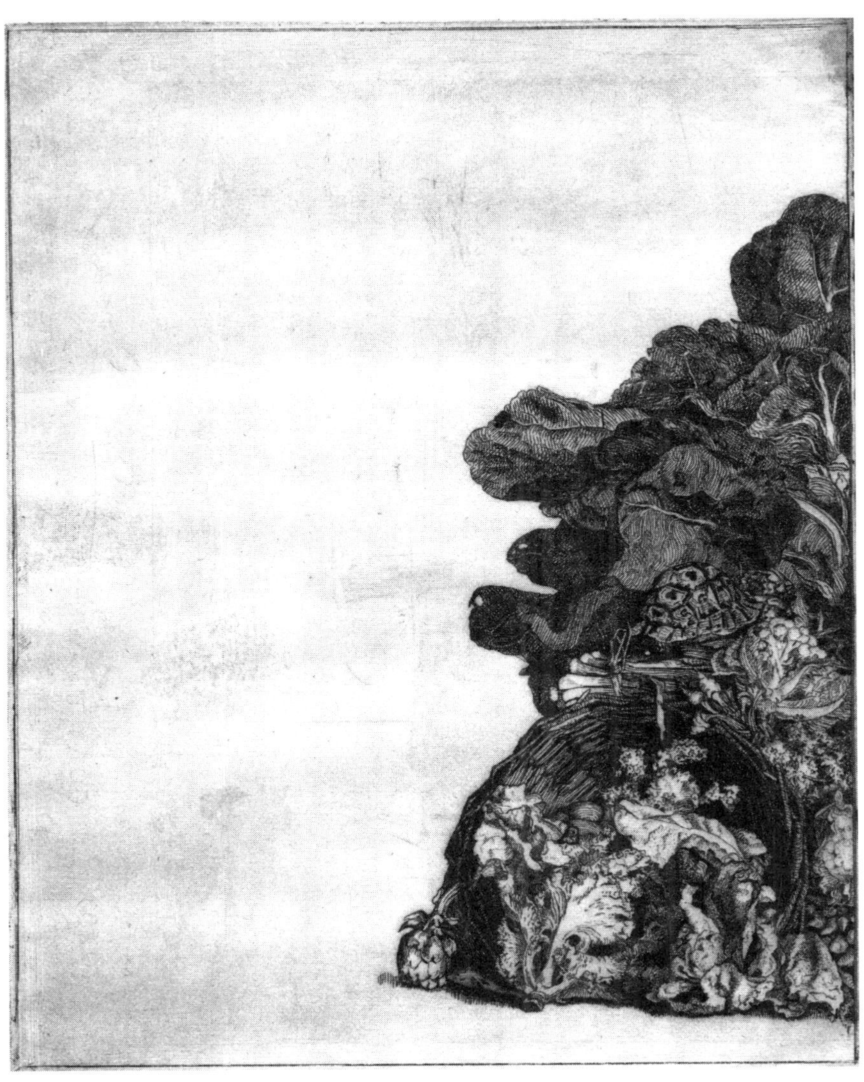

Figure C.1. Félix Bracquemond, *Le Panier des legumes* (c. 1854–55), etching with surface tone, British Museum, London. If realism is, among other things, belief in a world beyond (and including) the representation, this might be a great realist image.

such a collector. The monumental pile of chard, cauliflower, lettuce, and artichokes, dwarfing comically the little capsized wicker basket, houses a small tortoise that will perhaps soon begin feasting on the produce. Textural profusion — curls of leafage, basketwork, tortoise-shell — balances a vast expanse of plate tone, which covers the paper's leftmost half and reaches over the vegetables and under them. Besides the merest hint of shadow, no concession is made to the world around these things. This kind of willful, if not gratuitous, virtuosity, the print's lack of literary or anecdotal interest, even of the kind of moral value associated with the Dutch still lifes Bracquemond admired, must surely have pleased the amateurs. This has as much a right to be called an abstract print — one meant to be enjoyed in isolation from dramatic or intellectual considerations — as the perhaps incomplete *Saint Cloud*. But, being a rendering of concrete objects, it is also something more. There is more to the print than what we see of the vegetables, or what we don't see to their left, above, and immediately below them. What is there? All the vegetables on the other side of the right edge of the picture, without which the pyramid would fall. The picture is as much about them as it is about the vegetables we do see, and whose rightmost parts are cut off by the frame. A bunch of things one does not see: that is a pretty remarkable thing for a picture to be about.

Acknowledgments

In my previous book on Henry Fuseli, I did not include an acknowledgments page, hoping instead that scholars would find my tribute to them in the footnotes and index. I have outgrown that scruple, but I still despair of thanking everyone properly. In graduate school, my advisors Ewa Lajer-Burcharth and Henri Zerner nourished my interest in the nineteenth century, even while shepherding an eighteenth-century dissertation. They were abetted by a remarkable cast of scholars I met over the decade since my dissertation research, especially David Bindman, Juliet Bellow, Emmelyn Butterfield-Rosen, Whitney Davis, André Dombrowski, Mark Ledbury, Patrick McGuinness, Elizabeth Mansfield, Pierre Michel, Claire Moran, Natasha Ruiz-Gómez, Susan Siegfried, Alison Syme, Merel van Tilburg, and Barbara Vinken. Wen-Shing Chou, who shared an office with me at the Center for Advanced Study in the Visual Arts at the National Gallery, got the first big earful of my symbolism project, when it was still a James Ensor project.

In Austria and Switzerland, where I taught and my family lived for the better part of a decade, I am grateful for all kinds of support, intellectual as well as practical: in Vienna particularly from Wolfram Pichler, Sebastian Egenhofer, Ramón Reichert, and Werner Hanak, and from Theresa, Hans, and Gerheid Widrich. Curators and archivists in Basel, Vienna, Graz, and Munich were resourceful in helping me track down works of art and documents: I would like to thank Raphaël Bouvier, Ulf Höfer, Friedrich Polleross, Matthias Röschner, and Wolfgang Schinhan, as well as the public institutions and private collectors who allowed me to reproduce singular works of art. In Basel, where I began writing this book, I was particularly fortunate to count as my employer and friend Ralph Ubl, and to be surrounded by a talented team of modernist art historians and art theorists in the Kunsthistorisches Seminar and the eikones Center for the Theory and History of the Image, as it is now called, especially Markus Klammer, Inge Hinterwaldner, Rahel Villinger, and Vera Wolff, to say nothing of eikones founder Gottfried Boehm and, on his return from Paris, fellow Fuseli enthusiast Andreas Beyer. T. J. Clark, Michael Fried, Dario Gamboni, Catriona MacLeod, Richard Shiff, and Chris Wood, on their visits to Basel, were generous with erudition as well as sage advice.

On arriving at the University of Chicago, I was surrounded by amiable and authoritative scholars, of the late nineteenth century and beyond, that made the completion of the book a joy and a learning experience. Among the art historians, Ina Blom, Joyce Cheng, Susanna Caviglia, Jaś Elsner, Anne Leonard, Christine Mehring, W. J. T. Mitchell, Richard Neer, Joel Snyder, and Martha Ward deserve special mention, as do colleagues across Chicago, in particular Nina Dubin, Stephen Eisenman, James Elkins, Gloria Groom, Margaret MacNamidhe, Suzanne McCullagh, Jennifer Nelson, and Kevin Salatino. While working on the book, I was fortunate enough to be invited to present my work in progress by James Chandler at the University of Chicago's Franke Institute, by Bryan Garsten at Yale University, by Marty Powers in Ann Arbor, and by Eva Struhal and Matthew Hunter of the Université de Laval and McGill University respectively. In these places I fondly recall the constructive criticism and collegiality of Matthew Birro, Deborah Coen, Michèle Hannoosh, Chriscinda Henry, Jeehee Hong, Mary Hunter, Joan Kee, Alex Potts, and Nicola Suthor. My students at the University of Chicago are the unsung heroes of much of this book, on whom I have tried most of its arguments. Michelle Facos, whom I met near the end of editing, is a scholar of symbolism whose breadth and precision I strive to emulate. My colleagues in the Committee on Social Thought, notably Robert Pippin, Rosanna Warren, Thomas Pavel,

Lorraine Daston, David Nirenberg, Joel Isaac, and Nathan Tarcov, and beyond the Committee, Alison James, Florian Klinger, Michèle Lowrie, Robert Morrissey, Larry Norman, and above all Françoise Meltzer, whose vision of symbolism overlaps with mine in several ways, have encouraged me immensely.

Parts of the book, particularly those dealing with Frege, Russell, and Wittgenstein were enlivened by discussion and correspondence with philosophers. I thank Paul Benacerraf, Patricia Blanchette, James Conant, Roy Cook, David Egan, Solomon Feferman, Juliet Floyd, Warren Goldfarb, Michael Kremer, Gabriel and Jonathan Lear, Omar Nasim, Hilary Putnam, Mark Sainsbury, and Josef Stern, for their good will in what was no doubt an unequal exchange. Michael Dummett once noted how he could not affirm that none of his mistakes came from his interlocutors, since had he known what they were, he would have fixed them. I can add that in my own case some error is probably due to misunderstanding.

This book would not exist without Zone Books and the intellectual sympathy of Jonathan Crary. My editor Meighan Gale is responsible for many of the book's virtues, as is artist and designer Julie Fry, whose evocative cover and imaginative work throughout embodies the book's argument about the parallel developments constituting symbolism. Gregory McNamee and Alena Jones transformed the manuscript for the better early and late in the process, as did my research assistant and fellow student of symbolism Ena Gojak. A generous gift from Jerry and Lois Beznos to the Committee on Social Thought paid for her work and for image-related costs, and the thoughtful support of Anne Gamboa and my chair Robert Pippin helped the project materialize on time and in one piece. Mechtild Widrich has read more versions of this book, and done more to make it intelligible — and to make it exist — than I can properly express. She and our son Laurens Pop, my most consistent critic, are the dedicatees.

Notes

PREFACE: WHY SYMBOLISM?

1. Donald Preziosi, *The Art of Art History*, 2nd ed. (London: Oxford University Press, 2009), p. 7.

2. Mark Twain, *The Adventures of Huckleberry Finn* (London: Chatto & Windus, 1884), p. 124. The entire passage, beginning with the discussion of "King Sollermun," was the last part of the novel to be composed, after Twain had written the conclusion.

3. George Washington Cable wrote Twain on October 25, 1884: "When we consider that the programme is advertised & becomes cold-blooded newspaper reading I think we should avoid any risk of appearing — even to the most thin-skinned and super-sensative [*sic*] and hypercritical matrons and misses — the faintest bit gross." CU-MARK (University of California, Mark Twain Papers, The Bancroft Library, Berkeley, UCLC 42319). See Guy Cardwell, *Twins of Genius* (East Lansing: Michigan State College Press, 1953), p. 105. See also Walter Blair and Victor Fischer's edition of the *Adventures* (Berkeley: University of California Press, 1985), pp. 376–77. Cable was a noted novelist of New Orleans, dealing thoughtfully with race and opposing Jim Crow laws.

4. The fact that Jim wins the argument is recognized in the literature — e.g., the classic discussion of D. L. Smith, "Huck, Jim, and American Racial Discourse," in *Satire or Evasion? Black Perspectives on Huckleberry Finn* (Durham, NC: Duke University Press, 1992), p. 111 — but it is still routinely misunderstood. Howard Horwitz, "Can We Learn to Argue? *Huckleberry Finn* and Literary Discipline," *ELH* 70.1 (Spring 2003), p. 283, for instance, insists that "Huck and Jim argue in exactly the same way." This is as false as Horwitz's claim that neither hero examines his premises.

5. A strong statement of the generality and logical abstractness of scientific laws can

already be found in the final chapter of George Boole, *An Investigation of the Laws of Thought* (London: Macmillan, 1854). For an extended argument that the issue of perspectival knowledge should be distinguished from that of ontological realism, see Lorraine Daston and Peter Galison, *Objectivity* (New York: Zone Books, 2007). I reconnect the two in this book.

6. The sole published interpretation of this work links it to photography: "This 'cubification' of the eye's volume can be 'read' as a comment on the camera's mechanization of the human gaze." Raphaël Bouvier in *Odilon Redon* (Ostfildern: Hatje Cantz, 2014), p. 30. To "read" thus is a symbolic, allegorical process. On Redon's interest in biology and astronomy, see Barbara Larson, *The Dark Side of Nature: Science, Society, and the Fantastic in the Work of Odilon Redon* (University Park: Pennsylvania State University Press, 2005).

7. The eye inscribed in a triangle is an emblem of reason. See Albert Potts, *The World's Eye* (Lexington: University of Kentucky Press, 1982). The single eye was also a beloved romantic memento. See Hanneke Grootenboer, *Treasuring the Gaze: Intimate Vision in Late Eighteenth-Century Eye Miniatures* (Chicago: University of Chicago Press, 2012).

8. Odilon Redon, "Letter to Monsieur X," December 1911, in *To Myself*, translated by Mira Jacob and Jeanne L. Wasserman (New York: George Braziller, 1986), p. 98. For the original, see Redon, *À soi-même* (Paris: Henri Floury, 1922), p. 113. Cf. p. 26 (p. 30 in the original) on "putting the logic of the visible at the service of the invisible."

9. I cannot agree that "it was largely by absorptive dynamics that traditional painting had achieved its effects of 'significance' and 'meaning.'" Michael Fried, *Manet's Modernism or the Face of Painting in the 1860s* (Chicago: University of Chicago Press, 1996), p. 355. Meaning is a wider phenomenon than absorption, which it makes possible.

10. The major reconsideration of Bragdon, Jonathan Massey's *Crystal and Arabesque: Claude Bragdon, Ornament, and Modern Architecture* (Pittsburgh: University of Pittsburgh Press, 2009), connects his liberal rationalism to the "exoticizing symbolist movement," emphasizing the latter's backward-looking, medieval tendencies (p. 170).

11. Consider Stephen Eisenman's discussion in *The Temptation of Saint Redon* (Chicago: University of Chicago Press, 1992), p. 67, of Redon's sympathy with Pascal and with workers and peasants in light of their "apparent irrationality." Eisenman's good sense is manifest in the qualifying adjective: reason and intuition are, as Pascal insists, compatible.

12. The best treatments of science and painting remain John Gage, *Color and Culture* (Berkeley: University of California Press, 1999), covering many artists, and Jonathan Crary, *Suspensions of Perception* (Cambridge, MA: MIT Press, 1999), focusing intensely on a few. On the graphic practice of Victorian science, see Omar Nasim, *Observing by Hand: Sketching the Nebulae in the Nineteenth Century* (Chicago: University of Chicago Press, 2014).

CHAPTER ONE: SYMBOLISMS IN THE PLURAL

1. Caillebotte's address was in fact 30 Boulevard Haussmann.

2. One wonders whether Munch read this, agreed, and painted a scream for just this reason.

3. Letter to Schlabbrendorf, August 16, 1766, in Johann Joachim Winckelmann, *Briefe*, ed. Walther Rehm with Hans Diepolder (Berlin: Walter de Gruyter, 1956), vol. 3, p. 199.

4. On the context of this unusual drawing, see Andrei Pop, *Antiquity, Theatre, and the Painting of Henry Fuseli* (Oxford: Oxford University Press, 2015), ch. 3. Fuseli already challenged Winckelmann's interpretation of the Laocoon in his English translation of the *Reflections on the Painting and Sculpture of the Greeks* (London: Andrew Millar, 1765), p. 31, where the "sigh" becomes a "groan": a plausible reading of Winckelmann's authority, Jacopo Sadoleto, whose sixteenth-century Latin Laocoon emits a *gemitus*.

5. Do these sayings have anything to do with Laocoon? Maybe. Blake thought morals had been supplanted by commerce in much the same manner that the composition of the Laocoon group, supposedly inspired by God's statue on "Solomon's Temple," was copied "by three Rhodians & applied to Natural Fact or History of Ilium." For references, see Andrei Pop, "Laocoön in Art," in *The Virgil Encyclopedia*, ed. Richard F. Thomas and Jan M. Ziolkowski (Malden, MA: Wiley-Blackwell, 2013), pp. 715–18.

6. Richard Brilliant, *My Laocoon: Alternative Claims in the Interpretation of Artworks* (Berkeley: University of California Press, 2000).

7. J. J. Bernoulli, *Über die Laokoongruppe* (Basel: Schweighauserische Universitätsbuchdruckerei, 1863), p. 4. This Bernoulli, an art historian and archaeologist rather than a mathematician like most of his family, discusses among other issues the modern flexed-elbow restoration of Laocoon's right arm, also considered by Winckelmann.

8. Anselm Feuerbach, *Der Vaticanische Apollo: Eine Reihe Archäologisch-Ästhetischer Betrachtungen*, 2nd ed. (Stuttgart/Augsburg: Cotta, 1855), pp. 339–411; Jakob Wilhelm Henke, *Die Gruppe des Laokoon oder über den kritischen Stillstand tragischer Erschütterung* (Leipzig/Heidelberg: Winter, 1862). Henke begins his book with an epigraph from Nathaniel Hawthorne: "In any sculptural object, there should be a moral stillstand, since there must of necessity by a physical one. In Laocoon the horror of a moment grew to be the Fate of interminable ages." Both sentences are from *The Marble Faun*, 2 vols. (Boston: Houghton Mifflin, 1860), vol. 1, p. 25, and vol. 2, p. 200 (with no capitals on "fate"). Interestingly, Hawthorne's women painters, Hilda and Miriam, challenge these Lessingian precepts (vol. 1, p. 25). Henke's own debt to Lessing is at best an ironic one.

9. Hippolyte Taine, *Voyage en Italie*, vol. 1, *Naples et Rome* (Paris: Hachette, 1866), p. 199.

Taine considers Apollo Belvedere and Laocoon late works; more oddly, he calls them "feminine," as he does Euripides.

10. Hermann Lotze, *Geschichte der Ästhetik in Deutschland* (Munich: Cotta, 1868), p. 554.

11. *Ibid.*, pp. 554–55. Daumier had a more literal Laocoon caricature in *Charivari* in the same year Lotze published his book. It is possible that Lotze mixed this up with Gillray, which better fits his description.

12. Thus Brilliant, *My Laocoon*, p. 99: "Three figures struggling with snakes, or snakey forms, constitute the core identity of the motif, and they must be present — and recognizable — to ensure the success of the 'take-off.' "

13. Lotze, *Geschichte der Ästhetik*, p. 555. Emphasis in the original.

14. Gottlob Frege, "Der Gedanke," *Beiträge zur Philosophie des Deutschen Idealismus* 1.2 (1918), pp. 70–71, my translation. There are various English translations, e.g., "The Thought," *Mind* 1.259 (1956), pp. 303–304. An earlier version of the passage can be found in Frege's unpublished "Logik," composed no later than 1897. See Gottlob Frege, *Nachgelassene Schriften* (Hamburg: Felix Meiner, 1983), p. 156.

15. Salomon Stricker, *Studien über die Association der Vorstellungen* (Vienna: Wilhelm Braumüller, 1883), p. 23. Though Stricker might disavow it, a note of elitism creeps in, as it does in the "social definition of art" as whatever is called art by experts. Much important art — and science — is not initially accepted by experts.

16. A diagram of Hauck's perspectograph and a drawing produced with its aid can be found in Friedrich Dalwigk, *Vorlesungen über darstellende Geometrie* (Leipzig/Berlin: B. G. Teubner, 1914), vol. 1, p. 235.

17. Guido Hauck, *Die subjektive Perspektive und die horizontalen Curvaturen des dorischen Styls: Eine perspektivisch-ästhetische Studie* (Stuttgart: Konrad Wittwer, 1879); cf. Erwin Panofsky, *Perspective as Symbolic Form*, trans. Christopher Wood (New York: Zone Books, 1991), p. 33.

18. A refutation of Hauck is M. H. Pirenne, "The Scientific Basis of Leonardo da Vinci's Theory of Perspective," *British Journal for the Philosophy of Science* 1.10 (August 1952), pp. 169–85, which points out that *whatever subjective curvature we see*, it would be exaggerated, not "matched," by drawing curves instead of straight lines. Nevertheless, Hauck-like theses have appealed to thinkers as diverse and probing as Johannes Kepler and James Elkins. See the latter's "'Das Nüsslein beisset auf, ihr Künstler!' Curvilinear Perspective in Seventeenth Century Dutch Art," *Oud Holland* 102.4 (1988), pp. 257–76.

19. The relevant texts are Edmund Husserl, *Philosophie der Arithmetik: Psychologische und logische Untersuchungen* (Halle: Pfeffer, 1891); Ernst Mach, *Beiträge zur Analyse der*

Empfindungen (Jena: Gustav Fischer, 1886); Gustav Theodor Fechner, *Elemente der Psycho-physik*, 2 vols. (Leipzig; Breitkopf und Härtel, 1860); Wilhelm Wundt, *Logik: Eine Unter-suchung der Prinzipien der Erkenntnis und der Methoden wissenschaftlicher Forschung*, 3 vols. (Stuttgart: Enke, 1880–83); and Hermann von Helmholtz, "Zählen und Messen erkennt-nisstheoretisch betrachtet," *Philosophische Aufsätze. Eduard Zeller zu seinem fünfzigjährigen Doctor-Jubiläum gewidmet*, ed. Friedrich Vischer (Leipzig: Fues's Verlag, 1887), pp. 14–52.

20. Andrea K. Henderson, "Symbolic Logic and the Logic of Symbolism," *Critical Inquiry* 41.1 (Autumn 2014), pp. 78–101, draws interesting connections between Lewis Car-roll's symbolic logic and fantastic stories.

21. Helmholtz, "Zählen und Messen," p. 21.

22. Letter to William Graham, July 3, 1881, in *The Life and Letters of Charles Darwin Including an Autobiographical Chapter*, ed. Francis Darwin (London: John Murray, 1887), vol. 1, p. 316.

23. Stricker shrugs it off: "people who do not associate their experiences correctly *causaliter* do not find their way in the world" (*Studien*, p. 28). But Darwin's deeper point is that success in the world is not scientific justification.

24. William James, *The Principles of Psychology* (New York: Henry Holt, 1890), vol. 1, p. 269. There is another way to read James's diagram: as distinguishing subjective from objective thought.

25. *Ibid.*

26. Christine Ladd [Franklin], "Intuition and Reason," *The Monist* 3.2 (January 1893), p. 211.

27. *Ibid.*, pp. 211–12.

28. Griselda Pollock, "Modernity and the Spaces of Femininity," in *Vision and Difference: Femininity, Feminism and Histories of Art* (London and New York: Routledge, 1988), p. 56.

29. "L'homme y passe à travers des forêts de symboles / Qui l'observent avec des regards familiers." A retrospective gaze, and an "overlaying of disparate realities that remain, opposingly, in the poet's gaze" (p. 248), is the theme of Françoise Meltzer's very fine *Seeing Double: Baudelaire's Modernity* (Chicago: University of Chicago Press, 2011).

30. The text was a second, more considered response to a critique of "les poètes déca-dents" by Paul Bourde, in *Le Temps*, August 6, 1885. All these were printed, alongside an exchange with Anatole France, in a pamphlet designed to make of symbolism a great liter-ary scandal, *Les Prèmieres armes du symbolisme* (Paris: Vanier, 1889). In a letter to the editor Vanier printed as introduction to the volume, Moréas inveighs, interestingly, against the charge of obscurity: "We repudiate only the *Unintelligible*, that charlatan" (p. 10).

31. Moréas, "Le Symbolisme," *Les Prèmieres armes du symbolisme*, pp. 38–39. The concluding sentence is even clearer: "the symbolist novel will edify its work of *subjective deformation* in accordance with this axiom: that art will not seek in the *objective* any but a simple and extremely succinct point of departure" (*ibid.*).

32. Allison Morehead, *Nature's Experiments and the Search for Symbolist Form* (University Park: Pennsylvania State University Press, 2017), links symbolism to science (particularly psychology) above all through the practice of deformation of bodies, the painterly form of Moréas's "subjective deformation."

33. The point is emphasized by Bourde in "Les Poètes décadents," *Les Prèmieres armes du symbolisme*, p. 14. Bourde is also clear that the succession of important French poetic movements in the nineteenth century is "De Gautier à Baudelaire, de Baudelaire au Parnasse, du Parnasse au décadent" (p. 24).

34. Jules Huret and Stéphane Mallarmé, "Enquête sur l'évolution littéraire. Suite 1: Stéphane Mallarmé," *L'Écho de Paris*, March 14, 1891, p. 2. Emphasis in the original. Cf. Henri Peyre, *Qu'est-ce que le symbolisme?* (Paris: Presses Universitaires de France, 1974), pp. 75–76. Mallarmé goes on to provide his own definition: "The perfect use of that mystery is what constitutes symbolism: evoking an object little by little in order to show a state of mind, or inversely, choosing an object and setting off a state of mind through a series of decipherments."

35. Réne Ghil, *Traité du verbe avec avant-dire de Stéphane Mallarmé* (Paris: Giraud, 1886), p. 6: "Je dis: une fleur! et, hors de l'oubli où ma voix relègue aucun contour, en tant que quelque chose d'autre que les calices sus, musicalement se lève, idée rieuse ou altière, l'absente de tous bouquets."

36. Stéphane Mallarmé, "Crise de vers," *Divagations* (Paris: Bibliothèque-Charpentier, 1897), p. 245: "d'inclure au papier subtil du volume autre chose que par exemple l'horreur de la forêt, ou le tonnerre muet épars au feuillage: non le bois intrinsèque et dense des arbres."

37. Paul Valéry, *Degas Manet Morisot*, translated by David Paul (Princeton: Bollingen/Princeton University Press, 1971), p. 62. Cf. Ludwig Wittgenstein, *Tractatus Logico-Philosophicus* (London: Routledge and Kegan Paul, 1922), 49: objects I can only *name*. Signs represent them. I can only speak *of* them. I cannot *assert them*" (3.221).

38. A locus classicus is Jacques Derrida, "La Double séance," *Tel Quel* 41 (Spring 1970), p. 343, in *Dissemination* (Chicago: University of Chicago Press, 1981), pp. 173ff. In a brilliant article, Françoise Meltzer agrees that Mallarmé "is no Platonist" but adds that "Derrida is confusing the conscious absence of a referent with a rejection of Platonism," a philosophy

that requires abstract reference to ideal objects. See "Color as Cognition in Symbolist Verse," *Critical Inquiry* 5.2 (Winter 1978), p. 265. Meltzer's own subtler argument is that the wealth of color terms in Mallarmé privileges subjective image and Fregean sense over reference. This is convincing, but it does not have the consequences she attributes to it. For one, as we saw, Mallarmé believed (and as we will see in Chapter 4, Frege agreed) that art deals in sense not reference. An emphasis on sense yields a more sophisticated Platonism, wherein not every term need refer. But secondly, color words *do* refer. They stand for color concepts and are mainstays of Platonism: in Frege, but also in Plato's *Meno* (where Socrates defines color as what accompanies shape) and Plotinus's *Enneads* (Treatise 30 argues against materialism that we can build machines, but not their colors).

39. Meltzer's fine analysis of this poem encourages me, despite its emphasis on volatile referents like "ice, frost, and glaciers [which] can be melted by the sun" (p. 262). Meltzer even cites Wordsworth's sonnet "Mutability": "Truth fails not; but her outward forms that bear / The longest date do melt like frosty rime." What could be more Platonist?

40. Téodor de Wyzewa, *Mallarmé, Notes* (Paris: Publications de *La Vogue*, 1886), p. 10: "Une consciente logique a créé le thème, avec — mais rien au delà — son expansion nécessaire."

41. Wyzewa, *Mallarmé*, p. 8: "Et j'ai peur de penser (/ mourir) lorsque je couche seul."

42. "Vers, marbre, onyx, émail." In *Émaux et camées* (Paris: Poulet-Malassis et De Broise, 1858), p. 211. The moral, given in the first stanza, is that art must take on a recalcitrant subject, like the artisan: "Oui, l'oeuvre sort plus belle / D'une forme au travail / Rebelle." It is reinforced in the final stanza: "Sculpte, lime, cisèle; / Que ton rêve flottant / Se scelle / Dans le bloc résistant!" (p. 216). His "Fantaisies d'hiver" should be compared with Mallarmé's "Cygnes."

43. Wyzewa, *Mallarmé*, pp. 12–13, emphases in original. This self-reflexive concern with the means by which an art achieves meanings is the rationale for Wyzewa's repeated invocation of Mallarmé as a logician: "Mais M. Mallarmé se manifeste, ici même, un logicien et un artiste" (p. 10); "M. Mallarmé, logicien et artiste, cherchait, infatigablement, la rénovation logique de l'Art" (p. 12). Wyzewa, in order to credit Mallarmé with this innovation, controversially classes Verlaine as a Parnassian, the last and greatest of that movement, and the most successful at reducing poetry to pure sound and feeling. This was not the orthodox position, nor Mallarmé's; witness "Crise de vers," p. 237. Philip Stefan points out that as reviewer for *La Revue Indépendant* Wyzewa wrote about Mallarmé and Laforgue but never Verlaine. *Paul Verlaine and the Decadence, 1882–90* (Manchester: Manchester University Press, 1974), p. 171 n. 58.

44. Merel van Tilburg's forthcoming *Staging the Symbol: The Nabis and Symbolist Theatre in Paris* is particularly thorough on the intellectual forces flowing from painting to theater and vice versa in the Paris of the 1890s.

45. A good account of the colorful practices of this artistic community remains Agnès Humbert, *Les Nabis et leur époque, 1888–1900* (Geneva: Pierre Cailler, 1954). A sample of the confusion in the older literature is Charles Chassé, *Le Mouvement symboliste dans l'art du XIXe siècle: Gustave Moreau, Redon, Carrière, Gauguin et le groupe de Pont-Aven, Maurice Denis* (Paris: Librarie Floury, 1947). That heterogeneous list is in fact due in part to Aurier.

46. G.-Albert Aurier, *Oeuvres posthumes* (Paris: Edition du "Mercure de France," 1893), p. 208. Italics in the original. Aurier died of typhus at the age of twenty-seven in October 1892; his mother published his works a year later, with an introductory essay by Remy de Gourmont.

47. *Ibid.* Art must be "1) ideist 2) symbolist 3) synthetic 4) subjective and 5) (as a consequence) decorative" (pp. 215–16).

48. Aurier, *Oeuvres posthumes*, p. 297. Emphasis in the original. "Les Peintres symbolistes" originally appeared in the *Revue encyclopédique*, April 1, 1892, pp. 474–86.

49. "*Il n'y a jamais d'Art sans symbolisme.*" *Ibid.*, p. 298.

50. *Ibid.* Italics and capitals are Aurier's. "*Idéiste*" refers not to a group but to the mind itself — hence, eidetic.

51. Aurier, *Oeuvres posthumes*, p. 213.

52. *Ibid.*, p. 214, on Baudelaire's verses and "l'homme supérieur" who embodies them. Apropos Germans, Paul Sérusier adds, tongue in cheek: "Quelques-uns parlaient de Nietzsche, mais sans le connaître, et rien que pour pouvoir se croire surhommes" (*ABC de la peinture* [Paris: Librarie Floury, 1942], pp. 146–47).

53. Antonin Proust, *Édouard Manet: Souvenirs* (Paris: Librairie Renouard, 1913), p. 123. Proust's accuracy has been doubted. But the words that follow are plausibly Manetian: "Minerva is good. Venus is good. But the heroic image, the amorous image, will never prevail over the suffering image. It is the fount of humanity, it is the poem." Note the persistent equation of symbol with image. Also Manetian is the laughing disclaimer that doctors make him morbid.

54. Aurier, *Oeuvres posthumes*, p. 305, apropos Gauguin: "one could almost say that here is Plato interpreted plastically by a savage of genius." Sérusier agrees: "The movement to which we belong was anterior to the German influences. In philosophy we spoke about Plato, Aristotle, the Neo-Platonics and never about Kant." Letter to Maurice Denis, February 16, 1915, in Sérusier, *ABC de la peinture*, p. 146.

55. Other "Isolés" articles followed on the Belgian Wagnerian painter Henry de Groux, and the rather less symbolist Eugène Carrière and J.-J. Henner (*Oeuvres posthumes*, pp. 257–89). On these articles, see Geneviève Comès, "Le Mercure de France dans l'évolution des arts plastiques 1890–1895," *Revue d'Histoire Littéraire de la France* 92.1 (January–February 1992), pp. 40–55, and Erin M. Williams, "Signs of Anarchy: Aesthetics, Politics, and the Symbolist Critic at the *Mercure de France*, 1890–95," *French Forum* 29.1 (Winter 2004), pp. 45–68.

56. English translation from Linda Nochlin, ed., *Impressionism and Post-Impressionism 1874–1904* (Englewood Cliffs, NJ: Prentice Hall, 1966), p. 137. See *Oeuvres posthumes*, p. 262.

57. *Oeuvres posthumes*, pp. 262–63. The concern seems also to attach to Aurier: in his March 1892 article on Monet, he finds in that painter too a "mystic heliotheism . . . satisfied to love" (*Oeuvres posthumes*, p. 223).

58. Van Gogh's reply, his first letter to be printed, is in the *Oeuvres posthumes*, pp. 265–68. On the exchange, see Patricia Mathews, "Aurier and Van Gogh: Criticism and Response," *Art Bulletin* 68.1 (March 1986), pp. 94–104.

59. Nochlin, *Impressionism and Post-Impressionism 1874–1904*, p. 154. I have modified the translation of the last clause, "*mais j'en redoute le ridicule*," which Nochlin gave a speculative interpretation of: "*but I'm afraid of the preposterousness of it.*" Van Gogh seems more literally to have dreaded all the ridicule flung about by the partisan critics. *Oeuvres posthumes*, p. 267.

60. I have used the translation in H. R. Rookmaaker, *Synthetist Art Theories: Genesis and Nature of the Ideas on Art of Gauguin and His Circle* (Amsterdam: Swets en Zeitlinger, 1959), p. 1, except for the flowery ending: I have restored Aurier's "unhooked the stars" for Rookmaaker's "unhinged the constellations." *Oeuvres posthumes*, p. 293.

61. Letters to Theo van Gogh of February 12, 1890, and to Émile Bernard of early December 1889, cited after Nochlin, *Impressionism and Post-Impressionism 1874–1904*, pp. 152 and 149, respectively.

62. Michael Marlais, *Conservative Echoes in Fin-de-Siècle Parisian Art Criticism* (University Park: Pennsylvania State University Press, 1992), argues at length that "the radical, reactionary, Catholic movement and the avant-garde had something in common at this time, namely their hatred for the Academy and for the French scientific/positivist establishment" (p. 18; see also p. 126 on Aurier's "idealism"). See also Patricia Mathews, *Passionate Discontent: Creativity, Gender, and French Symbolist Art* (Chicago: University of Chicago Press, 1999), ch. 2, esp. p. 44, and Juliet Simpson, *Aurier, Symbolism, and the Visual Arts* (Berlin: Peter Lang, 1999).

63. Aurier, *Oeuvres posthumes*, p. 175.

64. Patricia Mathews's dissertation *Aurier's Symbolist Art Criticism and Theory* (Ann Arbor: UMI Research Press, 1986), pp. 19–23, though good on Aurier's critique of positivism, missed his admiration of mathematics (pp. 70, 74). Marlais discusses "Essai sur une nouvelle méthode de critique," and knows that Sérusier studied mathematics, yet quotes a sub-Nietzschean passage from the *Promenades philosophiques* of Remy du Gourmont, "who was close to Aurier," to the effect that "science is the need to know suffocating the need to live" (p. 71).

65. Martin Kusch, *Psychologism: A Case Study in the Sociology of Philosophical Knowledge* (London: Routledge, 1995), has revived study of the phenomenon, which Kusch believes is vindicated by modern cognitive science.

66. Mirbeau, it is true, once told Aurier himself that "the beginning of comprehension in painting is the hatred of symbolist painting!" Letter to Camille Pissarro of December 14, 1891, Octave Mirbeau, *Correspondance générale* (Lausanne: L'Age d'Homme, 2002), vol. 2, no. 962. In my "Ennemis de l'absolu?," forthcoming in *Cahiers Octave Mirbeau*, ed. Pierre Michel (2019), I argue that both Mirbeau's 1898 novel *Jardin des supplices* and Auguste Rodin's illustrations for it are nevertheless paradigmatically symbolist. On the politics of the symbolist writers, see Alain Pessin and Patrice Terrone, eds., *Littérature et anarchie* (Toulouse: Presses Universitaires du Mirail, 1998). On Mallarmé's own ties to radical politics, see Patrick McGuinness, "Mallarmé and the Poetics of Explosion," *Modern Language Notes* 124.4 (September 2009), pp. 797–824.

67. Cf. Huret and Mallarmé, "Enquête," p. 2: "Above all, [Hugo] lacked this indubitable notion: that in a society without stability, without unity, one cannot create a stable art, a definitive art. From this incomplete social organization, which explains also the disquiet of spirits, there is born the inexplicable desire for individuality, of which present literary manifestations are the direct reflection."

68. Denis expressed support for l'Action Française in his "Réponse à une enquête sur l'orientation de la peinture moderne," *La Revue du Temps Présent* (June 2, 1911), p. 569. See the "Réponse à l'enquête de Louis Dimier sur l'Art Chrétien," *Revue de l'Action Française* 23 (1912–13), pp. 260, 278–383, and Albert Marty, *L'Action française racontée par elle-même* (Paris: Nouvelles Éditions Latines, 1986), p. 247.

69. *Oeuvres posthumes*, "Paul Gauguin," p. 213. The "ciel des vérités" and "des idées pures" recurs in his sole novel, *Ailleurs* (1890). See *Oeuvres posthumes*, pp. 34, 64, respectively.

70. In his Gauguin article (*Oeuvres posthumes*, p. 205) Aurier did not cite Cousin's translation of Plato but instead that of the Catholic mystic Jean Nicolas Grou, *La République de Platon, ou dialogue sur la justice* (Paris: Humblot, 1765), vol. 2, p. 153; reprinted in Émile

Saisset, *Oeuvres completes de Platon* (Paris: Charpentier, 1873), vol. 7, pp. 340–41.

71. *Oeuvres posthumes*, p. 195. Cf. Nina Athanassoglou-Kallmyer, *Cézanne and Provence: The Painter in His Culture* (Chicago: University of Chicago Press, 2003), pp. 177–79, on the appeal of the new idealism to painters.

72. Claude Bernard, *Introduction à l'étude de la médicine expérimentale* (Paris: Baillière, 1865), p. 179. Bernard pursues the idea into the very body of the vivisected animal: "He no longer hears the animals' cries, he no longer sees the blood that flows, he sees only his idea and perceives only the organisms that hide the problems he wishes to solve" (p. 180).

73. Eder likely knew of the discovery from Viennese physicist Franz-Serafin Exner who received on January 1, 1896 an offprint of Wilhelm Conrad Röntgen's publication "Über eine neue Art von Strahlen," *Sitzungsberichte der phys.-med. Gesellschaft zu Würzburg* (1895/1896); through him the news was also printed in *Die Presse* for January 5, 1896. The earliest known x-rays by Eder, of the hand of a rachitic girl, are dated January 25, 1896. See Albertina Inv. No. FotoGLV2000/14616/29. On Eder's activities as experimenter, writer, and teacher, see Monika Faber, "Josef Maria Eder und die wissenschaftliche Fotographie 1855–1918," in *Das Auge und der Apparat: Die Fotosammlung der Albertina*, ed. Monika Wagner and Klaus Albrecht Schröder (Paris/Berlin: Seuil/Hatje Cantz, 2003), pp. 142–69.

74. Ernst Mach, *Einleitung in die Helmholtz'sche Musiktheorie* (Graz: Leuschner und Lubensky, 1866), p. 1.

75. Frege, often seen as outside this development, taught physics for most of his university career in Jena; during his studies there, his mentor was Ernst Abbe, one of the founders, with Carl Zeiss, of the Carl Zeiss Stiftung, then as now at the forefront of optical instruments. Lothar Kreiser, *Gottlob Frege: Leben — Werk — Zeit* (Hamburg: Felix Meiner, 2001), pp. 63–64, notes that Abbe and his colleague Carl Snell used in their courses the superior new Zeiss instruments.

76. Hermann Helmholtz, *Das Denken in der Medicin* (Berlin: August Hirschwald, 1877), p. 3.

77. *Ibid.*, p. 22. Medical students were treated gratis; instruments, on the other hand, they had to buy.

78. Helmholtz has anecdotes about older colleagues who found it tasteless to use a chronometer while taking a pulse, much less an ophthalmoscope, for "it is dangerous to throw harsh light into a diseased eye" (p. 20).

79. *Ibid.*, p. 15.

80. *Ibid.*, p. 31.

81. Helmholtz, "Zählen und Messen," p. 20. The essay by Leopold Kronecker in

Philosophische Aufsätze tended in the same direction, reducing irrational and imaginary numbers to psychological operations on the integers. The book as a whole, a Festschrift for Eduard Zeller, is a kind of high-water mark for mathematical psychologism. It also contains a wide-ranging review essay on "The Symbol" in aesthetics by Friedrich Vischer, editor of the volume.

82. *Ibid.*, p. 21.

83. A very readable account of these debates is Jeremy Gray, *Plato's Ghost: The Modernist Transformation of Mathematics* (Princeton: Princeton University Press, 2008). On modern and ancient efforts to make mathematics rigorous, see John Burgess, *Rigor and Structure* (Oxford: Oxford University Press, 2015).

84. Set theory, which serves as the foundation for most mathematics, began with Cantor's distinction between cardinal numbers (one, two, three: the "total" number of items in a group, irrespective of ordering) and ordinal numbers (first, second, third: their place in an ordering). The most readable introduction to set theory and its philosophy remains Bertrand Russell, *Introduction to Mathematical Philosophy* (New York: Macmillan, 1919).

85. Georg Cantor, "Mitteilungen zur Lehre vom Transfiniten," *Zeitschrift für Philosophie und philosophische Kritik* 91 (1887), pp. 88–89. The passage is quoted with approval by Frege in an otherwise critical review of Cantor: see *Zeitschrift für Philosophie und philosophische Kritik* 100 (1892), p. 272.

CHAPTER TWO: CRISES OF SENSE

1. *Les Prèmieres armes du symbolisme* (Paris: Vanier, 1889), p. 5. The year 1886 became enshrined as the birth of symbolism with a 1936 Bibliothèque Nationale exhibition commemorated in the catalogue *Cinquantenaire du symbolisme* (Paris: Éditions des Bibliothèques nationales, 1936).

2. I trace an earlier stage of this modern confrontation with the past, in eighteenth-century classicism, in my first book, *Antiquity, Theatre, and the Painting of Henry Fuseli* (Oxford: Oxford University Press, 2015).

3. Gauguin presented this second state to Mallarmé, and to Aurier (a copy inscribed "à l'ami Aurier; au Poëte," formerly in the Guardsmark Collection).

4. The success of "The Raven" led Poe reluctantly to publish his poems, many of which he felt were marred by dearth of leisure to polish them; see his preface to *The Raven and Other Poems* (New York: Wiley and Putnam, 1845).

5. On Poe as the prototype "damned poet," see Paul Verlaine, *Les Poètes maudits* (Paris: Léon Vanier, 1888), p. 53. Poe's English editor, John Henry Ingram, to whom Mallarmé was

close, fought off the more sensational libel spread by the "Memoir" appended to the 1850 edition by Poe's literary executor (and once-foe) Rufus Wilmot Griswold.

6. "Literary Notices," *The Knickerbocker, or New-York Monthly Magazine* 27 (January 1946), p. 70. Poe in fact asserts that "beauty is the sole legitimate province of the poem." "The Philosophy of Composition," *Graham's Magazine* 28.3 (March 1846), p. 164. Poe does allow ideas in poetry, but subservient to lyric form.

7. Poe, "The Philosophy of Composition," p. 165. Thomas Ollive Mabbott, in *Collected Works of Edgar Allan Poe*, vol. 1, *Poems* (Cambridge, MA: Belknap Press, 1969), pp. 353–54, discusses possible sources featuring a parrot or owl.

8. Gauguin to Émile Gauguin, October or November 1888, in *Correspondance de Paul Gauguin: Documents; Témoignages*, ed. Victor Merlhès, vol. 1 (Paris: Fondation Singer-Polignac, 1984), p. 270; Dario Gamboni, "The *Noa Noa Suite*: 'Veiled in a Cloud of Fragrance,'" in *Gauguin Paintings, Sculpture, and Graphic Works at the Art Institute of Chicago*, ed. Gloria Groom and Genevieve Westerby (Chicago: Art Institute of Chicago, 2016), paragraph 21 (cat. no. 51–60), links this advice with Manet's use of shadow in his *Raven*.

9. Stéphane Mallarmé, *Les Poèmes d'Edgar Poe* (Paris: Léon Vanier, 1889), p. 140.

10. Letter to Verlaine, November 16, 1885, in Stéphane Mallarmé, *Autobiographie: Lettre à Verlaine* (Paris: Échoppe, 1991), p. 13. On Baudelaire's creative use of Poe, see Françoise Meltzer, *Seeing Double: Baudelaire's Modernity* (Chicago: University of Chicago Press, 2011), pp. 98–102.

11. Alphonse Lemerre, editor of *Le Parnasse Contemporain*, rejected Mallarmé's "Après-midi d'un faune" later the same year, ending their relationship. Though he made excuses to Mallarmé in terms of his collection of textbooks, which would be compromised by including the Poe (letter of March 11, 1875), he wrote to Manet four days later that "le poëme de Mallarmé... offre de telles insanités qu'il est impossible à une maison sérieuse de le publier." See the invaluable collection *Edgar Poe Le Corbeau — Dossier réalisé par Michael Pakenham* (Paris: Séguier, 1994), pp. 12–13.

12. A good survey of the debate is Juliet Wilson-Bareau and Breon Mitchell, "Tales of a Raven: The Origins and Fate of *Le Corbeau* by Mallarmé and Manet," *Print Quarterly* 6.3 (September 1989), pp. 258–307. Pakenham's dossier, with its repeated references to lithography by Lesclide and his printers, has closed the case.

13. Arsène Houssaye, "Life in Paris: Letter from Arsene Houssaye," *New York Tribune*, August 9, 1875, p. 2 (Pakenham dossier, p. 60).

14. Thus Georges Mayrant in *Le Gaulois* (June 9, 1875), the unsigned note (possibly by Castagnary) in *Le Siècle* (June 13), Jules Claretie in *L'Indépendence Belge* (June 14), the note

in *La Liberté* (July 10), Gygès (*Paris-Journal* and *La Chronique des Arts et de la Curiosité*, both July 17), and Houssaye (see above). Philippe Burty, print connoisseur that he was, got the medium right: "five violent lithographs by M. Édouard Manet" (*The Academy*, July 24). The fifth is the ex libris of the bird with open wings.

15. The mistake about etchings was abetted by the fact that Lesclide printed the book under his Librairie de l'Eau-Forte imprint and that he had in fact printed a book with etchings by Manet, Charles Cros's *Le Fleuve*, a year earlier.

16. *L'Artiste* 5.10 (1 March 1853), p. 1. The first volume of tales, and critical articles in French, followed: see Célestin Pierre Cambiaire, *The Influence of Edgar Allan Poe in France* (New York: Stechert & Co., 1937), pp. 34–45.

17. The set contained nine scenes, but two were printed together on the last sheet. See Françoise Cachin, Charles S. Moffett, and Juliet Wilson-Bareau, *Manet 1832–1883* (New York: Metropolitan Museum of Art, 1983), p. 59 (nos. 7–9); and Carol Armstrong, *Manet/Manette* (Princeton: Princeton University Press, 2002), pp. 71–98.

18. Because there were rumors that Bracquemond etched the *Olympia*, art historians have overreacted in disclaiming influence. Thus Léon Rosenthal, in his *Manet aquafortiste et lithographe* (Paris: Le Goupy, 1925), p. 45, noting that Bracquemond added aquatint to Manet's *Torero mort*, asks rhetorically: "Dans l'œuvre si varié de Bracquemond peut-on montrer une seule page dont le métier, dont l'allure se rapprochent d'une eau-forte de Manet?"

19. On the source of the motto in Poulet-Malassis's letters to Bracquemond, and the probable date of the print, see Jean-Paul Bouillon, "Manet vu par Bracquemond," *La Revue de l'Art* 27 (1975), pp. 37–45.

20. Ewa Lajer-Burcharth, in "Modernity and the Condition of Disguise: Manet's *Absinthe Drinker*," *Art Journal* 45.1 (Spring 1985), pp. 18–26, argues that the sitter's anonymity, typical of the modern metropolis, infects also Manet's manner of painting him. If so, the third state of the etching takes this further, by other means.

21. Mayrant thought "he found a way to speak American in French, that is to say a French more Poe-like than Poe himself," a language "closer to American than to English, technical, practically telegraphic." Pakenham dossier, p. 40.

22. Poe himself was very interested in vignettes and their dreamlike effects: see the story "The Oval Portrait," in Thomas Ollive Mabbott, ed., *Collected Works: Tales and Sketches*, vol. 1, pp. 659–67. On Mallarmé's translations, see Dominique Jullien, "Translation as Illustration: The Visual Paradigm in Mallarmé's Translations of Poe," *Word & Image* 30.3 (September 2014), pp. 249–60. See also Haskell M. Block, "Poe, Baudelaire, Mallarmé and the Problem of the Untranslatable," in *Translation Perspectives: Selected Papers, 1982–1983,*

ed. Marilyn Gaddis Rose (Binghamton: SUNY-Binghamton, 1984), pp. 104–11.

23. In *Les Poèmes d'Edgar Poe*, only Mallarmé's "Le Tombeau d'Edgar Poe," which opens the volume, is in verse.

24. There is a letter to Nadar apologizing for having failed to sit for a photograph before, and asking to have one made for Whitman. See Mallarmé, *Correspondance*, vol. II, 1871–1885, ed. H. Mondor and J. P. Richard (Paris: Gallimard, 1965), p. 145. See also the letter to Whitman of March 31, 1877 (p. 149), where he complains of the resulting photograph.

25. Mallarmé, *Correspondance*, vol. XI, p. 120. Whitman added, and crossed out: "Don't tell M. Manet what I say about it." Caroline Ticknor gets closest to Whitman's reaction: "Manet's illustrations, done, it has been said, in his most 'intimidating' style, are startling productions. A glance at some of them reveals strange blotches of black ink apparently without form or meaning, but presently what has seemed merely a splash of ink proves to be a grotesque vision which takes hold on the imagination with a haunting persistency." *Poe's Helen* (New York: Scribner's, 1916), p. 272.

26. Rossetti adds: "A copy shd [*sic*] be bought for every hypochondriacal ward in Lunatic Asylums. To view it without a guffaw is impossible." *D. G. Rossetti and Jane Morris: Their correspondence*, ed. John Bryson and Janet Camp Troxell (Oxford: Clarendon, 1976), p. 174. Rossetti's copy had belonged to Arthur O'Shaughnessy, who had praised Mallarmé in *The Athenaeum* and had helped revise the English translation of his essay on Manet, discussed in the fourth chapter of this book (*Correspondance*, vol. II, p. 129).

27. Juliet Wilson-Bareau, in *Manet 1832–1883*, p. 385. Most art historians working on Manet more broadly have neglected these works: Melot does not even mention them in his essay on "Manet and the Print" in *Manet 1832–1883* (pp. 36–40). A notable exception, besides Léon Rosenthal (pp. 97–103), is James H. Rubin, *Manet's Silence and the Poetics of Bouquets* (Cambridge, MA: Harvard University Press, 1994), pp. 144–49.

28. "Literature, Science, Art and the Drama," *Civil Service Review* (June 26, 1875), reprinted in the Pakenham dossier, p. 47.

29. *Edgar Poe and His Critics* (New York: Rudd & Carleton, 1860). The book is not uncritical: Whitman takes Poe to task for his presumed atheism in *Eureka* (pp. 65–68), but this too is done with great nuance and sympathy.

30. Mallarmé, *Correspondance*, vol. II, p. 149.

31. I cannot endorse Jay Fisher's observation that "Manet departed slightly from the text" in the second print by having the "shutters open inward." Fisher, *The Prints of Édouard Manet* (Washington, DC: International Exhibits Foundation, 1985), p. 115. Manet has not drawn *shutters* at all, but windows, which quite routinely open inward. This meets David

Van Zanten's objection that "shutters **NEVER** open inward — so that Manet has them do so for some reason of meaning or composition" (Hollis Clayson, "Looking Within the Cell of Privacy," *The Darker Side of Light*, p. 161 n. 58), though I agree with Clayson that, qua composition, the open window "shifts the emphasis somewhat by staging the sleek black bird's arrival as a clear, even dramatic interruption or invasion of [Poe's poetic] space" (p. 79).

32. Wilson-Bareau and Mitchell give as titles "Once upon a Midnight Dreary," "Open here I Flung the Shutter," "Perched upon a Bust of Pallas," "That Shadow that Lies Floating on the Floor."

33. See the memorable description of this print in Étienne Moreau-Nélaton, *Manet raconté par lui-même* (Paris: Henri Laurens, 1926), vol. II, pp. 26–27: "l'apparition, enfin, dans la chambre où la lampe la projette, de la grande ombre noire qui y règne désormais en maîtresse."

34. It is interesting in this connection to note that Mallarmé regards the craftsmanship of "The Philosophy of Composition" as essentially dramatic in nature. *Poèmes d'Edgar Poe*, p. 139.

35. Poe himself was passionate about the aesthetics of furniture, professing an austere taste (for his time). See "The Philosophy of Furniture" (1840) in Mabbott, *The Collected Works: Tales and Sketches*, vol. 2, pp. 494–504.

36. Mallarmé rests the poet's head on a *housse* (slipcover) atop a *chaise à coussins*, Poe's cushioned seat.

37. *The Works of the Late Edgar Allan Poe* (New York: Redfield, 1850), "Marginalia: XVI," vol. III, pp. 494–96. The "Marginalia," a rubric of Poe's, have been renumbered and arranged in a bewildering variety of ways since 1850.

38. *Graham's Magazine* 28.3 (March 1846), p. 117 reprinted in Poe, *The Literati* (*Works*, vol. 3), ed. R. W. Griswold (New York: Redfield, 1850), pp. 494-96 and *The Works of Edgar Allan Poe*, ed. John H. Ingram, vol. 3 (Edinburgh: Charles and Adam Black, 1875), pp. 379-82, under the heading "Expression." Cf. Mallarmé, *Poèmes d'Edgar Poe*, p. 156.

39. "Marginalia," *Graham's Magazine*, p. 117.

40. "I have seen a person in a discussion on this subject strike himself on the breast and say: 'But surely another person can't have **THIS** pain!' — The answer to this is that one does not define a criterion of identity by emphatic stressing of the word 'this.' Rather, what the emphasis does is to suggest the case in which we are conversant with such a criterion of identity, but have to be reminded of it." Ludwig Wittgenstein, *Philosophical Investigations*, ed. G. E. M. Anscombe (Oxford: Basil Blackwell, 1986), p. 91e (§253). The argument is circular, for the emphatic *this* may be taken by others to state a criterion of identity available

to the speaker. Wittgenstein touches on this reply in his discussion of a clock by which one might record the occurrence of sensations, but he was stuck on the idea that any criterion must be shared, which he must argue, rather than assume, in order to show that there is no private language.

41. Gottlob Frege, *Die Grundlagen der Arithmetik* (Breslau: Koebner, 1884), pp. 35–36. There are in Wittgenstein's *Philosophical Investigations* apparently divergent statements of the concept: "the language which describes my inner experiences and which only I myself can understand" (p. 91e, §256) and "the individual words of this language are to refer to what can only be known to the person speaking; to his immediate private sensations. So that another person cannot understand the language" (p. 89e, §243). Are "inner experiences" or "private sensations" "described" or "referred to" in private language? This lack of clarity is typical, as critics like A. J. Ayer and Judith Jarvis noticed.

42. Gottlob Frege, "Der Gedanke," trans. Anthony and Marcelle Quinton as "The Thought: A Logical Inquiry," *Mind* 65.259 (July 1956), pp. 289–311, at p. 298. The essay has been retranslated often, sometimes under the (equally appropriate) title "Thoughts," sometimes under the (for Frege misleading) singular word "Thought."

43. Peter Geach, in *Mental Acts* (London: Routledge and Kegan Paul, 1957), pp. 117–21, declares even Descartes's *cogito* invalid because words are used to communicate and are thus useless for assuring oneself of one's own existence in a context wherein the existence of others is in doubt. Cf. his introduction to Frege, *Logical Investigations* (Oxford: Basil Blackwell, 1977), p. viii, which construes Frege as anticipating, but not fully grasping, Wittgenstein's argument.

44. This way of reading Frege is due to Saul Kripke, whose influential reading of Wittgenstein presents the private language argument as part of the larger problem as to how one may objectively follow a rule (the difficulty being *private* rule following). See Saul A. Kripke, "Frege's Theory of Sense and Reference: Some Exegetical Notes," *Theoria* 74.3 (September 2008), pp. 181–218, and for a position closer to Frege's, Guy Longworth, "Sharing Thoughts about Oneself," *Proceedings of the Aristotelian Society* 103 (2013), pp. 57–81.

45. Especially the lamp-lit first scene, which makes the figure's hair look fairer than Poe's, has been treated as a portrait of Mallarmé, notably in the 1991 Échoppe edition of the autobiographical *Lettre à Verlaine*.

46. Poe, "Philosophy of Composition," p. 166.

47. *Ibid.*, p. 167. Ségolène Le Men, "Manet et Doré: L'illustration du *Corbeau* de Poe," *Nouvelles de l'Estampe* 78 (December 1984), pp. 4–20, contrasts Manet in this respect with

the later version of Gustave Doré, who turns even "the Night's Plutonian shore" into a phantasmagoria of a woman floating over the waves.

48. Edgar Allan Poe, "The Philosophy of Composition," *Graham's Magazine* 28.4 (April 1846), p.167.

49. Juliet Wilson-Bareau, in *Manet 1832–1883* (1983), pp. 382–83.

50. Some did treasure the performance: Charles Algernon Swinburne wrote Manet of "ces pages merveilleuses où le premier poète américain se trouve deux fois si parfaitement traduit." Mallarmé, *Correspondance*, vol. XI, p. 91.

51. Frege, *Grundlagen der Arithmetik*, pp. 35–36. Frege is relying here on the nineteenth-century discovery of projective duality by Poncelet, Gergonne, and Plücker, a principle he applied at length, and creatively, in his dissertation, *Über eine geometrische Darstellung der imaginären Gebilde in der Ebene* (Jena: A. Neuenhahn, 1873).

52. On the reception of the *Caprichos* and its relevance to questions of subjectivity and meaning, see Andrei Pop, "Goya and the Paradox of Tolerance," *Critical Inquiry* 44 (Winter 2018), pp. 242–74.

53. Besides the delays and expenses incurred by the artists, and Lesclide's distraction by other projects, ill luck was involved: English and American interest did not translate into substantial sales or a new print run, and prominent supporters (Victor Hugo, for whom Manet had printed a sumptuous and expensive one-off copy on Japan paper; but also Hoschedé and Swinburne) did not write publicly in support of the project. To this must be added the work's initial difficulty and startling qualities, remarked on by virtually *all* critics writing on it, positive as many of them were. Lesclide lost so much money that a second Poe volume, *The City in the Sea*, was scrapped; his Librairie de l'Eau-Forte folded in 1876.

54. Apropos the poem's identification as an eclogue, as opposed to the freer "improvisation" (Mallarmé's first subtitle), see David J. Code, "The Formal Rhythms of Mallarmé's Faun," *Representations* 86.1 (Spring 2004), pp. 73–119, which argues that the structure of the text is very finely calibrated.

55. A similar ambiguity attends the nymphs, whose illustration goes inexplicably unmentioned on the title page, in strange contrast to the cursory *cul-de-lampe*, unless, once again, they are meant themselves to serve as fleurons.

56. Rosenthal adds: "avec une crainte d'appuyer tout à fait exceptionnelle, délicates et subtiles" (*Manet aquafortiste et lithographe*, p. 103). He may be parrying Béraldi's irritation that the prints "sont si peu faites qu'elles peuvent se regarder à l'envers, comme les vers se lire à rebours" (*Les Graveurs du xixe siècle*, vol. IX, p. 211). This needn't be a fatal objection: Horne in his review had said that the *Corbeau* illustrations are just as impressive turned upside down.

57. Too bad his maquette for *Le Corbeau*, which gave Lesclide's printer Lévy-Alcan so much trouble, has been lost.

58. On Mallarmé and Redon, see Léon Cellier, "Mallarmé, Redon, et Un coup de dés," *Cahiers de l'Association internationale des études françaises* 27.2 (1975), pp. 363–75. Cellier points out that the original title for the poem had a different word order, *Jamais un coup de dés n'abolira le hasard*. For an art-historical treatment of this collaboration, see Penny Florence, *Mallarmé, Manet, and Redon: Visual and Aural Signs and the Generation of Meaning* (Cambridge: Cambridge University Press, 1986), pp. 84–88.

59. Rosalind Krauss, "The Motivation of the Sign," in *Picasso and Braque: A Symposium*, ed. William Rubin (New York: Museum of Modern Art, 1992), pp. 261–87, and see esp. the discussion that followed (pp. 288–305). In particular, Christine Poggi questioned what she saw as an equation of Mallarmé and Picasso (p. 293), upon which Krauss backed off somewhat, insisting, "I wasn't really trying to line Picasso up with Mallarmé, or claiming a complete overlap in their thinking" (p. 294). Also interesting is Edward Fry's suggestion (p. 304) that a "permanent, absolute, latent truth" about art and language was discovered by Krauss's protagonists. Florence (p. 70 and passim), like Krauss, takes Julia Kristeva and her idiosyncratic understanding of symbolist poetry and structuralist linguistics as a key to Mallarmé.

60. Michèle Hannoosh, "From *Nevermore* to Eternity: Manet, Mallarmé and the Raven," in *Livres d'Artistes 1874–1999: The Dialogue Between Painting and Poetry*, ed. Jean Khalfa (Cambridge: Black Apollo Press, 2000), pp. 37–57, does not quite go so far, but sees in the work of Poe, Manet, Mallarmé, and Baudelaire modernist concerns with "negation, nothingness, chance" and their aesthetic productivity (p. 50). Accordingly, *Un coup* anchors the discussion.

61. T. J. Clark, *Picasso and Truth* (Princeton: Princeton University Press, 2013), p. 150. Krauss does not rely merely on a poststructuralist reading of Saussure; her *Passages in Modern Sculpture* (New York: Viking, 1977), already cited Wittgenstein's critique of private language in discussing the modernist rejection of heavy sculptural bodies (with their presumed "interiority") (p. 261).

62. On Mallarmé's many pictorial metaphors, see Gayle Zachmann, "Developing Movements: Mallarmé, Manet, the 'Photo' and the 'Graphic,'" *French Forum* 22.2 (May 1997), pp. 181–202.

63. See the editorial note to the author's preface, Stéphane Mallarmé, *Un coup de dés jamais n'abolira le hasard* (Paris: NRF, 1914), n.p.: "L'innovation principale établie par lui dans ce dernier 'état' de son oeuvre, pour reprendre le terme dont il se servit, nous semble consister en ceci qu'il n'existe pas de page recto ou verso, mai que la lecture se fait sur les

deux pages à la fois, en tenant compte simplement de la descente ordinaire des lignes." There remains *this* linguistic element, that the two-page landscape must be scanned from top to bottom per conventional lineation.

64. An edition reproducing the original spacing and font was published by Michel Pierson (Paris: Ptyx, 2002).

65. I should add that, in investigating this kind of misunderstanding, modern artworks, from Kafka's *Das Schloß* to Tarkovsky's *Stalker* and beyond, often rehearse symbolist themes and techniques.

66. Gottlob Frege, "On Sense and Reference," trans. Peter Geach and Max Black, *Mind* 57.3 (May 1948), pp. 209–30, at p. 209. I have added an "else" for clarity, though as a matter of fact a sign might stand for itself; but then it would not function as a sign as we conventionally understand the word so much as an ordinary object.

67. Rosemary Lloyd, "Mallarmé at the Millennium," *Modern Language Review* 95.3 (July 2000), p. 681.

68. But Chad Engelland's *Ostension: Word Learning and the Embodied Mind* (Cambridge, MA: MIT Press, 2014) interestingly defends Augustinianism (what he calls "the eavesdropping model") as the actual way we learn to speak.

69. I exclude as beyond the scope of this study the complex, at times whimsical and mystical, at other times pragmatic and even reductively sociological, study of the English language Mallarmé published as *Les Mots anglais* (Paris: Truchy, 1877). See Antonin Wiser, "D'un déplacement avantageux: *Les Mots anglais* de Mallarmé," *Littérature* 157 (March 2010), pp. 3–16, and the chapter in Mireille Ruppli and Sylvie Thorel-Cailleteau, *Mallarmé: La Grammaire & Le Grimoire* (Geneva: Droz, 2005), pp. 93–112.

70. Frege may have thought that, though all thoughts be public, there are different ways of dividing some into constituent senses, some of which may be private. In that case, that "He is in pain" can sometimes convey the same thought as "I am in pain" need not involve us in what Peter Geach calls the "cancelling-out fallacy" that "I" and "he" have the same sense. One way Frege might untangle this thicket would be to say that a unique way of being given to oneself is not the sense of the word, but a way of attaining that sense; the same sense may be attained by the public formula "the speaker or writer or thinker of this sentence." The means of achieving sense in the first case remain private: "Someone can have sympathy for me but my pain always belongs to me and his sympathy to him. He does not have my pain, and I do not have his sympathy" ("The Thought," p. 300).

71. See Alexandre Billon, "Why Are We Certain That We Exist?" *Philosophy and Phenomenological Research* 91.3 (November 2015), pp. 723–59; and in a more traditional mode,

Béatrice Longuenesse, *I, Me, Mine: Back to Kant, and Back Again* (Oxford: Oxford University Press, 2017).

CHAPTER THREE: WHERE DO WE COME FROM?

1. Vincent van Gogh to Willemien van Gogh, Arles, Sunday, September 9, and around Friday, September 14, 1888, Letter 678; Br. 1990: 681 | CL: W7, Amsterdam, Van Gogh Museum, inv. nos. b707 a–b V/1962.

2. Vincent van Gogh to Theo van Gogh, Arles, Friday, September 21, 1888, Letter 685; Br. 1990: 689 | CL: 540, Amsterdam, Van Gogh Museum, inv. no. b584 a V/1962.

3. Vincent van Gogh to Theo van Gogh, Arles, Tuesday, October 9 or Wednesday, October 10, 1888, Letter 685; Br. 1990: 705 | CL: 548, Amsterdam, Van Gogh Museum, inv. no. b592 V/1962.

4. Vincent van Gogh to Theo van Gogh, Arles, Monday, October 8, 1888, Letter 699; Br. 1990: 704 | CL: 547. Amsterdam, Van Gogh Museum, inv. nos. b591 V/1962 and b590 a–b V/1962.

5. Another text, available on the museum's homepage, attempts a technical explanation of the remarkable coloring: "This has partly to do with the specific pigments Van Gogh used (his reds have faded with time), but the artist also confessed his dissatisfaction with the work, writing to his sister, 'I don't like Mother's picture enormously' " (https://www.nortonsimon.org/collections/browse_title.php?id=M.1968.32.P; accessed September 29, 2015). The invocation of red paint seems a desperate expedient, given the reddish skin of Patience Escalier in a painting which hangs nearby in the museum. As for the letter to Willemien (of November 12, 1888, #720), both the context and the editors of the letters make it clear that it is the *photograph* that Vincent didn't like, not his own picture. See also Geoffrey Batchen, *Forget Me Not: Photography and Remembrance* (Princeton: Princeton Architectural Press, 2004).

6. Frege, "Der Gedanke," trans. Anthony and Marcelle Quinton as "The Thought: A Logical Inquiry," *Mind* 65.259 (July 1956), p. 299.

7. Wittgensteinians can make this argument their own: "such undetectable disagreement appears intelligible only if we already assume that the meanings of colour-terms are determined by their alleged application in private, phenomenal space." Michael Hymers, *Wittgenstein on Sensation and Perception* (London: Routledge, 2017), p. 169. Unfortunately, the Wittgensteinian solution favored by Hymers, which has it that color reports, like utterances of pain, are expressive rather than informative, can certainly not explain the divergence over the strawberry.

8. Maurice Denis, under the pseudonym Pierre Louis, "Notes d'art: définition du

néo-traditionnisme," *Art et Critique* 2.65 (23 August 1890), p. 540. English translation from Linda Nochlin, *Impressionism and Post-Impressionism* (Englewood Cliffs, NJ: Prentice Hall, 1966), p. 187. Nochlin observes that the manifesto "might more aptly have been entitled 'A Definition of Symbolism'" (p. 186).

9. Cf. Paul-Henry Kahnweiler: "Faithful to the teaching of Manet and Mallarmé, the cubist painter painted with 'oils and colors.' On their canvases they juxtaposed, not bottles and trees, but colored forms." "Mallarmé and Painting," in *From Baudelaire to Surrealism*, ed. Marcel Raymond (New York: Wittenborn, Schultz, Inc., 1950), pp. 359–63.

10. Paul Sérusier, *ABC de la peinture* (Paris: Librarie Floury, 1950), p. 7, in Nochlin, *Impressionism and Post-Impressionism*, p. 184. Though not printed until 1921, the *ABC* recalls a correspondence with Denis in 1890–91. Sérusier continues: "All these coefficients have acted upon the sensation to the point of transforming it into an image that we call a *mental image*" (*ABC*, p. 9; Nochlin, *Impressionism and Post-Impressionism*, p. 185).

11. Nochlin, *Impressionism and Post-Impressionism*, p. 187.

12. Alphonse Allais, *Album primo-avrilesque* (Paris: P. Ollendorff, 1897), n.p.

13. On Girodet's *Endymion: Effet de lune* (1791, Louvre), see Thomas Crow, *Emulation: David, Drouais, and Girodet in the Art of Revolutionary France* (New Haven: Yale University Press, 2006), pp. 136–37.

14. "Les Arts Incohérents, exposition du 15 Octobre au 15 Novembre" *La Libre Revue*, November 1, 1883, in Félix Fénéon, *Oeuvres plus que complètes*, ed. Joan U. Halperin (Geneva: Droz, 1970), vol. 1, p. 12. Fénéon also plays rhetorically with the exhibition's color sense in describing, with tongue in cheek, a scandalous work attributed to a famous courtesan: "Mlle Valtesse de la Bigne shows *Coherent lizards*. These two saurians are in a position that my pen cannot describe without taking on a chaste vermillion." Corinne Taunay, "Les incohérents," in *Impressionnisme et littérature*, ed. Gérard Gengembre, Yvan Leclerc, Florence Naugrette (Mont-Saint-Aignan: Presses Universitaires de Rouen et du Havre, 2012), pp. 213–24, stresses the avant-garde impact of the "incohérents": Manet, Renoir, and Pissarro visited (p. 215) and incohérent Jules Chéret exhibited in Le Barc de Boutteville's "Impressionistes et symbolistes" (p. 224).

15. In *Le Grelot au Salon*, reproduced in Françoise Cachin, Charles S. Moffett, and Juliet Wilson-Bareau, *Manet 1832–1883* (New York: Metropolitan Museum of Art, 1983), p. 219.

16. On Bilhaud, see *The Spirit of Montmartre: Cabarets, Humor, and the Avant-Garde, 1875–1905*, ed. Philip Dennis Cate and Mary Lewis Shaw (New Brunswick, NJ: Jane Voorhees Zimmerli Art Museum, 1996), p. 31. The role of racist humor and blackface in modernist experimentation has begun to be studied: see Angela Rosenthal, Adrian Randolph, and

David Bindman, eds., *No Laughing Matter: Visual Humor in Ideas of Race, Nationality, and Ethnicity* (Lebanon, NH: University Press of New England, 2016). But even such studies are thin on late nineteenth-century phenomena like Bilhaud or Arthur Burdett Frost's 1884 proto-comic book *Stuff and Nonsense*.

17. See the *Catalogue illustré de l'exposition des arts incohérents* (Paris: É. Bernard et cie, 1884), p. 7: "*La Nuit de Noces du brave charbonnier (grande composition tragico-comique)* (no. 53). Allais was represented by the red *Récolte de la tomate* (no. 3), and the proto-Cagean *Les grandes douleurs sont muettes. — Marche funèbre incohérente* (no. 5).

18. Sérusier's 1889 letter to Denis from Le Pouldu claims skill "shouldn't be bothered with," or "it should even be repressed." The critique of skill is not, however, a matter of eliminating subjectivity: Sérusier hopes on the contrary that "personality" will shine all the brighter for lack of skill, for as in handwriting, "if one pays no special attention to it, it will become so much more the more personal as it is maladroit." Nochlin, *Impressionism and Post-Impressionism*, p. 183.

19. Paul Smith, "The Neo-Impressionist Painter: Color, Facture, and Fiction," in *Neo-Impressionism and the Dream of Realities: Painting, Poetry, Music*, ed. Cornelia Homburg (New Haven: Yale University Press, 2015), p. 67. See also Christopher Riopelle's essay in the catalogue, as well as Arthur Danto, *Unnatural Wonders* (New York: Columbia University Press, 2005), p. 251, which insists on a distinction between Allais and the serious monochromes of the early twentieth century.

20. See Ivan Nechepurenko, "Examination Reveals a Mysterious Message on Malevich's 'Black Square' Painting," *New York Times*, November 18, 2015 (https://artsbeat.blogs.nytimes.com/2015/11/18/examination-reveals-a-mysterious-message-on-malevichs-black-square-painting; accessed 11 August 2017). Nechepurenko cites Konstantin Akinsha, a Malevich scholar, as asserting that "there is no doubt" Allais was Malevich's source—at least for the final state of the painting. A polychrome geometric painting was also discovered underneath. Noam Elcott, *Artificial Darkness: An Obscure History of Modern Art and Media* (Chicago: University of Chicago Press, 2016), p. 3, argues that the black monochrome is unique in negating the image itself; but the same could be said of Allais's snow image.

21. "Celebrities at Home: Mr. James Whistler at Cheyne Walk," *The World*, May 22, 1878, reprinted in Whistler, *The Gentle Art of Making Enemies*, ed. Sheridan Ford (Paris: Delabrosse, 1890), pp. 68–70; the text is reprinted, with various small revisions, in Whistler's own edition of *The Gentle Art of Making Enemies* (London: Heinemann, 1892), pp. 126–28.

22. Personal communication, August 23, 2016. I have not found the term in his works. Fredric Jameson, in *The Political Unconscious: Narrative as a Socially Symbolic Act* (Ithaca,

NY: Cornell University Press, 1981), p. 106, views genres as "social institutions, or social contracts between a writer and specific public, whose function it is to specify the proper use of a particular cultural artifact." This is suggestive, but too rigid to explain either genre or individual artwork; the point is that one can expect to *find* some shared sense in an artwork, not that it is prescribed in advance.

23. Whistler, by contrast, retitled his piece "The Red Rag" in his own edition of *The Gentle Art of Making Enemies*.

24. Ironically, the key B-sharp major is sonically indistinguishable from C major (the two keys are enharmonic), which may be part of the joke. This title was inscribed (not by Whistler) on the back of a Whistler canvas. See "Titles a Key to My Work," *The Athenaeum* (November 22, 1873), reprinted in the Ford edition of *The Gentle Art*, p. 55, concerned especially with combatting the association of his notorious 1861 *Girl in White* with Wilkie Collins's popular 1859 novel *The Woman in White*.

25. Hermann Hankel, *Vorlesungen über die Complexen Zahlen und ihre Functionen* (Leipzig: Leopold Voss, 1867), vol. 1, p. 10, under the heading "Principle of the Permanence of Formal Laws."

26. Frege, "Die Unmöglichkeit der Thomaeschen formale Arithmetik aufs Neue nachgewiesen," *Jahresbericht der Deutschen Mathematiker-Vereinigung* 17 (1908), pp. 52–55, at p. 55 n. 1.

27. Charles Baudelaire, "L'Oeuvre et la vie d'Eugène Delacroix," in *L'Art romantique* (Paris: Michel Lévy, 1868), p. 6.

28. *Ibid.*, p. 19.

29. *Ibid.*, p. 5.

30. E. T. A. Hoffmann, "The Sandman," *Tales of Hoffmann,* ed. and trans. Leonard J. Kent and Elizabeth C. Knight (Chicago: University of Chicago Press, 1972), p. 104. The original is stormier: "Hast du, Geneigtester! wohl jemals etwas erlebt, das Deine Brust, Sinn und Gedanken ganz und gar erfüllte, Alles Andere daraus verdrängend? Es gährte und kochte in Dir, zur siedenden Gluth entzündet sprang das Blut durch die Adern und färbte höher Deine Wangen... und nun wolltest Du das innere Gebilde mit allen glühenden Farben und Schatten und Lichtern aussprechen und mühtest Dich ab, Worte zu finden, um nur anzufangen." *Nachtstücke* (Berlin: Realschulbuchhandlung, 1817), vol. 1, p. 31.

31. Hoffmann was curiously identified with visual art by nineteenth-century German critics, who routinely called him "Callot-Hoffmann" in reference to his first work, the anonymous *Fantasiestücke in Callot's Manier* (Bamberg: Kunz, 1814). See, e.g., Wolfgang Menzel, *Deutsche Dichtung von der ältesten bis auf die neueste Zeit* (Stuttgart: Adolph Krabbe, 1859), vol. 3, pp. 359–60. Reinhold Grimm, "From Callot to Butor: E. T. A. Hoffmann and

the Tradition of the Capriccio," *MLN* 93.3 (April 1978), p. 404, cites a September 1813 letter to Hoffmann's publisher emphasizing that the author had "thought deeply" about the titular phrase *in Callot's Manner.*

32. Michael Fried, *Courbet's Realism* (Chicago: University of Chicago Press, 1990), p. 61.

33. *Ibid.,* p. 62.

34. In this sense, Poe's speculation about his poem's "effects" in the "Philosophy of Composition" are psychologistic, which is remarkable, since in his futuristic story-essay "Mellonta Tauta" (reprised in *Eureka*) he takes John Stuart Mill to task for a psychologistic inconsistency, namely for violating his own dictum that "ability or inability to conceive is in no case to be received as a criterion of axiomatic truth."

35. Another promise that was also a threat; as we have known since Jonathan Crary's *Techniques of the Observer* (Cambridge, MA: MIT Press, 1990), new visual technologies were attended by intrusive regimentation of human bodies.

36. T. J. Clark, "Preface," *The Painting of Modern Life*, rev. ed. (Princeton: Princeton University Press, 1999), p. xxx.

37. The most thorough reading of this psychologization of late nineteenth-century French art is Jonathan Crary, *Suspensions of Perception* (Cambridge, MA: MIT Press, 1999). Particularly the chapter on Paul Cézanne may be read with profit together with this chapter.

38. This study cannot unfortunately chart the fluctuating fortunes of Kant and his "Copernican revolution" through the nineteenth century, but suffice it to say here that the "return to Kant" went hand in hand with psychologism, as well as with various efforts by thinkers who admired Kant (Bolzano, Frege, Cantor) to evade psychologism.

39. Thomas Hardy, preface to *Jude the Obscure* (New York: Harper and Brothers, 1896), p. iv.

40. Alois Riegl, "Objective Aesthetik," *Neue Freie Presse* (13 July 1902), p. 34.

41. In "Die Stimmung als Inhalt der modernen Kunst," *Die Graphischen Künste* 22.2 (1899), pp. 47–56, Riegl discusses with sympathy the subjectivization of art, which may reassure psychically oppressed modern subjects.

42. Arthur Parsey, *Perspective Rectified; or, The Principles and Application Demonstrated* (London: Longman, Rees, Orme, Brown, Green, and Longman, 1836), p. 82.

43. *Ibid.,* p. 79.

44. *The Athenaeum* 1899 (March 19, 1864), p. 409, reprinted in Augustus De Morgan, *A Budget of Paradoxes* (London: Longmans, 1872), p. 177. De Morgan's criticism of Parsey is interesting: "What is called the picture is not the picture from which the mind takes its

perception; that picture is on the retina. The *intermediate* picture, as it may be called — the human artist's work — is itself seen perspectively. If the tower were so high that the sides, though parallel, appeared to meet in a point, the picture must also be so high that the *picture-sides*, though parallel, would appear to meet in a point." Cf. Nelson Goodman, *Languages of Art* (Oxford: Oxford University Press, 1968), p. 17.

45. *The Westminster Review* LXXI (October 1841), pp. 228–29. Lewes signed his reviews "L": see Edgar W. Hirshberg, *George Henry Lewes* (New York: Twayne Publishers, 1970), p. 219. Lewes, like De Morgan and Aaron Scharf after him, is under the mistaken impression that Parsey had already defended his new system in the 1836 book.

46. *The Works of John Ruskin: Early Prose Writings, 1834 to 1843*, ed. E. T. Cook and Alexander Wedderburn (London: George Allen, 1903), pp. 215–34. W. G. Collingwood, the father of the philosopher, in his *Life of John Ruskin* (Cambridge: Riverside Press, 1900), pp. 62–64, recounts the controversy, adding, if possible, to the confusion by observing that "if you draw what you really *see*, you would draw the top of a tower *greater* than its base, owing to the structure of the lens of the eye, as discussed, with curious experiment and improved knowledge of optics, by P. H. Emerson and Mr. Goodall in a recent tract."

47. Ruskin, "Remarks on the Convergence of Perpendiculars," *Architectural Magazine* 5.48 (February 1838), pp. 94–96, reprinted in *Early Prose Writings*, p. 216.

48. Parsey criticizes this metaphor in *Perspective Rectified*, pp. 44–46; Ruskin defends it in *Early Prose Writings*, pp. 226–27.

49. Of the roof of a cathedral he was sketching, Ruskin notes: "The head would be turned upward in looking at it; and it consequently cannot be represented in the drawing . . ." (*ibid.*, p. 217).

50. *Ibid.*, 219.

51. "The new optical laws of the camera obscura, or daguerreòtype." Cf. Parsey, *Science of Vision*, pp. xvi, 34–35. A good overview of Ruskin's own enthusiasm and final disenchantment with photography is still to be found in Michael Harvey, "Ruskin and Photography," *Oxford Art Journal* 7.2 (1985), pp. 25–33.

52. Thomas Sutton, *Unconventional* (London: Sampson Low, Son, & Marston, 1866), vol. 1, p. 159. The second and third volumes, more aptly called *Unending*, deal with more serious photographic sins: forging of banknotes (vol. 2), and photographic pornography, made under duress (vol. 3). On Sutton's life and career, see the obituary notes in the *British Journal of Photography* XXII.147 (May 12, 1875) and pp. 210–12 (April 30, 1875).

53. It is striking that there are no such prominent distortions in the recently redis-covered daguerreotypes Ruskin collected between 1849 and 1855, when he was writing

The Stones of Venice. See Ken and Jenny Jacobson, *Carrying off the Palaces: John Ruskin's Lost Daguerreotypes* (London: Bernard Quaritch, 2015). Joel Snyder, comparing one such daguerreotype (of a portal of San Marco) with Ruskin's drawing after it, in fact found Ruskin correcting the recession of the photograph! My thanks to him for his eye, intellect, and generosity.

54. There is a critical account of the three-lens system in the *BJP* (April 30, 1875), p. 211.

55. *BJP* (May 12, 1876), p. 228, and *BJP* (April 30, 1875), p. 211. On the Sutton camera in the Museum of the History of Science at Oxford, see the museum's newsletter: Willem Hackmann, "Sphere No. 8: Thomas Sutton Panoramic Camera Lens," *Sphaera* 8 (Autumn 1998), at http://www.mhs.ox.ac.uk/sphaera/index.htm?issue8/artic17.

56. This conceit was literalized in Bob Shaw's story "Light of Other Days," *Analog* 77.6 (August 1966), pp. 121–28.

57. De Morgan, *Athenaeum*, p. 409 (*Budget of Paradoxes*, p. 177).

58. Louis-Edmond Duranty, *La nouvelle peinture* (Paris: E. Dentu, 1876), p. 27. Modern editions include artist names taken from marginal notes in a copy belonging to Duranty. See Oscar Reuterswärd, "An Intentional Exegete of Impressionism: Some Observations on Edmond Duranty and His 'La nouvelle peinture,'" *Konsthistorisk tidskrift* 18.1–4 (1949), pp. 111–16, and Carol Armstrong's correlation of Duranty with Degas in *Odd Man Out: Readings of the Work and Reputation of Edgar Degas* (Chicago: University of Chicago Press, 1991), pp. 73–100.

59. Barbara Wittmann, *Gesichter geben: Édouard Manet und die Poetik des Portraits* (Munich: Wilhelm Fink, 2004), thinks Duranty's moments "escape the profaning objective of the photographer" (164). But the "pose sans but, sans signification devant l'objectif du photographe" (Duranty, p. 28) belongs to the studio hack, not photography as such, or as presented in Duranty's thought experiment. Cf. Marcel Crouzet, *Un Méconnu du réalisme: Duranty (1830–1880)* (Paris: Librairie Nizet, 1964), pp. 333, 603.

60. Duranty, "De l'excentricité (à propos de roman et de théâtre)," *Réalisme* 1 (15 November 1856), p. 10.

61. Duranty, "Confession d'un peintre âgé de quarante-cinq ans," *Réalisme* 1 (15 November 1856), p. 10.

62. "By these means one would arrive among the public with documents (I call these documents) of excessive interest, because they are familiar to the public, and can be appreciated by it, diversified by individual points of view" (*ibid.*, p. 11).

63. Winckelmann, in his *Versuch einer Allegorie* (Dresden: Walther, 1766), pp. 30–31, praises Guido Reni's penitent Magdalen for, among other things, thematizing her purity

by the detail of a couple of radishes she will be eating. I hope to show elsewhere that this tradition, elaborated by Poe, issues in the "real allegories" of Courbet and Melville.

64. Michael Fried, "Caillebotte's Impressionism," *Representations* 66 (Spring 1999), pp. 1–51, makes a spirited case that Duranty, though associated with Degas's faction among impressionists, also greatly appreciated Caillebotte.

65. Kirk Varnedoe and Thomas Lee, eds., *Gustave Caillebotte: A Retrospective Exhibition* (Houston, TX: Museum of Fine Arts, 1978), p. 157, and *Gustave Caillebotte: Urban Impressionist* (New York: Abbeville Press, 1987), p. 167. Cf. Michael Marrinan, *Gustave Caillebotte: Painting the Paris of Naturalism, 1872-1887* (Los Angeles: Getty Research Institute, 2016), p. 229.

66. James Clerk Maxwell, "Experiments on Colour, as Perceived by the Eye, with Remarks on Colour-Blindness," *Transactions of the Royal Society of Edinburgh* 21.2 (1855), pp. 275–98, first describes the photographic experiment with color mixing (pp. 283–84: the blue filter is described as violet, probably in deference to Newton). Cone cells had first been described by Leeuwenhoek as early as 1684, but they were not distinguished from rod cells until 1866.

67. James Clerk Maxwell, "On the Theory of Compound Colours, and the Relations of the Colours of the Spectrum," *Philosophical Transactions of the Royal Society of London* 150 (1860), p. 61.

68. R. M. Evans, "Maxwell's Color Photograph," *Scientific American* 205:5 (November 1961), pp. 118–28. See the essays of John Reid and Malcolm Longair in *James Clerk Maxwell: Perspectives on His Life and Work*, ed. Raymond Flood, Mark McCartney, and Andrew Whitaker (Oxford: Oxford University Press, 2014), and Jordi Cat, *Maxwell, Sutton, and the Birth of Color Photography* (New York: Palgrave Macmillan, 2013). Even Ducos du Hauron's and Charles Cros's three-color prints made a decade and a half later betrayed a curious materiality; at their edges, monochrome green, red, and blue overlap imperfectly, giving away the illusion of a colored world as an effect of printing.

69. As a matter of fact, in 1879 the group was called Les Indépendants in deference to Edgar Degas, who did not approve the Impressionistes label that graced, for the first time, the 1877 exhibition. See Joel Isaacson, "The Crisis of Impressionism," in *The Crisis of Impressionism 1878–1882* (Ann Arbor: University of Michigan Museum of Art, 1980), p. 4.

70. Four-plate color printing was invented by the German Jakob Cristof Le Blon in the 1720s, but the process came to be associated with the French aquatintist Philibert Louis Debucourt between 1785 and 1800, especially during the Directory. See Anthony Griffiths, *Prints and Printmaking: Introduction to the History and Techniques* (Berkeley: University of

California Press, 1996), pp. 118–19. Griffiths notes that polychrome printing was revived by Mary Cassatt — and, we might add, by her younger American and Japoniste contemporary Helen Hyde.

71. Bracquemond's role in founding Japonisme was first explored in Yvonne Thirion, "Le japonisme en France dans la seconde moitié du XIXe siècle à la faveur de la diffusion de l'estampe japonais," *Cahiers de l'Association internationale des études françaises* 13 (1961), pp. 117–30. See also Jean-Paul Bouillon, "*À gauche*: Note sur la Société du Jing-Lar et sa signification," *Gazette des Beaux-Arts* 91 (March 1978), pp. 108–18; Gabriel Weisberg, "Les albums ukiyo-e de la collection de Camille Moreau: Source nouvelle pour le japonisme," *Nouvelles de l'Estampe* 23 (1975), pp. 18–21; Martin Eidelberg, "Bracquemond, Delâtre, and the Discovery of Japanese Prints," *Burlington Magazine* 123 (April 1981), pp. 221–27; and Pamela Genova, *Writing Japonisme: Aesthetic Translation in Nineteenth-Century French Prose* (Evanston, IL: Northwestern University Press, 2016).

72. Jean-Paul Bouillon, "Au Jardin d'acclimatation" and "La Nuée d'orage," in *The Crisis of Impressionism*, pp. 52, 56.

73. The watercolor-and-pen sketch for the first state of the print in the New York Public Library, though essentially laying out the color scheme for the color print, attempts more brilliant color effects, particularly in the birds' plumage. The brown tint was also abandoned as a separate color, being achieved in the final print by admixture.

74. Henri Béraldi, *Les Graveurs du XIXe siècle. Tome III-Bracquemond* (Paris: L. Conquet, 1885).

75. Bracquemond in fact illustrated the *Jardin des plants* for Lacroix's 1867 *Paris Guide*, edited by Victor Hugo and Philippe Burty, *Paris guide par les principaux écrivains et artistes de la France. Prémiere partie: La science — L'art*, 2nd ed. (Paris: Librairie Internationale, 1867), p. 260.

76. *The Crisis of Impressionism*, pp. 52–53.

77. There are studies of both single figures (Béraldi, *Les Graveurs du XIXe siècle*, pp. 755–56) wherein the direction of the gaze is far less determinate.

78. Frank Weitenkampf, "Félix Bracquemond: An Etcher of Birds," *Arts and Decoration* (June 1912), p. 298; and *Prints and Their Makers: Essays on Engravers and Etchers Old and Modern*, ed. Fitzroy Carrington (New York: Century Co., 1912), p. 223.

79. The etching, with four others, appeared with a lithographic title page in an 1887 portfolio published by Lemercier (Béraldi, *Les Graveurs du XIXe siècle*, p. 810), under the title *Canards surpris*. The 1900 magazine printing shortened this to *La Surprise*.

80. I quote French sources close to Bracquemond (who, by his 1876 inclusion of a portrait of August Comte in the first Independent Exhibition, declared himself close

to positivism), but it should be emphasized that similar ideas were advanced by the very different Austro-German philosopher and psychologist Franz Brentano (*Psychologie vom empirischen Standpunkte*, 1874), and by the (more Taine-like) British psychologists Bain and Mill.

81. Hippolyte Taine, *De L'Intelligence* (Paris: Hachette, 1870), vol. 2, p. 6 (under the rubric "Idées qui composent l'idée de corps").

82. *Ibid.*, vol. 1, p. 411, quoted from Hippolyte Taine, *On Intelligence*, trans. T. D. Haye (New York: Holt, 1875), vol. 1, p. 224.

83. Félix Bracquemond, *Du Dessin et de la couleur* (Paris: Charpentier, 1885), p. 42.

84. *Ibid.*, 118.

85. There are such prints: Béraldi 1885–92, *Inventaire du Fonds Français: Bibliothèque Nationale, Département des Estampes* (Paris, 1930), p. 456; British Museum 1917, 1119.5. Bracquemond's first view of *St. Cloud* is dated 1856.

86. Vertical landscape compositions of this sort, with a hollow surrounded by tree trunks, are popular among the Fontainebleau painters. See, e.g., Narcisse Díaz de la Pena's *Forest Path* in the National Museum of Art, Bucharest.

87. Fragment of a draft letter of February 26, 1882, sent to Durand-Ruel by Renoir's brother Edmond while the latter was ill to indicate his motives for not wishing to exhibit. In Lionello Venturi, *Les Archives de l'impressionisme*, 2 vols. (Paris and New York: Editions Durand-Ruel, 1939), vol. 1, p. 122: "Débarrassez-vous de ces gens-là et présentez-moi des artistes telles que Monet, Sisley, Morisot, etc., et je suis à vous, car ce n'est plus de la politique, c'est de l'art pur."

88. John Rewald, *The History of Impressionism* (New York: Museum of Modern Art, 1961), p. 449. Cf. the 1881 letters to Pissarro in Marie Berhaut, *Caillebotte, sa vie et son oeuvre* (Paris: Bibliothèque des Arts, 1978), pp. 245–46.

89. The use of drypoint alongside aquatint is typical of Edgar Degas, with whom Cassatt collaborated on the print journal *La Jour et la nuit*.

90. Manet had written a letter to Rossetti to warm him up for an article concerning his art. See Nancy Locke, "Manet and His London Critics," *Burlington Magazine* 152.1293 (December 2010), pp. 780–82.

91. W. M. Rossetti, "The Deschamps Gallery," *The Academy* 208 (April 29, 1876), p. 416. Rossetti prefers an "aesthetic" realism: "*A Portrait* by Bastien-Lepage is also, but in a different way, an example of realism — an old *bourgeois* with whitened-yellow beard, cap, and snuffbox, his blue-checked handkerchief laid over his grey-trowsered knee, seated in a green alley which we might expect to identify in the Bois de Boulogne; excellently true,

and, though not particularly beautiful (the physique of the sitter militates against that), still executed with all self-respecting refinement."

92. P. Burty, "Paris Letter," p. 417. Burty, who had defended *Le Linge* earlier that year in his "Paris Letter," is no sycophant of Manet; in April, he attacks him for keeping aloof from the "Intransigeants," i.e., the impressionists.

93. On Pattison, best remembered for her political career as Emilia, Lady Dilke, and her polemical engagement with her early mentors John Ruskin and Walter Pater, see Elizabeth Mansfield, "Articulating Authority: Emilia Dilke's Early Essays and Reviews," *Victorian Periodicals Review* 31.1 (Spring 1998), pp. 75–86.

94. E. F. S. Pattison, "The Salon of 1876 (*First Notice*)," *The Academy* 209 (13 May 1876), pp. 463–64.

95. It is no accident that the hero of the review is Puvis de Chavannes (p. 463), followed by Gustave Moreau, of whose *Salomé* Pattison gives a fascinated, extended, if not entirely uncritical, account (p. 464).

96. *Ibid.*, p. 464. Of *Le Linge* she makes a similar point, again deploring the lack of finish: "We feel that an artist who can give so much with undoubted power and truth could certainly give much more." This concern echoes Fried's investigation of the finish of the "*tableau*" in contemporaneous art criticism, though the English do not use the word.

CHAPTER FOUR: WHAT ARE WE?

1. Hippolyte Taine, *On Intelligence*, trans. T. D. Haye (New York: Holt, 1875), vol. I, pp. 35–36. Cf. Hippolyte Taine, *De L'Intelligence* (Paris: Hachette, 1870), vol. I, p. 77.

2. Taine must have known the Salon des Refusés of 1863 (containing Whistler's *Girl in White* and Manet's *Déjeneur sur l'herbe*). See the discussion in Thomas H. Goetz, *Taine and the Fine Arts* (Madrid: Playor, 1973), pp. 59–60, 141.

3. "The participation of a hitherto ignored people in the political life of France is a social fact that will honour the whole of the close of the nineteenth century. A parallel is found in artistic matters, the way being prepared by an evolution which the public with rare prescience dubbed, from its first appearance, Intransigeant, which in political language means radical and democratic," Stéphane Mallarmé, "The Impressionists and Édouard Manet," *The Art Monthly Review and Photographic Portfolio* 1.9 (September 30, 1876), pp. 117–22, in *The New Painting: Impressionism 1874–1886*, ed. Charles S. Moffett (San Francisco: Fine Arts Museums of San Francisco, 1986), this quote, p. 33.

4. Mallarmé, "Manet," p. 31. See Jean C. Harris, "A Little-Known Essay on Manet by Mallarmé," *Art Bulletin* 46.4 (December 1964), pp. 559–63, and Isabella Checcaglini,

"Mallarmé en anglais: The Impressionists and Édouard Manet," in *Le Texte étranger* 5 (2004), pp. 116–30.

5. *Ibid.* Accordingly, Manet uses instead "that artistic perspective which we learn from the extreme East —Japan for example" (*ibid.*).

6. Mallarmé thinks sketchiness suggests mobile seeing, or, as he puts it at the end, the painting of "the Aspect" (p. 34).

7. Mallarmé, "Manet," p. 31.

8. *Ibid.*

9. Jules Antoine Castagnary, "L'Exposition du boulevard des Capucines: Les impressionistes," *Le Siècle* (April 29, 1874), 3, quoted in *Impressionism: A Centenary Exhibition*, ed. Anne Dayez, Michel Hoog, and Charles S. Moffett (Paris: Éditions des Musées Nationaux, 1974), p. 153. [Achille Piron], *Eugène Delacroix: Sa Vie et ses oeuvres* (Paris: Jules Claye, 1863), p. 421. I quote here in its entirety the French aphorism, collected by Piron among Delacroix's notes on "Le beau, l'idéal et le réalisme": " — Sur l'importance du sujet. — Tous les sujets deviennent bons par le mérite de l'auteur. Oh! jeune artiste, tu attends un sujet? Tout est sujet ; le sujet c'est toi-même; ce sont tes impressions, tes émotions devant la nature. C'est en toi qu'il faut regarder, et non autour de toi." Castagnary's text and its implications are discussed well in Richard Shiff, *Cézanne and the End of Impressionism* (Chicago: University of Chicago Press, 1984), pp. 3, 29.

10. Ludwig Wittgenstein, *Philosophical Investigations*, ed. G. E. M. Anscombe (Oxford: Basil Blackwell, 1986), p. 128e (§401). Cf. Judith Jarvis Thomson, "Private Languages," *American Philosophical Quarterly* 1.1 (January 1964), pp. 20–31, especially at the end.

11. The bird's-eye view was popularized in J. J. Grandville, *Un autre monde* (1844), and picked up, perhaps from Grandville's 1845 German edition *Eine andere Welt* by the painter Adolph Menzel, among others; but this vantage point, while suggestive of a spectator looking down, by no means specifies the viewer, as in McCay or Caillebotte.

12. I date the first drawing as 1868–69, "approximately 17 years" before the date of Mach's preface, November 1885. Mach later made another, more cursive drawing, substantially like the diagram, likely in preparation for the book.

13. Jonathan Crary, *Suspensions of Perception* (Cambridge, MA: MIT Press, 1999), p. 162, compares Mach with Georges Seurat: "What Mach and Seurat both renounce is any attempt to given an objective picture of the world; instead they each seek provisional and practical ways of representing various complexes of elements that have no essential substance or permanence." Yve-Alain Bois, "On Matisse: The Blinding," *October* 68 (Spring 1994), pp. 60–121, regards a drawing of Matisse resembling Mach's in the inclusion of the

artist's left hand and leg as an attempt to "verify whether it is really possible to be an Impressionist ... to be nothing but an eye" (p. 74). Though Bois quotes Mach at length, he assimilates his project to Maurice Merleau-Ponty's, seeing in the drawing "the indissociability of the physical and psychical domains, the interdependence of the self, the world, and one's own body, as well as the permeability of the senses" (p. 75). This astounding indifference to the actual context depends on Bois's understanding of the drawing as a "solipsistic vision that Mach ridicules" (p. 77). It would be hard to find better evidence for solipsism, at least among historians. See also the discussion in Horst Bredekamp, "Denkende Hände: Überlegungen zur Bildkunst der Naturwissenschaft," in *Von der Wahrnehmung zur Erkenntnis — From Perception to Understanding*, ed. Monika Lessl and Jürgen Mittelstraß (Berlin: Springer, 2005), pp. 123–26, and the first chapter of Karl Clausberg, *Neuronale Kunstgeschichte: Selbstdarstellung als Gestaltungsprinzip* (Vienna: Springer, 1999).

14. Alois Riegl, "Objective Aesthetik," *Neue Freie Presse* July 13, 1902, p. 34. His 1899 lecture "Die Stimmung als Inhalt der modernen Kunst" and his work on historic preservation emphasized mood (German *Stimmung*) as a paradigmatic modern, subjective value, especially in the treatment of nature. See Mechtild Widrich, "The Willed and the Unwilled Monument," *Journal of the Society of Architectural Historians* 72.3 (September 2013), pp. 382–98.

15. Ernst Mach, "Wozu hat der Mensch zwei Augen?" *Zwei Populäre Vorträge über Optik* (Graz: Leuschner & Lubensky, 1867), pp. 28–29. Cf. Mannfred Sommer, *Evidenz im Augenblick: Eine Phänomenologie der reinen Empfindung* (Frankfurt: Suhrkamp, 1987).

16. "We do not explain pictures: we explain our remarks about pictures — or rather, we explain pictures only in so far as we have considered some verbal description or specification." Michael Baxandall, *Patterns of Intention: On the Historical Explanation of Pictures* (New Haven: Yale University Press, 1985), p. 1. Frege rejects a view of truth as resemblance by pointing out that it covertly relies on such descriptions; to say that a picture of Cologne Cathedral is true is to say that the sentence "this picture matches the cathedral" is true. Gottlob Frege, "Der Gedanke," *Beiträge zur Philosophie des Deutschen Idealismus* 1.2 (1918), pp. 59–60; "The Thought: A Logical Inquiry," *Mind* 65.259 (July 1956), p. 291.

17. Ernst Mach, *Beiträge zur Analyse der Empfindungen* (Jena: Gustav Fischer, 1886), pp. 13–14. The picture is on p. 14, the omnibus anecdote in the long footnote on p. 3. Mach's text reads remarkably like that wherein John Stuart Mill argues that we conceive "other minds" by analogy with our own: "I look about me, and though there is only one group [of possibilities of sensations] (or body) which is connected with all my sensations ... I observe that there is a great multitude of other bodies ... closely resembling this particular one,

but whose modifications do not call up, as those of my own body do, a world of sensations in my consciousness." *An Examination of Sir William Hamilton's Philosophy* (London: Longman, Green, Longman, Roberts & Green, 1865), p. 209.

18. This feature impressed Ernst Bloch, who, despite his justifiable suspicion of Mach's "philosophical goal" in the drawing, declared the "headless representation of the drawer a *novum* in the history of the self-portrait." Ernst Bloch, "Selbstportrait ohne Spiegel," in *Verfremdungen* (Frankfurt: Suhrkamp, 1962), p. 14, and Clausberg, *Neuronale Kunstgeschichte*, p. 12.

19. Clausberg, *Neuronale Kunstgeschichte*, p. 15. Cf. Karl Clausberg, "Selbstschauung 'Ich' als Bild: Von Karl Christian Friedrich Krause zu Johannes Müller und Ernst Mach," in *Repraesentatio Mundi: Bilder als Ausdruck und Aufschluss menschlicher Weltverhältnisse*, ed. S. Blasche, M. Gutmann, and M. Weingarten (Bielefeld: Transcript, 2004), pp. 109–59, and Rudolf Haller and Friedrich Stadler, eds., *Ernst Mach: Werk und Wirkung* (Wien: Hölder-Pichler-Tempsky, 1988).

20. Karl Christian Friedrich Krause, *Abriss des Systemes der Philosophie* (Göttingen: Dieterich, 1828), p. 8: "*Task*: To complete the self-perception *I*. *Note*: Whoever lacked this perception could not get it through communication. We find that we have to attribute it to everyone. As instruction, it should suffice to take and grasp oneself pure and whole, without putting in anything that isn't there already. *Solution*: Each sees him- or herself as I, or: the I sees itself, and indeed right off, without looking on anything specific or special that the I might have or be. . . . For this reason the content of the I can only be pronounced: I. But not 'I am,' 'I am spirit,' 'I am human,' 'I act,' 'I think,' etc." Mach may have gotten the gist of Krause after all; he certainly imitated, drolly, Krause's problem-set style.

21. Johannes Müller, *Handbuch der Physiologie des Menschen* (Coblenz: J. Hölscher, 1840), vol. 2, p. 356, in a section significantly titled "Pictures of One's Own Body in the Visual Field."

22. Salomon Stricker, *Studien über die Association der Vorstellungen* (Vienna: Wilhelm Braumüller, 1883), p. 76. Significantly, this passage opens ch. 16, on "Mental Images of Numbers." Stricker's further point, according to which counting requires muscular efforts (for each point must be focused on), and thus images, is inconclusive; as Frege would reply, what makes a count correct is the fact that such efforts (acts of focusing) are put in a one-to-one relation with every one of the points in the figure. This is a logical, not a muscular, achievement.

23. Gottleb Frege, *Die Grundlagen der Arithmetik* (Breslau: Koebner, 1884), p. vi, and Stricker, *Studien*, p. 9. Husserl's *Philosophie der Arithmetik* denies that 0 and 1 are numbers.

24. Frege, *Grundlagen*, p. 71. He concludes: "Even if, as seems to be the case, thinking without images is impossible for us humans, their connection with what is thought is often very superficial, arbitrary, and conventional." This is so because, though thought is image-laden, "we are oft led by thinking beyond the imaginable (*das Vorstellbare*)." Very similar points were made in Bernard Bolzano, *Beyträge zu einer begründeteren Darstellung der Mathematik* (Prague: Caspar Widtmann, 1810), appendix, §9 (pp. 148–50): that in imagining triangles, different subjects may imagine substantially different shapes (an acute, a right, an obtuse triangle), and that such mental images, if inevitable, are by no means required for proofs, indeed that geometry abounds in propositions, like the infinite extension of straight lines, that are unvisualizable. Bolzano, though admired by Goethe, was unknown in the late nineteenth century, with the notable exception of Cantor, who cites his posthumous *Paradoxien des Unendlichen*.

25. These ideas are found in Frege's "Der Gedanke" (1919), going back to unpublished manuscripts as early as 1880.

26. "Thinking is grasping thoughts." "Einleitung in die Logik," in Gottlob Frege, *Nachgelassene Schriften* (hereafter *NS*), eds. Hans Hermes, Friedrich Kambartel, and Friedrich Kaulbach (Hamburg: Felix Meiner, 1969), p. 201.

27. Frege's contemporary Emily Dickinson knew this aesthetic desire and its limits. *The Poems of Emily Dickinson, 2nd series*, ed. T. W. Higginson and Mabel Loomis Todd (Boston: Roberts Brothers, 1892), p. 36: "The thought beneath so slight a film / Is more distinctly seen; — / As laces just reveal the surge, / Or mists the Apennine." Cf. Houghton Library, Harvard, Emily Dickinson Papers, Poems, Packet XXXVII, fascicle 10, 1860–1861.

28. With one important exception, noted by Frege at the end of "The Thought": humans and any other thinkers there may be *act* on them ("Der Gedanke," pp. 76–77).

29. Alexius Meinong, *Über Annahmen* (Leipzig: Barth, 1902), p. 42 (for the Siegfried example) and vii–viii for the credit to Radaković, whose work he dates to 1899. Meinong read Frege later (as the 1910 ed. of *On Assumptions* testifies). See also Mila Radaković, "Metaphysische Konsequenzen aus dem Persistenzgedanken Meinongs," in *Meinong-Gedenkschrift* (Graz: "Styria" Steirische Verlagsanstalt, 1952), pp. 103–12.

30. See "Der Gedanke," p. 69, and, on the objectivity rather than existence of thoughts, "Über das Trägheitsgesetz," *Zeitschrift für Philosophie und philosophische Kritik* 98 (1981), pp. 145–61 (esp. p. 157), translated by Rose Rand as "About the Law of Inertia," *Synthese* 13.4 (December 1961), pp. 350–63.

31. Gottlob Frege, "Über Sinn und Bedeutung," *Zeitschrift für Philosophie und philosophische Kritik* 100 (1892), p. 33 n. The original reads: "Es wäre wünschenswerth, für Zeichen,

die nur einen Sinn haben sollen, einen besondern Ausdruck zu haben. Nennen wir solche etwa Bilder, so würden die Worte des Schauspielers auf der Bühne Bilder sein, ja der Schauspieler selber wäre ein Bild." See Gottlob Frege, "Sense and Reference," *Philosophical Review* 57.3 (1948), p. 216, and *Translations from the Philosophical Writings of Gottlob Frege*, ed. Max Black and Peter Geach (New York: Philosophical Library, 1952), p. 63.

32. Note that here too we are distinguishing aesthetic *use* from research: literary and art historians of course inquire into the facts *concerning* artworks as well as those which artworks may contain, but like Heinrich Schliemann's investigations of the historical Troy, which Frege certainly knew about, this is plausibly a "scientific" activity.

33. Frege, "Über Sinn und Bedeutung," p. 33 n. English translators have ignored to the point of inaccuracy Frege's interest in pictures: Max Black and Peter Geach render Frege's *Bild* as "representation." This turns Frege's striking image of the actor being an image into a banality. The only author I know to have drawn attention to Frege's usage is James Elkins, *The Domain of Images* (Ithaca, NY: Cornell University Press, 1999), p. 64.

34. Marián Zouhar, "Frege on Fiction" in *Fictionality — Possibility — Reality*, eds. P. Koťátko, M. Pokorný, and M. Sabatés (Bratislava: Aleph, 2010), pp. 103–20, canvases two suggestions: (1) science and fiction are two homophonic languages, (2) a thought shifts from science to fiction or vice versa. The first suggestion, besides being incredible (one would, on first reading fiction, suddenly learn a whole language), fails to explain the scientific hypothesis: how could Poe's explanation of why the sky is dark at night in *Eureka: A Prose Poem* turn out to be correct? See Edward Harrison, *Darkness at Night: A Riddle of the Universe* (Cambridge, MA: Harvard University Press, 1989). The second suggestion, pace Zouhar, requires no change in truth-value, but only in the thinker: the step from thinking to judging.

35. See the examples involving Wilhelm Tell and Don Carlos in Frege's unpublished 1897 "Logik," which suggest that two names, one fictional, one real, are involved in respective fictional and asserted thoughts (*NS*, pp. 141–42). I do not find this discussion really supportive of the "two languages" thesis; it is worth noting that Frege was unhappy with the aesthetics in the 1897 *Logik*, as shown by the fact that he crossed out the paragraph contrasting truth with beauty (*NS*, p. 143 n.).

36. *Logik* (1897), *NS*, p. 142, makes clear that he has painting in mind as well: "Poetry, like for instance painting, is concerned with appearance." After the Don Carlos example, we are told: "The same happens in a history painting."

37. Frege, "Über Sinn und Bedeutung," p. 33.

38. Sentences like "Hamlet is a fictitious prince" seem true precisely if there is no

Hamlet. Michael Dummett, *Frege: Philosophy of Language*, 2nd ed. (London: Duckworth, 1980), p. 426, presses this point against Frege. The philosophy of fiction has proposed a bewildering variety of accounts of this usage, some more or less consistent with Frege's. His own theory gives a very commonsensical explanation: on what do we base the truth of the assertion that Hamlet is a fictitious prince? There "being" no Hamlet, such truths rely on uses of the name "Hamlet." But then the sentence itself is about the name, and not about the (nonexistent) prince, as it superficially appears.

39. Frege, "Über Sinn und Bedeutung," p. 34.

40. Frege, *Logik*, p. 142.

41. This is from the late unpublished text "Meine grundlegende logischen Einsichten" [1915], *NS*, 271.

42. Letter to Hugo Dingler, February 6, 1917, in Frege, *Wissenschaftlicher Briefwechsel*, ed. G. Gabriel, H. Hermes, F. Kambartel, C. Thiel, and A. Veraart (Hamburg: Felix Meiner, 1976), p. 34 (hereafter *WBW*). Dingler, a formalist, believed there was no essential difference between $x > y$ and $3 > 2$ (Letter to Frege of February 26, 1917, *WBW*, 38).

43. Draft of a letter to Philip Jourdain, January 1914, *WBW*, 127. Around this time, Saussure began, under the rubric of general linguistics, to present all language and sign use more or less in terms of conventional (railroad) signals.

44. Gottlob Frege, *Begriffsschrift: Eine der arithmetischen nachgebildete Formelsprache des reinen Denkens* [1879] (Hildesheim and Zürich: Georg Olms, 1964), p. 5. Frege's "Bedingtheit" or "dependence" (pp. 5–13) has entered logic permanently as material implication.

45. Caption to figure 5.2, titled "Pure Thought," in Lorraine J. Daston and Peter Galison, *Objectivity* (New York: Zone Books, 2007), p. 274. The quotation on p. 273 cites the figure.

46. Cf. "the joke about the teacher who says, 'Suppose there are x pounds of sugar in a box' and the pupil who puts up his hand and says 'But sir, suppose there aren't?'" G. E. M. Anscombe, "Ludwig Wittgenstein," *Philosophy* 70.273 (July 1995), p. 400. The moral is that x does not have meaning alone, but only as part of a conditional truth.

47. It is too bad that Daston and Galison's presentation of Frege in their fine book *Objectivity* (pp. 265–73) is less reliable than, say, their treatment of Helmholtz. The content and judgment stroke do *not* represent mental image and judgeable content respectively (p. 271). Only judgeable content can have a content stroke (—). " — House" is nonsense, because "House" is not something that could be true, though "There are houses" is. See *Begriffsschrift*, p. 2.

48. An early historical text on Leibniz's contemporary, the inventor Denis Papin, stresses his lack of precise tools. See Gottlob Frege, "Über der Briefwechsel Leibnizens

und Huygens' mit Papin," *Jenaische Zeitschrift für Naturwissenschaft*, 15 Suppl. (1881), pp. 29–32. As the concept-script evolved, he borrowed all sorts of notations, from prosody to phonetics of various languages, and the IPA (International Phonetic Alphabet), first published in 1888. See J. J. Green, Marcus Rossberg, and Philip Ebert, "The Convenience of the Typesetter: Notation and Typography in Frege's *Grundgesetze der Arithmetik*," *Journal of Symbolic Logic* 21.1 (March 2015), pp. 15–27.

49. Daston and Galison, *Objectivity*, pp. 263 and 265.

50. Gottlob Frege, *Über eine geometrische Darstellung der imaginären Gebilde in der Ebene* (Jena: A. Neuenhahn, 1873), p. 3.

51. Frege, *Logik*, p. 156.

52. Letter to O. H. Mitchell, 21 Dec. 1882, in *Writings of Charles S. Peirce — A Chronological Edition, vol.4: 1879–1884*, ed. Christian J. W. Kloesel et al (Bloomington and Indianapolis: Indiana University Press, 1989), p. 394.

53. *Zeitschrift für Mathematik und Physik: Historisch-literarische Abtheilung* 25 (1880), p. 90. The review ends with a bibliography of works that Schröder thinks Frege failed to consult. Frege's response appeared in the less-read *Jenaische Zeitschrift für Naturwissenschaft* 16 (1882), pp. 97–106; a much longer reply was refused publication.

54. This account necessarily simplifies, since at the time of publication of the *Begriffsschrift* (1879), Frege had not made the sense/reference distinction. The mature version of the script, in *Grundgesetze der Arithmetik*, 2 vols. (Jena: Hermann Pohle, 1893 and 1903), includes a proof that every correctly introduced formula has a reference as well as a sense, a proof rendered fallacious by the discovery of Russell's Paradox. It is best then to stick to sense.

55. Frege, "Logik in der Mathematik" (1914), *NS*, p. 233.

56. *Ibid.*, p. 235.

57. Frege, *Grundlagen*, p. 110 (§100).

58. It is often said that Frege is combating idealism. Against this, see Frege's response to Wittgenstein in a letter of April 3, 1920: "Perhaps I did not intend at all to combat idealism in the sense you mean. I most certainly did not use the expression 'idealism.' Take my sentences just as they are written, without foisting on me an intention that was perhaps alien to me." "Frege-Wittgenstein Correspondence," ed. Juliet Floyd and Burton Dreben, in *Interactive Wittgenstein: Essays in Memory of Georg Henrik von Wright*, ed. E. de Pellegrin (Dordrecht: Springer, 2011), p. 62.

59. Frege, "Der Gedanke," pp. 69–70.

60. I heartily concur with their estimate of Frege's motives: "The battle against subjectivity was not based in Platonic contempt for appearances or Cartesian distrust of bodily

sensations but was rooted in the struggle to transcend the privacy and individuality of representations and intuitions" (p. 273).

61. Frege, "Der Gedanke," pp. 71–72.

62. *Ibid.*

63. Compare the discussion of a Kandinsky painting with an irregular border in terms of Mach's self-exploration in Clausberg, *Neuronale Kunstgeschichte*, pp. 25–27.

64. The reprint is in *The Mathematical and Philosophical Works of the Right Rev. John Wilkins* (London, 1802).

65. See Lothar Kreiser, *Gottlob Frege: Leben — Werk — Zeit* (Hamburg: Felix Meiner, 2001), pp. 4–5 and Dale Jacquette, *Frege: A Philosophical Biography* (Cambridge: Cambridge University Press, 2019), pp. 20-30.

66. John Wilkins, *An Essay Towards a Real Character and a Philosophical Language* (London: Royal Society, 1668).

67. Boole, by contrast, dedicates a chapter of the *Laws of Thought* to finding fallacies in old metaphysical arguments.

68. On the former's attempt to correlate colors and the directions of lines, the indispensable source remains Barbara Maria Stafford, *Symbol and Myth: Humbert de Superville's Essay on Absolute Signs in Art* (Newark: University of Delaware Press, 1979). On Klee's pedagogy of signs in relation to his painting, see Annie Bourneuf, *Paul Klee: The Visible and the Legible* (Chicago: University of Chicago Press, 2015), esp. pp. 141–82.

69. James, though he coined the term "radical empiricism," made it clear in his *Essays in Radical Empiricism* (New York: Longmans, 1912) that one should avoid slipping from "logical" to "physical" or "psychological" points of view (pp. 102, 108, 111). James commented that his own position "may be regarded as somewhat eccentric in its attempt to combine logical realism with an otherwise empiricist mode of thought." *Some Problems of Philosophy* (New York: Longmans, 1911), p. 106, cited at *Essays in Radical Empiricism*, p. 16.

70. Gottlob Frege, *Function und Begriff* (Jena: Hermann Pohle, 1891), and "Was ist eine Funktion?" in *Festschrift Ludwig Boltzmann gewidmet zum sechzigsten Geburtstage* (Leipzig: Johann Ambrosius Barth, 1904), pp. 656–66.

71. The doctrine is found in "On Concept and Object" ("Begriff und Gegenstand," 1892), and also in an 1882 letter to Carl Stumpf (*WBW*, p. 164). The treatment of totalities of output values (*Werthverlaufe*) as objects made Frege's system vulnerable to Russell's Paradox, as he discovered in 1902; most scholars agree these difficulties can be met.

72. Gottlob Frege, "Über Begriff und Gegenstand," *Vierteljahrschrift für wissenschaftliche Philosophie* 16 (1892), pp. 192–205; the inutility of determining subject and predicate

grammatically is argued already in *Begriffsschrift*, pp. 2–4. The second interpretation involves the higher-order concept "exemplify," which takes saturated first-order concepts as objects.

73. "Sinn und Bedeutung," p. 27 (Aristotle), pp. 42–43 (the Chinese man on European history), both in footnotes.

74. Frege, "Der Gedanke," p. 65.

75. Dummett, in his wittily titled chapter "Original *Sinn*," defends Frege against Quine's skeptical argument that one may attach different senses to the same sensation. Dummett says finally that sense is "an ideal — a goal we strive towards" (p. 625), which is equally true of visual pictures. Cf. *Grundgesetze*, p. xxi, on "falling" into fiction.

76. *Grundlagen*, §60–2, and p. x, where "meaning" is given as the noun *Bedeutung*. The best discussion remains Dummett, "Nominalism," *Philosophical Review* 65.4 (October 1956), pp. 491–505. Dummett notes that Frege seldom stated the context principle after his 1884 book, though it is implicit. I think he cites it explicitly in "Was ist eine Funktion?," where he says of a conditional sentence with variables that "only the whole has a sense" (p. 659).

77. Recall the puzzling claim that "Hamlet is fictional": here, accepting the truth of the claim not only fixes a reference for the word "Hamlet" (the name in a work of literature) but indeed points us to *that* sense of the word rather than the sense we grasp in agreeing or disagreeing with Gertrude that "Hamlet is fat" (by which she may have meant sweaty). This complexity suggests that the context principle, though applicable to reference, does so *through* fixing an appropriate sense that accords both with the component reference and the truth conditions of the whole thought: one could select this reference for Hamlet (the name itself rather than the prince) *without* being sure of fictionality.

78. Frege, *Logik*, p. 151.

79. "Resonant side thoughts" (*mitanklingende Nebengedanken*) is from Frege's last essay, "Gedankengefüge," *Beiträge zur Philosophie des Deutschen Idealismus* 3 (1923), pp. 36–51, at p. 42. The concept was first introduced in "Sinn und Bedeutung": strongly suggestive sentences (like "He didn't have any wine at lunch," which implies that he does often indulge) make other thoughts "resonate" (*mitklingen*, literally "sound with" or "resound," as a tuning fork sets a guitar abuzz) though they are not asserted. Such thoughts arise in the hearer according to psychological laws. "It can thus come to be that we have more simple thoughts than sentences" (p. 46).

80. John Ruskin, *Modern Painters*, vol. 3, *Of Many Things* (New York: Wiley & Halsted, 1859), p. 48 (ch. IV, §8). The concept of assertion, if not the word, is used already by Sir

Philip Sidney in his *Apologie for Poetrie*, ed. Edward Arber (London: Murray, 1868), p. 52: "Now, for the Poet, he nothing affirmes, and therefore neuer lyeth. For, as I take it, to lye, is to affirme that to be true which is false." The ideas are at times strikingly Ruskinian: "The Poet neuer maketh any circles about your imagination, to coniure you to beleeue for true what he writes" (*ibid.*). Ruskin nowhere mentions the *Apologie*, but he surely knew it, having edited Sidney's psalms in *Rock Honeycomb* (1877).

81. Frege, *Logik*, p. 142.

82. Draper Hill, *Mr. Gillray the Caricaturist* (London: Phaidon, 1965), p. 150.

83. Allan Ingram, *Cultural Constructions of Madness in Eighteenth-Century Writing: Representing the Insane* (New York: Palgrave Macmillan, 2005), pp. 198, 199.

84. *Ibid.*, p. 200.

85. James Elkins, "Marks, Traces, *Traits*, Contours, *Orli* and *Splendores*: Nonsemiotic Elements in Pictures," *Critical Inquiry* 21.4 (Summer 1995), pp. 822–60, is a pioneering effort to say how such pictorial elements are "neither written marks . . . nor inarticulable, inchoate mutterings" (p. 824), as is Nicola Suthor's work on virtuoso paint handling, *Bravura: Virtuosität und Mutwilligkeit in der Malerei der Frühen Neuzeit* (Munich: Fink, 2010).

86. In a letter to the mathematician David Hilbert, Frege contrasted natural language and scientific symbolism quite lyrically: "Where the tree lives and grows, it must be soft and juicy. But if the juicy did not harden over time, no significant height could be reached. If on the other hand all that is green is turned to wood, growth ceases." Letter to Hilbert of October 1, 1895, in *WBW*, p. 59. This letter, which Hilbert admired so much that he read it aloud to his students, should dispose of vulgar accounts of Frege's wanting to replace ordinary language with a kind of Newspeak; as the letter shows, Frege would agree with Orwell that doing so across the board would bring thought to a standstill.

87. Frege, "Der Gedanke," p. 67, considers the possibility of thought transfer. Still, Frege concludes, "the question would remain unanswerable, whether the mental image [transferred from one brain to another] were one and the same."

88. In "Der Gedanke" (p. 59), Frege mentions but sets aside notions of "the truth of an artwork," which he equates with authenticity or "true feeling," as being a different sense of "true" from the logical one applying to thoughts: had he pursued his valuable reflections on images as thoughts, I believe he would have seen this to be one possible mode that truth in the logical sense can play a role in artworks, fictions, and even dreams and other modes of thought.

89. Gottlob Frege, "Über die wissenschaftliche Berechtigung einer Begriffsschrift," *Zeitschrift für Philosophie und philosophische Kritik* 81 (1882), p. 49. Philosophers like D. H.

Mellor, who think Frege "speciest" for attributing thoughts only to beings with language, have ignored this account of the emergence of objectivity. Cf. Tyler Burge, *Origins of Objectivity* (Oxford: Oxford University Press, 2010), which works out a philosophical and biological story about the development of objective representation in animals with much fascinating experimental evidence.

CHAPTER FIVE: WHERE ARE WE GOING?

1. I have discussed the politics and religion of Belgian symbolists in "Masks, Modernity, and Egoism: Theatrical Practice in James Ensor and Maurice Maeterlinck," in *The Art of Theatre: Word, Image, and Performance in France and Belgium, c.1830–1910*, ed. Claire Moran (Oxford and Berlin: Peter Lang, 2013), pp. 287–305.

2. This is also the lesson of the last and greatest of the religious symbolists, G. K. Chesterton, in his Father Brown tales. Father Brown's instistence on logic and reason and its compatibility with faith and a sensuous delight in the natural and urban world are paradigmatically symbolist, for all that he pokes fun at "blue birds" and mystic gurus.

3. See Walter Thijs, *De Kroniek van P. L. Tak: Brandpunt van Nederlandse Cultuur in de Jaren Negentig van de vorige Eeuw* (Amsterdam: Wereld-Bibliotheek, 1956), p. 130.

4. G.-Albert Aurier, "Les Peintres symbolistes," *Oeuvres posthumes* (Paris: Edition du "Mercure de France," 1893), p. 308. He briskly lists on the same page mystical influences: Porphyry, Plotinus, Santa Theresa, Saint Bonaventure, and Ruysbroeck. But the technique is the unifying thing.

5. *Ibid.* He cites Gauguin "above all" among the artists to whom this applies; interestingly, he excepts Redon.

6. This is also the lesson of Dario Gamboni's study of Gauguin's interest in perceptual psychology, *Paul Gauguin: The Mysterious Center of Thought* (London: Reaktion, 2014).

7. Georges Rodenbach, *Bruges-la-Morte* (Paris: Flammarion, 1892), p. ii. The 1900 Paris edition of the firm L. Carteret substituted for the photographs wood engravings of the same views by Henri Paillard; the 1904 Flammarion reissue kept the photographs but supplemented them with awkward engravings of the action (strangling included) by H. Delavelle.

8. [Émile Verhaeren], "Un peintre symboliste," *L'Art moderne* (Bruxelles) 7.17 (24 April 1887), p. 130.

9. The doctrine of self-aware thought is enjoying a renaissance in metaphysics: see Sebastian Rödl, *Self-Consciousness and Objectivity: An Introduction to Absolute Idealism* (Cambridge, MA: Harvard University Press, 2018).

10. There is a large literature on Wittgenstein's picture theory, unlike Frege's. A good starting point is the publication of the 33rd International Wittgenstein Symposium, *Bild und Bildlichkeit in Philosophie, Wissenschaft und Kunst*, 2 vols., ed. Elisabeth Nemeth, Richard Heinrich, and Wolfram Pichler (Kirchberg am Wechsel: Österreichische Ludwig Wittgenstein Gesellschaft, 2010) (hereafter *BB*). A fine, brief introduction to the picture theory remains Eva Cassirer, "On Logical Structure," *Proceedings of the Aristotelian Society* 64 (1964), pp. 177–98.

11. Jaś Elsner, "Art History as Ekphrasis," *Art History* 33.1 (February 2010), pp. 10–27; these quotes are at p. 12. The echo of iconoclastic critiques of images, and defenses of images, is hardly unintentional, since Elsner is a historian of such debates. See esp. his "Iconoclasm as Discourse: From Antiquity to Byzantium," *Art Bulletin* 94.3 (September 2012), pp. 368–94.

12. Elsner, "Art History as Ekphrasis," p. 12

13. William Blake, "To The Accuser who is The God of This World," *The Complete Poetry and Prose of William Blake*, ed. David B. Erdman (Berkeley and Los Angeles: University of California Press, 2008), p. 269.

14. Frege's discussion of the fundamentally unasserted nature of pictures offers one explanation why this is so.

15. Elsner, "Art History as Ekphrasis," p. 22.

16. Heinrich Wölfflin, *Das Erklären von Kunstwerken* (Leipzig: E. A. Seemann, 1921), p. 3, somewhat misquoting Schiller, whose exact words, in a letter to Goethe of February 27, 1798, are: "The relation of general concepts and the language built upon it . . . is for me always a precipice that I cannot comtemplate without vertigo." *Briefwechsel zwischen Schiller und Goethe, vol. 2, 1797–98*, ed. Heinz Amelung (Berlin: Deutsche Bibliothek, [1919]), p. 226.

17. Ludwig Wittgenstein, *Prototractatus: An Early Version of Tractatus Logico-Philosophicus*, ed. Brian McGuinness, T. Nyberg, and G. H. von Wright (Ithaca, NY: Cornell University Press, 1971), p. 3 (Sentence 2.1).

18. Ludwig Wittgenstein, *Tractatus Logico-Philosophicus* (London: Kegan Paul, Trench, Trubner & Co., 1922), 2.1. I give my own English from the German, using Wittgenstein's sentence numbers thus: "*T #*."

19. In the preface to the *Tractatus*, Wittgenstein says he owes much of his inspiration to "Frege's great works" (p. 28). There is the long-running debate between "metaphysical" readings (popular especially in the postwar period) and "resolute" (as in, "the book is resolutely nonsensical") readings, due especially to more recent American intepreters, notably Cora Diamond and James Conant. A recent blast from the metaphysical side is

Jaakko Hintikka, "What Does the Wittgensteinian Inexpressible Express?" *Harvard Review of Philosophy* 11 (2003), pp. 9–17. An answer with very little polemic is Juliet Floyd, "Wittgenstein and the Inexpressible," in *Wittgenstein and the Moral Life: Essays in Honor of Cora Diamond*, ed. Alice Crary (Cambridge, MA: MIT Press, 2007), pp. 177–234. Other interpreters seek a middle ground, e.g., Marie McGinn, "Between Metaphysics and Nonsense: Elucidation in Wittgenstein's *Tractatus*," *Philosophical Quarterly* 49.197 (October 1999), pp. 491–513.

20. See Matthew Ostrow, *Wittgenstein's Tractatus: A Dialectical Interpretation* (Cambridge: Cambridge University Press, 2002), p. 35: "It is important to note at once the emphasis here on picturing as an activity: we make pictures of facts and to ourselves, for our own purposes." Cf. Arley Moreno, "Bild: From Satz to Begriff," *BB*, vol. 1, pp. 73–104.

21. This source has not yet been located, and it sounds remarkable for a French publication of August or September 1914, when war news predominated. Perhaps Wittgenstein was remembering something he had read earlier.

22. Ludwig Witgenstein, *Notebooks 1914–1916*, 2nd ed., ed. G. H. von Wright and G. E. M. Anscombe (Chicago: University of Chicago Press, 1979), p. 7 (entry of September 29, 1914). Only the "experimentally" is retained in the *Tractatus*: "In the sentence it's as if a state of affairs is experimentally put together" (*T* 4.031).

23. *Notebooks*, p. 7. The English translation (p. 8) is careless: instead of "man" standing for a "person," we get a "figure representing a man" (p. 8). Wittgenstein is careful to indicate his male figures don't determine the gender of his referent, while the translators, missing the maleness of the stick figures, assign masculinity to the referent!

24. *Ibid.* Why not a three-dimensional script, like his Parisian courthouse dolls? Wittgenstein seems more interested in the concept of picturing as such than in giving rules à la Lessing (whom he admired) for different kinds of pictures.

25. *T* 4.014. In *T* 4.0141, Wittgenstein discusses the rule for translating notes to record grooves. See also *T* 3.1431, which compares a sentence to the furniture of a room. The notebooks had compared a tune to a tautology.

26. Ogden, the official translator of the *Tractatus*, followed Wittgenstein's formula for *T* 3.001 but kept "we make to ourselves pictures of facts" for *T* 2.1. Whitney Davis, in *A General Theory of Visual Culture* (Princeton: Princeton University Press, 2011), uses the verb "to image" similarly, distinguishing "human subjects who image" from pictures, which "can be imaged."

27. As consequence: "We cannot think anything illogical, for then we would have to think illogically" (*T* 3.03).

28. *T* 4.011. Henceforth, *Tractatus* citations not requiring further comment will be listed in the body text.

29. *T* 2.182: "Jedes Bild ist auch ein logisches. (Dagegen ist z. B. nicht jedes Bild ein räumliches.)" Joachim Schulte, *Wittgenstein: An Introduction* (Albany: State University of New York Press, 1992), p. 54, thinks there needn't be "photographic" but only conventional resemblance, as in the fairy tale. This is dubious, since the record's grooves are as mechanical as any photograph. Rather, "logical" resemblance encompasses both causal and conventional.

30. *T* 3.1432. (The English translation botches the quotation marks.) The emphases are all Wittgenstein's.

31. See the letter to Russell of September 19, 1919: "The main point is the theory of what can be expressed (*gesagt*) by prop[osition]s — i.e., by language — (and, which comes to the same, what can be *thought*) and what can not be expressed by prop[osition]s, but only shown (*gezeigt*); which, I believe, is the cardinal problem of philosophy." Wittgenstein, *Cambridge Letters*, ed. Brian McGuinness and G. H. von Wright (Oxford: Basil Blackwell, 1997), p. 124.

32. *T* 4.1212: "Was gezeigt werden *kann*, *kann* nicht gesagt werden." The emphasis ought to have covered the "nicht."

33. Of course, we might fail to define our terms because they are so basic that everything else is stated in terms of them. Frege, "Concept and Object," p. 193, offers that in such cases "gestures" (*Winken*) would have to suggest to a sympathetic reader what is meant but cannot be said. How *that* is done is precisely Wittgenstein's interest.

34. Frank Plumpton Ramsey, *Philosophical Papers*, ed. D. H. Mellor (Cambridge: Cambridge University Press, 1990), p. 146. The observation does not appear in Ramsey's review of the *Tractatus* in *Mind* 32.128 (October 1923), pp. 465–78, although there too Ramsey is critical of the showing/saying distinction.

35. *Notebooks*, p. 33 (November 26, 1914).

36. It is a shame that Wolfram Pichler's *Bildnegation*, which deals in detail with such negative pictorial modes, remains in manuscript. Some sense of Pichler's ideas can be found in the first part of Ralph Ubl and Wolfram Pichler, *Bildtheorie zur Einführung* (Hamburg: Junius, 2014).

37. *T* 4.064. Wittgenstein missed the point, emphasized by Frege since the *Begriffsschrift*, that negation is part of a thought's content and not an act parallel to assertion. For another blunder concerning Frege's theory of assertion, see *T* 4.442. G. E. M. Anscombe, *An Introduction to Wittgenstein's Tractatus* (London: Hutchinson University Library, 1959), deplores Wittgenstein's unreliability in his comments on Frege.

38. "Der Satz *zeigt* seinen Sinn. Der Satz *zeigt*, wie es sich verhält, *wenn* er wahr ist. Und er *sagt*, *dass* es sich so verhält." The relative pronoun *es* designates the common object of saying and showing (presumably logical form).

39. The best efforts to illuminate Wittgenstein's notion of "formal concepts," which according to him render phrases like "1 is a number" nonsensical, are Michael Kremer, "The Purpose of Tractarian Nonsense," *Noûs* 35 (2001), pp. 39–73, and Warren Goldfarb, "Das Überwinden: Anti-Metaphysical Readings of the Tractatus," in *Beyond the Tractatus Wars*, ed. Rupert Read and Matthew Lavery (New York: Routledge, 2011), ch.1, both reacting largely to the profound and original papers collected in Cora Diamond, *The Realistic Spirit: Wittgenstein, Philosophy, and the Mind* (Cambridge, MA: MIT Press, 1991).

40. The example is from Hilary Putnam, *Realism with a Human Face* (Cambridge, MA: Harvard University Press, 1990), pp. 96–104, and John R. Searle, *The Construction of Social Reality* (New York: Free Press 1995), pp. 160–67.

41. Cf. Anscombe's *Introduction to Wittgenstein's Tractatus*, pp. 163–66. Wittgenstein used similar arguments around 1930: "If I speak of 5 persons, I can represent them through strokes. But the fiveness [*die Fünfzahl*] of the persons is not represented, but is depicted in the fiveness of the strokes. Here we grasp the numerals directly as a picture." *Ludwig Wittgenstein und der Wiener Kreis*, ed. Brian McGuinness (Frankfurt: Suhrkamp, 1967), p. 225.

42. The cube was first discussed by Swiss crystallographer Albert Louis Necker in "Observations on Some Remarkable Optical Phaenomena Seen in Switzerland; and on an Optical Phaenomenon Which Occurs on Viewing a Figure of a Crystal or Geometrical Solid," *London and Edinburgh Philosophical Magazine and Journal of Science* 1.5 (1832), pp. 329–37. Wittgenstein's advice for switching aspects (by fixing on the corners labeled *a* or *b*) has been recently confirmed experimentally: see W. Einhäuser, K. Martin, and P. König, "Are Switches in Perception of the Necker Cube Related to Eye Position?" *European Journal of Neuroscience* 20.10 (2004), pp. 2811–18.

43. The duck-rabbit, "D-R head" for short (*H-E Kopf*), is in Ludwig Wittgenstein, *Philosophische Untersuchungen / Philosophical Investigations*, 4th ed., trans. G. E. M. Anscombe, ed. P. M. S. Hacker and Joachim Schulte (Oxford: Basil Blackwell, 2009), pp. 204–207 (II: §118–37). Wittgenstein got the picture from Joseph Jastrow, *Fact and Fable in Psychology* (Boston: Houghton Mifflin, 1900), pp. 292, 295. Jastrow, who discusses the shift in terms of feelings of surprise, got the picture from the October 1892 *Fliegende Blätter*. The cube also appears in the *Philosophical Investigations*. Note that Wittgenstein's *Hase* is in fact hare, but the original *Fliegende Blätter* title, *Kaninchen*, indeed means rabbit.

44. Cf. Sybille Krämer, "'The Mind's Eye': Visualizing the Non-visual and the 'Episte-mology of the Line,'" *BB*, vol. 2, p. 283.

45. *T* 4.461. Senseless (*sinnlos*), not "nonsense" (*unsinnig*), because contradictions and tautologies do not determine a way the world is. Perhaps it would be better to say they determine the way it is (or isn't) *anyway*. A psychologistic residue may have led Wittgenstein to underrate their sense, together with elucidations like "'1' is a number."

46. *T* 5.5423: Note the psychological tone ("perceive"), which leads to dubious empirical speculations: how much must we perceive of the parts to perceive the complex? We perceive a speckled hen, but not the number of speckles.

47. For a discussion and attempted drawing of a "negative fact," see *Notebooks*, p. 30 (entry of November 14, 1914).

48. There is a grotesque mistaking of this passage in Max Black, *A Companion to Wittgenstein's Tractatus* (Cambridge: Cambridge University Press, 1964), p. 301, as the claim on Wittgenstein's part that there just is no soul. But the "namely" clearly links this claim to the "superficial psychology" just ridiculed. One can see, however, how Black came to this, since the epistemologists criticized just one aphorism earlier (Russell and G. E. Moore) seem to assert unity ("*A* believes that *p*") where Wittgenstein sees a complex. But the result is meant to be an absurd consequence of this and the superficial psychology that makes the soul a collection of such beliefs. In his review of the Pears and McGuinness translation of the *Tractatus* in *Philosophical Review* 72.2 (April 1963), p. 265, Peter Geach insists that *Seele* here should be mind (cf. *Seelenleben*, mental life). This is pertinent, but the proffered alternatives ("subject, etc.") show that Wittgenstein was willing to equate a variety of terms for the mental core (person, will, agent, but certainly not "memory"). He didn't shy away from the theological connotations of the simplicity of the soul; a good restatement is Roderick Chisholm, "On the Simplicity of the Soul," *Philosophical Perspectives* 5 (1991), pp. 167–81, whose stated source Wittgenstein may have known: Bolzano's *Athanasia oder Gründe für die Unsterblichkeit der Seele* (Sulzbach: Seidel, 1827), a book of great philosophical penetration.

49. *T* 5.633: "*Wo in* der Welt ist ein metaphysisches Subjekt zu merken? Du sagst, es verhält sich hier ganz, wie mit Auge und Gesichtsfeld. Aber das Auge siehst du wirklich *nicht*. Und nichts *am Gesichtsfeld* läßt darauf schließen, daß es von einem Auge gesehen wird." On Wittgenstein's view of the subject, see José Zalabardo, *Representation and Reality in Wittgenstein's Tractatus* (Oxford: Oxford University Press, 2015), ch. 3.

50. The first known version of the diagram is in a diary entry of August 12, 1916. Judging by Frege's letter to him of August 28 (in response to a postcard of August 16), Wittgenstein was in Olmütz in a school for artillery officers. A month earlier he was on the Russian

front. See Gottlob Frege, "Briefe an Ludwig Wittgenstein aus den Jahren 1914–1920," *Grazer philosophische Studien* 33/34 (1989), p. 13.

51. Does any philosopher actually think this? Fichte might: "Intelligence, as such, *gazes on itself.*" (*Erste Einleitung in der Wissenschalftslehre*, §6); Black, in *A Companion to Wittgenstein's Tractatus*, p. 310, cites Stendhal. But Schopenhauer is closer, and avidly read by Wittgenstein: "'The world is my mental image' — this is a truth valid with reference to every living and knowing being . . . he knows no sun and no earth; but always only an eye, that sees a sun, a hand, that feels the earth; that the world, which surrounds him, is only there as a mental image, that is, only in relation to an other, the imagining, which is himself. If any a priori truth may be pronounced, this is it." *Die Welt als Wille und Vorstellung* (Leipzig: Brockhaus 1819), p. 3.

52. Ludwig Wittgenstein, "Notes on Logic," ed. Harry T. Costello, *Journal of Philosophy* 54.9 (1957), p. 234. (The passage reappears almost verbatim in *Notebooks*, p. 100).

53. The entire discussion is probably inspired by "Über Sinn und Bedeutung," 30: Frege compares sense to the picture seen through a telescope, which can with mirrors be made available to several viewers, whereas each viewer has unique retinal pictures. Even were these retinal pictures to be made available to others, they would not be had in the same way the first viewer had them (one would have a retinal picture of a retinal picture).

54. I am of course simplifying. Kant's argument in its mature form can be found in *Prolegomena zu einer jeden künftigen Metaphysik, die als Wissenschaft wird auftreten können* (Riga: Hartknoch, 1783), §13, pp. 56–59.

55. Russell, *The Principles of Mathematics* (Cambridge: Cambridge University Press, 1903), p. 418: "In itself, the fact would be no more puzzling than the distinction between the stretches AB and BA, which are metrically indistinguishable."

56. Bertrand Russell, *An Essay on the Foundations of Geometry* (Cambridge: Cambridge University Press, 1897), pp. 154–55. The idea is due perhaps to the popularizer of "the fourth dimension," C. H. Hinton, who in "Many Dimensions?" *Scientific Romances, Second Series* (London: Sonnenschein, 1896), p. 38, presented the rotation of figures as evidence of higher dimensions.

57. Honor Brotman, "Could Space Be Four Dimensional?," *Mind* 61.243 (July 1952), pp. 317–27. For discussion, see Richard Swinburne, *Space and Time* (London: Palgrave Macmillan, 1968), p. 154.

58. Arthur Conan Doyle, *A Study in Scarlet* (London: Ward, Lock, Bowden & Co, 1892), p. 6.

59. Joseph Kosuth, "Art after Philosophy, I" (1969), in *Conceptual Art: A Critical Anthology*, ed. Alexander Alberro and Blake Stimson (Cambridge, MA: MIT Press, 1999),

p. 171. Like the late Wittgenstein, these artists were often inclined to think the problem essentially linguistic, foregoing the effort, explored in this book, to correlate the logical with the subjective.

60. Bertrand Russell, "Obituary: Ludwig Wittgenstein," *Mind* 60.239 (July 1951), p. 298, and see below.

61. Russell, *The Principles of Mathematics*, p. 449.

62. *Ibid.*, p. v.

63. J. M. Synge, *The Playboy of the Western World: A Comedy in Three Acts* (Dublin: Maunsel & Co., 1907), p. vii.

64. *Principles*, pp. v–vi. The one fact bothered him as much as the other.

65. Russell, *The Principles of Mathematics*, 2nd ed. (London: Unwin, 1937), pp. ix–x. Russell did not simply recant his Platonism as a direct result of the set-theoretical paradoxes. Though by 1905 he had a theory of notation (the theory of types) to avoid paradox, and a theory of descriptions to avoid attributing existence to abstract nouns, he is comfortable calling Plato's theory of ideas "one of the most successful attempts hitherto made" and endorses it in his chapter "Universals" in *The Problems of Philosophy* (New York: Henry Holt, 1912), pp. 142–43. Cf. *An Inquiry into Meaning and Truth* (London: George Allen and Unwin, 1940), p. 340.

66. This influence was to a large degree personal, but Russell also took to heart Moore's criticisms of his first book of theoretical philosopy, *An Essay on the Foundations of Geometry*, in *Mind* 8.31 (July 1899), pp. 397–405. It is telling that both began as Kant experts, Russell specializing in his theory of space and Moore in his ethics.

67. On Russell and Bloomsbury (he was closest to the salonnière Lady Ottoline Morrell), see his *Autobiography, vol. 1, 1872–1914* (London: Allen and Unwin, 1967) and *The Bloomsbury Group: A Collection of Memoirs and Commentary*, ed. S. P. Rosenbaum (Toronto: University of Toronto Press, 1995). For comparisons with Woolf, see Jaakko Hintikka, "Virginia Woolf and Our Knowledge of the External World," *Journal of Aesthetics and Art Criticism* 38.1 (Autumn 1979), pp. 5–14; Timothy Mackin, "Private Worlds, Public Minds: Woolf, Russell, and Photographic Vision," *Journal of Modern Literature* 33.3 (Spring 2010), pp. 112–30; and Ann Banfield, *The Phantom Table: Woolf, Fry, Russell and the Epistemology of Modernism* (Cambridge: Cambridge University Press, 2000).

68. In a letter of October 19, 1913, to Bryn Mawr professor Lucy Donnelly, Russell confesses: "I feel sure learned aesthetics is rubbish, and that it ought to be a matter of literature and taste rather than of science." *The Basic Writings of Bertrand Russell*, ed. Robert E. Egner and Lester E. Denonn (London: Routledge, 2009), p. ix.

69. Virginia Woolf, "Pictures," *Nation and Athenaeum* 37.4 (April 25, 1925), pp. 101–102.

70. Russell, "Appearance and Reality," *Problems of Philosophy*, pp. 11–12. See also Banfield, *The Phantom Table*, 44.

71. *Ibid.* He makes the same point about shape: "We are all in the habit of judging as to the 'real' shapes of things, and we do this so unreflectingly that we come to think we actually see the real shapes. But, in fact, as we all have to learn *if we try to draw*, a given thing looks different in shape from every different point of view" (p. 12). My emphasis.

72. Or, for that matter, the psychological phenomenon of "lightness constancy," which allows us to identify variously shaded areas as possessing uniform local color. See Mark Sainsbury, "A Puzzle about How Things Look," in *Perspectives on Perception*, ed. Mary Margaret McCabe and Mark Textor (Frankfurt: Ontos, 2007), pp. 7–17.

73. *Roman Art: Some of Its Principles and Their Application to Early Christian Painting* (London: Heinemann, 1900), originally the introduction to *Die Wiener Genesis*, ed. Wilhelm von Hartel (Vienna: Tempsky, 1895).

74. Russell, *An Inquiry into Meaning and Truth*, p. 15. Of course, if physics assumes naïve realism, it too is false.

75. Russell, "Meinong's Theory of Complexes and Assumptions (I.)," *Mind* 13.50 (April 1904), p. 205.

76. That term, already used by Arthur Schopenhauer and Eduard Hartmann, was popularized in English by Dickinson R. Miller, "Naïve Realism: What Is It?" in *Essays, Philosophical and Psychological, in Honor of William James* (London: Longmans, 1908), pp. 231–61, where we are told that philosophically speaking "there is no such theory." I owe thanks to the late Hilary Putnam for correspondence about the origin and significance of the term.

77. Their view is not unlike that presented more recently in Nancy Cartwright, *The Dappled World: A Study of the Boundaries of Science* (Cambridge: Cambridge University Press, 1999), esp. ch. 1.

78. Plato introduced many of these themes, from the dream to the stick seen in water (which looks bent), which were to remain mainstays in the theory of knowledge from Descartes to the present — or at least, until the savage attack on the epistemologist's use of such chestnuts in J. L. Austin, *Sense and Sensibilia* (Oxford: Oxford University Press, 1962).

79. Bertrand Russell, "The Relation of Sense-Data to Physics," in *Mysticism and Logic* (hereafter *M&L*) (London: Longmans, Green, & Co, 1917), pp. 178–79, originally printed in *Scientia* 16.4 (1914), p. 27. Strikingly, Russell feared the onset of mental illness his whole life, for it had afflicted members of his family, as it was to do his son. See Ray Monk, *Bertrand*

Russell: The Spirit of Solitude, 1872–1921 (New York: Free Press, 1996), pp. xix, 303, 317–18, and passim.

80. Bertrand Russell, "The Ultimate Constituents of Matter," *The Monist* 25.3 (July 1915), p. 413 (*M&L*, p. 140).

81. Russell, "The Relation of Sense-Data to Physics," improving on lengthier arguments in *Our Knowledge of the External World* (Chicago: Open Court, 1914). On the eventful evolution of the doctrine of sense-data in Russell and Moore, and especially of the assertion that sense-data are physical, see Omar Nasim, *Bertrand Russell and the Edwardian Philosophers: Constructing the World* (Basingstoke: Palgrave Macmillan, 2008).

82. I do not mean to suggest that Russell was a Poe devotee; he was, however, an admirer of Poe's contemporary Nathaniel Hawthorne (from whom he borrowed the title "Sketches from Memory"), whose essay-tale "The Haunted Mind," in vol. 2 of *Twice-Told Tales* (Boston: James Munroe & Co., 1842) explores the whole gamut of what Poe called "psychal" states.

83. This probably also motivated his insistence on bivalence (truth or falsity) of statements about nonexistent objects in "On Denoting," *Mind* 14.56 (October 1905), pp. 479–93. For Frege, empty terms rendered thoughts fictional; for Russell, assertion of existence is implicit in definite descriptions, so that empty terms make sentences false.

84. William James, "Does 'Consciousness' Exist?," *Journal of Philosophy, Psychology, and Scientific Methods* 1.18 (September 1, 1904), pp. 477–91, reprinted in *Essays in Radical Empiricism*, pp. 1–38. James knew and admired Mach, whose priority claim is relative; in reprints of the *Analyse der Empfindungen* (Jena: Gustav Fischer, 1918), p. 253, Mach cites an early text by James, "The Sentiment of Rationality," *Mind* 4.15 (July 1879), pp. 317–46, which he however claims to have seen after the publication of his own book.

85. There are some disagreements. For Mach and James, it is a virtue of the theory that the same experience may be both mind and matter according to its relations. Russell notes that he can think of no case of a constituent belonging both to mind and matter, and offers to defend Cartesian dualism in "The Ultimate Constituents of Matter."

86. With tongue in cheek, Russell called objectless biographies "official" ("The Ultimate Constituents," p. 414). In the same vein, in his revision of the system of *Our Knowledge of the External World* in "The Relation of Sense-Data to Physics," he introduced under the name of *sensibilia* (singular: *sensibile*) sense-data that go unperceived.

87. That aspects of a thing attribute to it contradictory properties and are illusory was a mainstay of British idealism. Cf. Russell, "The Relation of Sense-Data to Physics," p. 7 (*M&L*, p. 153).

88. "The Ultimate Constituents," pp. 402–403 (*M&L*, pp. 128–29).

89. Though Keaton's *Seven Chances* (1927) is perhaps the *beau idéal* of the chase sequence (featuring all of Russell's desiderata), this was a fixture of film in the 1910s and was experienced by film writers as disjointedly as by Russell. Compare for instance this précis of the climax of the 1915 political thriller *The Precious Packet*: "After a thrilling chase and the precipitation of an automobile over a cliff, resulting in the death of the Frenchman, the packet is opened and found to contain an order for the Englishman to marry pretty Jacqueline, whose marriage with one of his nationality will thwart the plan of her countrymen to place her on a throne in Canada." *The Moving Picture World* 27.8 (Feb 26, 1916), p. 1311.

90. Women, too, led chases, e.g., in *The Girl Telegrapher's Nerves*, wherein "Helen follows [Nelson the crook] on another locomotive and after a thrilling chase overhauls him, couples the two engines together, and then crawls on the swaying coupling to the other locomotive to bring Nelson to bay." "Stories of the Films," *The Moving Picture World* 27.9 (March 11, 1916), p. 1698.

91. *Principles of Mathematics*, 347 (no. 327). Russell returns to these points about Zeno in "The Philosophy of Bergson," *The Monist* 22.3 (July 1912), pp. 321–47, reprinted in *A History of Western Philosophy*.

92. This discontinuous way of seeing time gained an intuitive-psychologistic spokesman in Gaston Bachelard's 1932 book *L'Intuition de l'instant*. See Jennifer Wild, *The Parisian Avant-Garde in the Age of Cinema, 1900–1923* (Berkeley: University of California Press, 2015), ch. 5.

93. "The Relation of Sense-Data to Physics," p. 9 (*M&L*, p. 155), where it is "the supreme maxim in scientific philosophizing." Cf. *Our Knowledge of the External World*, pp. 112, 116, and passim.

94. *Our Knowledge of the External World*, pp. 140–41. For a stern rebuke of Russell's "fictionalism," see H. A. Pritchard, "Mr. Russell on our Knowledge of the External World," *Mind* 24.94 (April 1915), pp. 145–85; for a more recent defense of it, Mark Sainsbury, "Russell on Constructions and Fictions," *Theoria* 46.1 (1980), pp. 19–36.

95. Russell's behaviorism dates to his "On Propositions: What They Are and How They Mean," *Proceedings of the Aristotelian Society, Suppl. Vol.* 2 (1919), pp. 1–43, and *The Analysis of Mind* (London: George Allen & Unwin, 1921).

96. Sainsbury is unsure whether Russell is right, but concludes: "If this issue can be resolved in a way which favours Russell's eliminative intentions, his metaphysics will have a breathtaking austerity, for with the elimination of classes he eliminates also numbers, physical objects, public space and time, and last but not least, our selves" (p. 36).

97. Peter Geach, in "Class and Concept," *Philosophical Review* 64.4 (October 1955), pp. 561–70, opposes this reduction: "Given that we know what sense-data are, we can treat the extension of a predicate that is true only of certain sense-data as identical with a certain physical object. But this does not reduce the physical object to a logical construction out of the sense-data; no more than I am reduced to a logical construction out of certain undergraduates, if the object that is the extension of a predicate applying to the undergraduates is taken to be myself."

98. Russell, "Sensation and Imagination," *The Monist* 25.1 (January 1915), pp. 28–44, here at p. 42.

99. I am in sympathy with the classic readings of Seurat's subject-matter by Meyer Schapiro, T. J. Clark, and Linda Nochlin, though Martha Ward's warning bears reflection that the mixing of classes on which Clark dwells was very rarely noted in the period criticism. I have benefited from reading Emmelyn Butterfield-Rosen's unpublished work on posture and the human body in Seurat.

100. See Inge Fiedler, "A Technical Evaluation of the *Grande Jatte*," *Art Institute of Chicago Museum Studies* 14.2 (1989), p. 178: "Dots composed mainly of zinc yellow, which were originally a bright yellow, have become a yellow-brown that is almost ocher; zinc yellow mixed with vermilion, which had been orange, has turned to warm brown; and yellow-green dots, composed mainly of zinc yellow with varying amounts of emerald green, have shifted to olive. The alteration of the zinc yellow was probably caused by oxidative changes in the chromium."

101. Paul Signac, "Les Besoins individuels et la peinture," in *Encyclopédie française*, ed. A. de Monzie, vol. 16: *Arts et littératures dans la société contemporaine* (Paris: Société des Gestion de l'Encyclopédie Française, 1935), p. 395. John House, whose English translation I reproduce, quotes an 1894 diary entry from Signac complaining of the same thing, but counters that most of the critics he cited discuss all these possibilities. "Reading the *Grande Jatte*," *Art Institute of Chicago Museum Studies* 14.2 (1989), pp. 114–16.

102. On the dating of the campaigns, see Fiedler, "Technical Evaluation," p. 175. The third and last campaign of 1889 or later resulted mainly in the painted "frame," which required re-stretching the canvas to produce a larger surface area. Staple holes from the prestretched canvas in its original form, which Seurat filled with gesso, are observable, especially at upper left.

103. Fiedler, "Technical Evaluation," p. 175, attributes these "few, essentially continuous, linear features" to the second campaign.

104. Jules Laforgue, "L'impressionnisme," in *Textes de critique d'art*, ed. Mireille Dottin (Lille: Presses Universitaires de Lille, 1988), p. 170, insists that daylight is "no dead

whiteness, but a thousand vibrant combats, of rich prismatic decompositions." These optic combats are "infinite and infinitesimal" (p. 172). From this complexity, Laforgue concluded that the painter paints not what he sees, but only a "compte-rendu" of it.

105. Michelle Foa, *Georges Seurat: The Art of Vision* (New Haven: Yale University Press, 2015), pp. 48–49. A link with Helmholtz (albeit only with his acoustics) was drawn by Michael F. Zimmerman, *Seurat and the Art Theory of His Time* (Brussels: Fonds Mercator, 1991), pp. 262–63. Foa, by contrast, emphasizes "On the Relations of Optics to Painting," translated in French in 1867 and popularized by Taine.

106. Foa, *Georges Seurat*, pp. 91–92. Foa quotes Helmholtz falling prey to what I call the doubling fallacy, indeed demanding that painters do so too: "Subjective phenomena of the eye must be objectively introduced into the picture, because the scale of color and of brightness is different upon the latter" (p. 91).

107. A recent defense of this kind of "irrealism" about objects is Jody Azzouni, *Ontology without Borders* (Oxford: Oxford University Press, 2017), which argues, e.g., that we arbitrarily regard two pieces of cloth stitched together as one but not two humans stitched together (p. 153). The fallacy should be obvious by now: the question is not one about "worldly" boundaries but about concepts. One body can be distinguished from two persons; ditto for pieces of cloth.

108. Perhaps an "arbitrary object" of this sort is the woman sitting with an umbrella grasped in her hands at far left, who seems to have no eyes! But more likely the woman is closing her eyes, or the shadow of her hat's brim hides them.

109. Foa, *Georges Seurat*, p. 74. Foa says "the little girl and her outward gaze make explicit and visible the ways that linear perspective posits a single, ideal vantage point, part of Seurat's broader pulling apart of the constitutive components of the classical tableau." It is not clear to me that exposing perspective's ordering functions necessarily undermines them.

110. Geometric and pragmatic reasons why figures "look after you" are set forth in Felix Exner, "Über das sogenannte 'Nachschauen' von Bildern," in *Festschrift Ludwig Boltzmann* (Leipzig: Ludwig Ambrosius Barth, 1904), pp. 652–55.

111. This point is pressed in John Russell, *Seurat* (New York: Praeger, 1965), pp. 146–47.

112. Russell, *Analysis of Mind*, pp. 160–61. Cf. pp. 104, 130, and the introduction of the analogy at 99–102.

113. Sigmund Exner, "Das Netzhautbild des Insectenauges," *Sitzungsbericht der Kaiserlichen Akademie der Wissenschaften in Wien, Mathematische-naturwissenschaftliche Classe* 98.III (1889), pp. 13–65.

114. Albertina Inv. No. FotoGLV2000/3586, and the later enlargement on silver gelatin paper, whose note insists it is "the direct photograph of a retinal image in the eye of the firefly" (Inv. No. FotoGLV2000/8411). This note gives the animal's genus as *Lampyris*, referring presumably to *Lampyris noctiluca*, the common European glowworm. Eder also published the image in his *Jahrbuch für Photographie und Reproduktionstechnik* 4 (1890), p. 50.

115. Russell, "My Mental Development," in *The Philosophy of Bertrand Russell*, ed. Paul Arthur Schilpp (Evanston, IL: Library of Living Philosophers, 1946), p. 16 (*Basic Writings*, p. 46): "thus, broadly speaking, minimum vocabularies are more instructive when they show a certain kind of term to be indispensable than when they show the opposite."

CONCLUSION: BEYOND SYMBOLISM

1. Friedrich Nietzsche, "Über Wahrheit und Lüge im außermoralischen Sinne" (1873), first printed with the *Unzeitgemäße Betrachtungen* (Leipzig: Alfred Kröner, 1921), p. 3.

2. *Ibid.*, p. 8. Robert Pippin, *Nietzsche, Psychology, and First Philosophy* (Chicago: University of Chicago Press, 2010), argues that Nietzsche is both a critic of traditional philosophical psychology and believes such a critique is the "path to the fundamental problems" (p. xii, quoting Nietzsche). I agree, but I believe Nietzsche did not extricate himself from all the traps of philosophical psychology, notably the abstraction theory of concept formation.

3. *Ibid.*, p. 7. Note that this is the first part of Frege's point about snow being white. Had Nietzsche thought further, he might have concluded that we attribute to the stone the power to cause this subjective effect in others as well.

4. *Ibid.*, p. 10. Nietzsche does admit tautologies as "real" truths, but calls them, traditionally, "empty husks" (p. 7).

5. *Ibid.*, pp. 8–9. I capitalized "leaf" and "honesty" where Nietzsche has, literally, "the leaf," "the honesty," definite article suggesting definite thing. As the last sentence makes clear, this usage is highly ironic. This linguistic play is not evident in the excellent Cambridge Texts in the History of Philosophy translation by Ronald Speirs, with its single quotes: see *The Birth of Tragedy and Other Writings* (Cambridge: Cambridge University Press, 1999), p. 145.

6. Even if he did mean that, why call them leaves? There are no words, nor things, without corresponding concepts. See *Parmenides* 135c, in Plato, *Complete Works*, ed. John M. Cooper (Indianapolis: Hackett, 1997), p. 369.

7. First printed in 1929 (Leipzig: Hadl) with the "Truth" essay. Interestingly, it was the body that nature gives humans access to only through the "imposter" (or "juggler,"

fairground entertainer) consciousness. The essay contains, near the end, a draft of the opening passage of the "Truth" essay, concerning the "animals that invented knowing."

8. "Wahrheit" (1921), p. 15. His case against natural laws is less crude: Nietzsche sees in them only regularities, that is, "relations to other natural laws," which are thus self-referential and do not afford any essential knowledge of nature.

9. Gottlob Frege, *Die Grundlagen der Arithmetik* (Breslau: Koebner, 1884), p. x.

10. Félix Bracquemond, *Du Dessin et de la couleur* (Paris: Charpentier, 1885), pp. 266–67.

Index

Zone Books series design by Bruce Mau
Typesetting by Meighan Gale
Image placement and production by Julie Fry
Printed and bound by Maple Press